fP
T0119282

THE CASE FOR MARS

The Plan to Settle the Red Planet and Why We Must

25TH ANNIVERSARY EDITION

ROBERT ZUBRIN

with RICHARD WAGNER

Free Press

New York London Toronto Sydney New Delhi

Free Press
An Imprint of Simon & Schuster, Inc.
1230 Avenue of the Americas
New York, NY 10020

This Free Press trade paperback edition February 2021

Frontispiece map: NASA Goddard Spaceflight Center

Manufactured in the United States of America

10 9 8 7 6 5 4 3 2 1

Library of Congress Cataloging-in-Publication Data

Zubrin, Robert.
The case for Mars : the plan to settle the red planet and why we must / Robert Zubrin with Richard Wagner ; foreword by Arthur C. Clarke.—Rev. and updated.
p. cm.
Includes bibliographical references and index.
1. Mars (Planet). 2. Life on other planets. 3. Space colonies.
I. Wagner, Richard, 1953– II. Title.
QB641.Z83 2011
919.9'2304-dc22 2011005417

ISBN 978-0-684-82757-5
ISBN 978-1-9821-7292-3 (pbk)
ISBN 978-1-4391-3521-1 (ebook)

To Linda; My sister, and true lifelong friend

Henceforth I spread confident wings to space

I fear no barrier of crystal or of glass;

I cleave the heavens and soar to the infinite.

And while I rise from my own globe to others

And penetrate even further through the eternal field,

That which others saw from afar, I leave far behind me.

—Giordano Bruno

"On the Infinite Universe and Worlds," 1584

CONTENTS

PREFACE TO THE 25TH ANNIVERSARY EDITION

It has be twenty-five years since I penned the first edition of *The Case for Mars*. A lot has happened since then. A lot more is going to happen very soon.

When spaceflight began in the Sputnik era, the price to reach orbit was astronomical. Under pressure of the space race, the basic technology of expendable space launch was perfected so that by the time of the 1969 Moon landing, the cost of delivery to orbit had declined to $10,000 per kilogram—and there it stayed for the next forty years.

But then, ten years ago, things began to change. The entrepreneurial SpaceX company entered the game, reducing the cost of launch, first by introducing cheaper ways of doing business, and then, more profoundly, in December 2015, by starting to make its Falcon 9 medium-lift boosters reusable. Then, on February 6, 2018, the SpaceX Falcon Heavy took flight, demonstrating a capacity to lift 60 tons to low Earth orbit while playfully sending a Tesla Roadster on a trajectory that is taking it beyond the orbit of Mars. To add to the coup, two of the Falcon's three booster stages flew back to land gracefully together at the Cape, while the third barely missed pulling off a recovery landing on a drone ship stationed downrange.

To understand how extraordinary this accomplishment was, let us recall that in 2009 the Obama administration's blue-ribbon review committee headed by former Lockheed Martin CEO Norm Augustine declared that NASA's Moon program had to be cancelled, because the development of the necessary heavy lift booster would take 12 years and 36 billion dollars.

Yet SpaceX did it, in half the time and at a thirtieth of the cost. And, to cap it all, the launch vehicle is three-quarters reusable.

In consequence, the cost of launch to orbit today is now $2,000 per kilogram, one-fifth of what it was a decade ago. Moreover SpaceX has not just introduced some new launch capabilities. They have proven the power of creative entrepreneurial approaches to spaceflight. As a result, dozens of other companies, developing not only launch vehicles but spacecraft, instruments, communications, power systems, and all sorts of other necessary technologies are now receiving investment and growing rapidly.

This is a revolution. Promethean fire has been unleashed. But it is only the beginning.

As I write these lines, a shipyard is rapidly expanding in Boca Chica, Texas, whose purpose is to build a fleet of interplanetary transports to support the exploration and settlement of Mars. In February 2020, my

The author and Elon Musk during their meeting at the Starship development facility in Boca Chica, Texas. (Photo Hope Zubrin)

wife, Hope, and I travelled to there to meet with SpaceX CEO Elon Musk. While we talked inside the SpaceX onsite headquarters, a mariachi band played outside, providing entertainment for long lines of people queued up to apply for multiple categories of jobs building craft to take humans to Mars. Hundreds were already hired and at work in the complex. By the time you read this there will be thousands.

Musk calls his design the "Starship." It's a methane-oxygen-driven, stainless steel, two-stage to orbit rocket with a payload capacity equal to the Saturn V booster that sent Apollo astronauts to the Moon. The Saturn V, however, was expendable, with each unit destroyed in the course of a single use. Starship will be fully reusable, like an airliner, and thus promises a radical reduction in payload delivery costs.

Furthermore, the Starship is being designed to be refueled in space, which means it won't be limited to delivering payloads to low Earth orbit, but, after orbital refilling, will be able to fly all the way to the Red Planet. We can make methane-oxygen propellant out of local materials on Mars, so once it lands, a Starship will be able to refuel there for the return flight to Earth.

Prototypes of Starship have already flown on low-altitude hops, but orbital flight has yet to be demonstrated. Yet here was Musk, building not only the first experimental ship to prove the concept but also—as we witnessed touring the place the next day—a shipyard and a fleet. Is he mad? According to conventional aerospace industry thinking he certainly is. But there is a method to his madness.

Both the Mars Direct strategy I first set forth in this book and Musk's Starship plan use direct flights from Earth orbit to the surface of Mars, with direct return from the surface to Earth using methane/oxygen propellant made on the Red Planet from local materials. Both shun any need for orbital infrastructure, orbital construction, interplanetary motherships, specialized small landing craft, or advanced propulsion. Both involve long-duration stays on Mars starting from the very first mission. For both, the central purpose of the mission is not flying to Mars but accomplishing something serious on Mars.

But there is a difference. Starship is about ten times more massive than the Earth Return Vehicle I proposed to use in Mars Direct. As a result, it will need about ten times as much propellant, making the

Mars surface power and other base requirements needed to support Starship operations factor of ten higher than that needed to implement Mars Direct.

So a large base will need to be built in advance, by sending several Starships loaded with lots of base equipment one-way to Mars—ten football fields' worth of solar panels and robots to set it all up—before the first crew-carrying Starship can arrive. This makes the system suboptimal for exploration. But exploration is not what Musk has in mind.

Mars Direct is designed to support an exploration program that could evolve toward base building and then settlement; Musk's plan is like D-Day. He needs a fleet. So he's creating a shipyard to build a fleet. But why build a fleet before you've even tested one ship? There are several reasons. The first is that he wants to be ready to take losses. By the time the first Starship is ready for its maiden test flight, he'll have three or four more already built and on deck, ready to be modified to fix whatever caused the first to fail. Launch, crash, fix, repeat—until it works—then keep launching, improving payload and cutting turnaround time, advancing performance, flight by flight, ferociously.

But there is another reason to build a fleet. It's to make Starships cheap. NASA built five Space Shuttles over a twelve-year period, each one costing several billion dollars. Musk is creating a shipyard designed to ultimately mass-produce Starships at a rate of fifty or more—he says hundreds—per year. As I write these lines, they are already rolling out prototypes at a rate of one per month.

So Starship is coming. But when?

Musk says they will reach orbit in 2021. That's possible. But Musk is frequently optimistic in his schedule prediction. So maybe it won't be until 2022, or even 2023. But surely Starships will be flying regularly to Earth orbit by 2024. That will create a very interesting situation.

Because while NASA and the other government space agencies may have gone adrift in the space-launch and human-spaceflight arenas since Apollo, they have delivered extraordinary results in the fields of space science. A string of incredibly successful NASA orbiters and rovers launched since the mid-1990s has revealed Mars to be a world rich in all the resources needed to support life and therefore future technological civilizations. A few years ago, methane—which can only exist on Mars as a product of life or of hydrothermal environments that can support life—was detected by the Curiosity rover. Then, in 2018, scientists using the MARSIS ground penetrating radar on the European Mars Express Orbiter announced the discovery of an underground lake of liquid salt water near the Martian South Pole. That same year the SHARAD ground penetrating radar team on NASA's Mars Reconnaissance Orbiter announced the discovery of massive ranges of glaciers on Mars, covered by only a few meters of dust, extending down from the poles to latitudes as far as 38 North (the same latitude as San Francisco), and containing an amount of water equal to 10 percent of the fresh water supply of Earth. Then, in the fall of 2020, MARSIS scientists announced the discovery of three more underground salt-water lakes on Mars. On the basis of such data, it is becoming increasingly likely that we will discover not only the remains but even living survivors of ancient microbial life on the Red planet. As humble as such Martian microbes might be, the implications drawn from their existence are spectacular: the processes that lead to the origin of life are not peculiar to the Earth.

If we combine such a finding with the Kepler space telescope mission's discovery that most stars have planets, and that virtually every star has a region surrounding it—near or far depending upon the brightness of the star—that can support the type of liquid-water environments that gave birth to life on Earth and Mars, the conclusion would have to be that a very large number of stars currently possess planets that have given rise to life.

But there is even more to it than that. We know from fossil evidence dating back at least 3.5 billion years that life appeared on Earth virtually as soon as it could. This means that either life evolves quickly and spontaneously from chemistry, or that life is being spread in microbial form across interstellar space and readily takes hold as soon as it finds a habitable environment. Either way, the conclusion is that life must be plentiful throughout the universe. In particular, since the thickly CO_2-enshrouded early warm and wet Mars was very similar to the early Earth, it means the Red Planet almost certainly once had life. We need to go to Mars, drill, bring up samples of subsurface water, and see what is there.

The key question is not whether there is life on Mars, but *what is its nature*? At the biochemical level, all life on Earth is the same. Whether bacteria, mushrooms, grasshoppers, or people, we all use the same DNA/RNA genetic alphabet. That's because we all share a common ancestor. But what about the Martians? If we both came from a common source, our alphabets will resemble each other, as the English alphabet does that of French. But if each biosphere originated locally, they could be as different as English and Chinese.

The necessary program of drilling, sample taking, culturing, biochemical analysis, and related observations is far beyond the ability of robotic rovers. It will require human explorers on the surface of Mars to carry out such a quest. But it will be worth the cost and risk involved, because it would not only once again astound the world with the daring creative genius of freedom, it would provide answers to fundamental questions about the prevalence and potential diversity of life in the universe that thinking men and women have wondered about for thousands of years.

In short, unlike NASA's human spaceflight program current lunar venture, dubbed project Artemis, which frankly is being done simply to have something to do, *sending human explorers to Mars would have a truly compelling purpose*. Once Starship is flying to orbit, achieving that purpose will be clearly within reach—not as something to be done in twenty or thirty years, but before this decade is out, and at a cost well within current space program budgets.

In short, by making human missions to Mars practical, SpaceX is going is going to make such a program sellable.

That is where you come in.

What SpaceX is doing is great, and we can all celebrate it. But we need to do more. We can't expect SpaceX the bear the full cost of a humans-to-Mars program. They are creating the transportation system, which is the single largest task. But there are numerous other technologies that need to be developed, including those for power generation, propellant making, spacesuits, surface vehicles, robots, greenhouse agriculture, construction equipment, and many others. SpaceX can develop Starship because it has many other commercial applications that will pay for it, ranging from satellite launch to rapid intercontinental travel on Earth. But many of these other technologies don't offer that kind of short- to medium-term payback. They will need government support. Furthermore, some of them, such as surface nuclear power, would be very difficult for the private sector to develop regardless of cost, as they require access to controlled materials, such as highly enriched uranium.

So we are going to need a public-private partnership. We need to get the U.S. and allied governments to meet SpaceX and the other entrepreneurial space companies halfway.

History is not a spectator sport. Musk, his team, and others like them are doing their part. We need to do ours.

If you have the technical or business smarts to sign up with one of the entrepreneurial space teams or start one of your own, I urge you to do so. If you have political, organizational, artistic, or literary abilities, use your talent to spread the vision.

Victor Hugo once said that nothing can stop an idea whose time has come. That is true, provided the idea has messengers who can help it recruit to its banners the forces necessary for its victory. The poet Percy Shelley said, "poets are the legislators of mankind." That is also true, because culture is upstream of politics.

Mars needs engineers. Mars needs messengers. Mars needs poets.

Mars needs you.

All of human history up to this point, from the trek out of our African birthplace to the settling of the continents and then the linking together of the disparate branches of humanity through first long-distance sailing ships, then telegraphs, telephones, radio, television, satellites, and the internet, has been a process of our rise from a local Kenyan Rift Valley biological curiosity to a full-fledged global civilization. That transition is now nearly complete, and we stand at the beginning of a new history—our rise to become an interplanetary species capable of measuring itself against the challenge of the stars.

I am convinced that someday humans will live on millions of worlds, scattered throughout this region of the galaxy. They will not remember the political, business, and geostrategic struggles that preoccupy most people today. But they will look back with wonder on the first human landing on Mars and those who made it happen.

Contrary to some prestigious commentators, we are not living at the end of history.

We are living at the beginning of history.

What a grand time to be alive.

Robert Zubrin
Golden, Colorado
October 2020

PREFACE TO THE 15TH ANNIVERSARY EDITION

Our doubts are traitors
And make us lose the good we oft might win
By fearing to attempt.

—William Shakespeare, *Measure for Measure*

A lot has happened in the fifteen years since *The Case for Mars* was first published. A string of robotic missions were launched to the Red Planet, including *Mars Pathfinder* and *Mars Global Surveyor* in late 1996; *Mars Polar Lander* and *Mars Climate Orbiter* in 1999; *Mars Odyssey* in 2001; *Spirit, Opportunity,* and *Mars Express* in 2003; *Mars Reconnaissance Orbiter* in 2005; and *Phoenix* in 2007. With the exception of the 1999 flights, all of these missions have been brilliantly successful. As a result, our knowledge of the planet has greatly increased.

We now know for certain that Mars was once a warm and wet planet, possessing not only ponds and streams but *oceans* of water on its surface, and continued to have an active hydrosphere for a period on the order of a billion years—a span *five times* as long as the time it took for life to appear on Earth after there was liquid water here. Thus, if the theory is correct that life is a natural phenomenon emerging from chemistry wherever there is liquid water, various minerals, and a sufficient period of time, then life must have appeared on Mars.

Furthermore, we know that much of that *water remains on the planet today* as ice or frozen mud, with the soil of *continent-sized* regions of the planet assessed as being more than 60 percent water by weight. Not only that, we have discovered that *Mars has liquid water,* not on the surface, but underground, where geothermal heating has warmed it to create environments capable of providing a home for life on Mars today. We have found places where water flowed out of the underground water table and down the slopes of craters *within the past ten years.* Indeed, we have detected methane emissions characteristic of subterranean microbial life emerging from vents in the Martian surface. These are either the signatures of Martian life or the proof of subsurface hydrothermal environments fully suitable for life. Either way, they identify exactly the places where astronauts could go, drill, and bring up water samples whose contents would reveal to us the truth about the nature, prevalence, and potential diversity of life in the universe.

Beyond that, we have mapped the mineral content and topography of the planet from orbit and photographed it in sufficient detail to be able to see and guide our small robotic rovers, as well as to identify ideal landing sites and travel routes for future human explorers.

So now we know why we should go, and where we should go. But are we on our way? Not yet. In startling contrast to the brilliant and continuing success of the robotic Mars exploration program, over the fifteen years since the first publication of this book, NASA's human spaceflight program has made no progress whatsoever. The point requires emphasis. Aside from the information returned by the robots, *NASA today is no better prepared to send humans to Mars than it was in 1996.*

How can that be? The most frequent answer is lack of money. If only NASA had the kind of funding it did during the Apollo era, it is claimed, we would see great accomplishments in human spaceflight. This excuse, however, is completely false. The fact of the matter is that in today's dollars the average NASA budget between 1961 (when President Kennedy gave his speech announcing the Apollo program) and 1973 (when the final Apollo-Skylab mission was flown) was $19 billion per year, nearly exactly the same as NASA's budget is today, and has been in round numbers, since about 1990.

Nor is it the case that the Apollo-era NASA was able to accomplish more in the human spaceflight area because it did so at the expense of robotic exploration. In fact, during that period the unmanned exploration program was more active than it has been over the past fifteen years, with some forty lunar and planetary probes launched. In fact, if we extend our baseline to fifteen years, matching the 1961 to 1975 period against 1996 to 2010, we find that the earlier NASA launched ten Mars probes with eight successes, nearly identical (but slightly superior) in flight rate and batting average to the modern NASA's track record of nine Mars probes with seven successes.

Yes, it is true that the NASA budget during the 1960s got a larger share of federal outlays, however that is not because NASA was richer but because the nation was smaller and poorer. During the 1960s, America's population was 60 percent what it is today, and its GNP was 25 percent as great. These were hardly advantages for Apollo.

Furthermore, the technology available to America a half century ago was vastly inferior to that of today. The men who designed Apollo did their calculations on slide rules capable of performing, at most, one calculation per second, not on computers doing billions. Yet in eight years they solved all the problems necessary to take us from nearly zero human spaceflight capability to landing men on the moon and returning them to Earth.

As this book will show in detail, from a technological point of view, *we are much better prepared to send humans to Mars today than they were to get men to the Moon in 1961.* Yet they got there in eight years. We've gone nowhere in the past three and a half decades.

So, the question is, what did NASA have then that it doesn't have now?

The answer is *Resolution.*

By resolution I mean that quality associated with being able to determine what it is you truly want to accomplish, committing to that objective, creating a plan to achieve it, and then doing what is necessary to actually implement that plan.

During the Apollo period, that is how America's human spaceflight program operated. The objective was clear—get men to the moon and back *by the end of the decade*—and the commitment to it was absolute. Accordingly, a plan was devised to achieve that goal in accord with that schedule, vehicle designs were created to implement that plan, technologies were developed to enable those vehicles, then the vehicles were built and the missions were flown.

The robotic space program also operated in that manner at that time, and continues to do so today. That is why it continues to deliver ever greater achievements.

It is not the fact that the unmanned exploration program employs robots that has made it a success. Rather it owes its success to the fact that the people running it are using their brains.

In contrast, NASA's human spaceflight program has abandoned this rational approach entirely. Instead of designing things to implement plans, it develops things and then tries to find some use for them. It created the space shuttle without any clear idea of what it would be for, and thus it has proved to be of very limited value for supporting human space exploration.

The International Space Station (ISS) was conceived of for the purpose of giving the Shuttle something to do, but requiring that the station be built by the shuttle has vastly increased the station program costs and risks, overcomplexified its design, and limited its size while burdening it with a nightmare twenty-year assembly launch sequence. In contrast, the simpler yet bigger *Skylab* was designed and built in four years, and launched in one day. Moreover, the ISS itself has no rational purpose commensurate with its cost, risk, or multidecadal preoccupation of the agency's time. The fact that this dismal assessment of the Station's value, while unacknowledged, is generally understood was made amply clear by the sequel to the February 1, 2003, *Columbia* disaster. Coming down harshly on the space agency, the accident review committee chairman Admiral Harold Gehman pronounced that "if we are to accept the costs and risks of human spaceflight, we need to have goals worthy of those costs and risks." In response, the Bush administration did not even attempt to make the case that the ISS program met that standard. Instead it launched a new initiative to give NASA human spaceflight program something worthwhile to do, specifically a return to the Moon by 2020.

While it is true that flying to the Moon is certainly a more interesting activity than hanging out in a space station in low Earth orbit, creating urine and stool samples so that guinea pig scientists can catalog still more data on the progressive deterioration of human physiology in zero gravity (which is completely unnecessary, since any competent Mars mission designer would employ artificial gravity aboard his interplanetary spacecraft in order to avoid such effects—unless, of course, he was mutilating his design in order to provide justification for space station research), it still fails the test of rationality. We have, after all, been to the Moon six times. Over 300 kilograms of lunar material has been returned to Earth, and few people show any active interest in it. The big picture regarding the nature of lunar geology is already understood, with further work largely a matter of filling in details. Moreover, the whole subject is of limited interest anyway, trivial, in fact, in comparison with the questions of the origins and fundamental nature of life that would be addressed by the human exploration of Mars. And as to the matters of national brilliance and glory, self and world image, and reassertion of our will as a people to embrace and meet new challenges, one wonders what it says about America if the highest aspiration of our space program is to repeat a mission it accomplished a half-century before.

Notwithstanding the above, an even bigger problem with the Bush administration's goal of returning to the Moon was that it was not a real goal at all. Rather it was an attempt to create sizzle without the steak, since, as proclaimed in 2004 for achievement by the year 2020, it did not actually require NASA to do anything toward its fulfillment during the administration's time in office, even assuming a second term. Thus five more Bush years went by, without any Moon mission hardware being built, after which the putative program was handed off to the Obama administration, which had no stake in it.

Thus orphaned, without political protection, without any valid or compelling reason for existence, and without any material progress to show for itself, the program was predictably cancelled. In its place, the Obama administration put first a "flexible path" concept without even a pretense of purpose. Then, when that was found too absurd for even Congress to bear, a pseudogoal of reaching a near-Earth asteroid by

2025 (i.e., beyond the time horizon requiring any action by the world of the present) was duly proclaimed and ignored. However, since there are, after all, twenty-seven swing electoral votes in Florida, the administration set forth a fanciful assortment of new projects, including spending several billion dollars to refurbish the shuttle launch pads after the shuttle stops flying, developing a high-power electric thruster without the very large space nuclear reactor required to drive it, building an orbiting refueling station to service interplanetary spaceships that do not exist, and creating a space capsule that can fly astronauts down from orbit but not up.

None of these strange projects serve any useful purpose, nor could any other alternative random set, not merely because they don't fit together into any functional combination, but because, *in the absence of a goal, there is no useful purpose for them to serve.* Without question, they'll all be cancelled when Obama leaves office, if not before, without producing anything useful. And after spending another 40 or 80 billion dollars and wasting another four to eight years, we'll be back to square one once again.

Where there is no vision, the people perish.

The American people want and deserve a space program that really is going somewhere. But no goal can be sustained unless it can be backed up, and not by "rationales," but by *reasons.*

There are real and vital reasons why we should venture to Mars. It is the key to unlocking the secret of life in the universe. It is the challenge to adventure that will inspire millions of young people to enter science and engineering, and whose acceptance will reaffirm the nature of our society as a nation of pioneers. It is the door to an open future, a new frontier on a new world, a planet that can be settled, the beginning of humanity's career as a spacefaring species with no limits to its resources or aspirations as it continues to push outward into the infinite universe beyond.

For the science, for the challenge, for the future; that's why we should go to Mars.

The only meaningful counterargument against launching a humans to Mars initiative is the assertion that we cannot do it. This claim, however, is completely false.

We would need a heavy lift launch vehicle (HLV), which we lack, say the opponents, and it would take vast sums and extended periods of time to create one—$36 billion and twelve years, according to the Obama administration's blue-ribbon human spaceflight review panel. This is nonsense. We flew our first heavy lift vehicle, the Saturn V, in 1967, following a five-year development program during which we had to invent it as we went along. Today we know exactly what to do. As to cost, SpaceX company president Elon Musk testified directly to the panel that he would be willing to develop a 100 tonne to orbit class HLV for a fixed-price contract of $2.5 billion. This claim is very credible, since SpaceX recently developed and flew a 10 tonne to orbit medium lifter for a total program cost of $300 million. Indeed Lockheed Martin, the aerospace giant formerly led by panel chairman Norm Augustine, has designs for HLVs whose development it prices at $4 billion.

A human Mars lander would require a huge parachute, the opponents say, much bigger than anything we have used. A large parachute? Please, give me a break. If we could send men to the Moon, we can certainly make a large parachute. Or if we didn't care to do so, we could just use a more modest-sized parachute system and complete the landing deceleration using rockets.

It takes too long to get to Mars, they say, so we have to delay launching the initiative until we can develop radically more advanced types of space propulsion capable of getting us there much faster. Wrong. Using existing chemical propulsion, we can go from Earth to Mars in six months, and in fact the *Mars Odyssey* spacecraft did exactly that in 2001. Trips of this duration are quite manageable by humans. In fact, it's the standard tour that scores of astronauts and cosmonauts have already performed aboard Russian space station *Mir* and the ISS.

We would need a nuclear reactor to power our base on the Martian surface, they say, and we don't have one. True. But we fielded our first practical nuclear reactor in this country, the one that powered the submarine *Nautilus*, in 1952, and the laws of physics haven't changed much since. We had nuclear power before we had color TV, passenger jets, or push-button telephones. Nukes are 1940s technology. We can certainly build the little one needed to power a Mars base.

Cosmic rays, solar flares, zero-gravity health effects, psychological factors, dust storms, life support systems, excessive cost—the list of alleged showstoppers put forward by the naysayers goes on and on. They're wrong on every point.

In this book I will prove that to you. I will lay out in detail a plan for a near-term human Mars exploration that negates or solves every single one of these difficulties, accomplished using technology that we possess today.

The human exploration of Mars is not a task for some future generation. It is a task for ours.

We hold it in our power to begin the world anew.

Let's do it.

Golden, Colorado
March 9, 2011

FOREWORD BY ARTHUR C. CLARKE

The planet Mars is where the action will be in the next century. It is the only world in the solar system on which there is a strong probability of finding Life Past, and perhaps even Life Present. Also, we can reach it—and survive on it—with technologies which are available today, or which we can acquire in the very near future.

Robert Zubrin's book—which is often very amusing and contains asides which will not endear him to NASA—is the most comprehensive account of the past and future of Mars that I have ever encountered. It explains why we should go there, how we may go there—and, perhaps most important of all, how we may "live on the land" when we get there.

Personally, I am delighted to think that—if Dr. Zubrin's persuasive arguments are accepted—the first expedition to Mars may leave shortly before my ninetieth birthday. Meanwhile, if all goes well, the Russian Mars Lander will be leaving just before my seventy-eighth, carrying a message I have videoed for the colonists of the next century:

MESSAGE TO MARS

My name is Arthur Clarke, and I am speaking to you from the island of Sri Lanka, once known as Ceylon, in the Indian Ocean, Planet Earth.

It is early spring in the year 1993, but this message is intended for the future. I am addressing men and women—perhaps some of you already born—who will listen to these words when they are living on Mars.

As we approach the new millennium, there is great interest in the planet which may be the first real home for mankind beyond the mother world. During my lifetime, I have been lucky enough to see our knowledge of Mars advance from almost complete ignorance—worse than that, misleading fantasy—to a real understanding of its geography and climate. Certainly we are still very ignorant in many areas, and lack knowledge which you take for granted. But now we have accurate maps of your wonderful world, and can imagine how it might be modified—terraformed—to make it nearer to the heart's desire. Perhaps you are already engaged upon that centuries-long process.

There is a link between Mars and my present home, which I used in what will probably be my last novel, *The Hammer of God.* At the beginning of this century, an amateur astronomer named Percy Molesworth was living here in Ceylon. He spent much time observing Mars, and now there is a huge crater, 175 kilometers wide, named after him in your southern hemisphere. In my book I've imagined how a New Martian astronomer might one day look back at his ancestral world, to try and see the little island from which Molesworth—and I—often gazed up at your planet.

There was a time, soon after the first landing on the Moon in 1969, when we were optimistic enough to imagine that we might have reached Mars by the 1990s. In another of my stories, I described a survivor of the first ill-fated expedition, watching the Earth in transit across the face of the Sun on May 11—1984! Well, there was no one on Mars then to watch that event—but it will happen again on November 10, 2084. By that time I hope that many eyes will be looking back towards the Earth as it slowly crosses the solar disk, looking like a tiny, perfectly circular sunspot. And I've suggested that we should signal to you then with powerful lasers, so that you will see a star beaming a message to you from the very face of the sun.

I too salute to you across the gulfs of space—as I send my greetings and good wishes from the closing decade of the century in which mankind first became a space-faring species, and set forth on a journey that can never end, so long as the universe endures. Doubtless in many of its details, Dr. Zubrin's book—like my own exercise in terraforming Mars, *The Snows of Olympus*—will be bypassed by future advances in technology. However, it demonstrates beyond all reasonable doubt that the first self-sustaining human colony beyond Mother Earth lies within the grasp of our children.

Will they seize the opportunity? It is almost fifty years since I ended my first book, *Interplanetary Flight*, with these words:

> The choice, as Wells once said, is the Universe—or nothing. . . . The challenge of the great spaces between the worlds is a stupendous one; but if we fail to meet it, the story of our race will be drawing to its close. Humanity will have turned its back upon the still untrodden heights and will be descending again the long slope that stretches, across a thousand million years of time, down to the shores of the primeval sea.

Arthur C. Clarke
1 March 1996

PREFACE TO THE FIRST EDITION

> We choose to go to the Moon! We choose to go to the Moon in this decade and do the other things, not because they are easy but because they are hard, because that goal will serve to organize and measure the best of our energies and skills, because that challenge is one that we are willing to accept, one we are unwilling to postpone, and one which we intend to win. . . . This is in some measure an act of faith and vision, for we do not know what benefits await us. . . . But space is there and we are going to climb it.
>
> —John F. Kennedy, 1962

The time has come for America to set itself a bold new goal in space. The recent celebrations of the twenty-fifth anniversary of the Apollo Moon landings have reminded us of what we as a nation once accomplished, and by so doing have put the question to us: Are we still a nation of pioneers? Do we choose to make the efforts required to continue as the vanguard of human progress, a people of the future, or will we allow ourselves to be a people of the past, one whose accomplishments are celebrated only in museums? When the fiftieth anniversary arrives, will our posterity honor it as the touchstone of a frontier pushing tradition that they continue? Or will they look upon it much as a seventh century Roman may have once gazed upon the aqueducts and other magnificent feats of classical architecture still visible among the ruins, saying to himself in amazement, "We once built that?"

There can be no progress without a goal. The American space program, begun so brilliantly with Apollo and its associated programs, has spent most of the subsequent twenty years floundering without direction. We need a central overriding purpose to drive our space program forward. At this point in history, that focus can only be the human exploration and settlement of Mars.

Mars is the fourth planet from the Sun, about 50 percent farther out than Earth, making it a colder place than our home planet. While daytime temperatures on Mars sometimes get up to 17° centigrade (about 63° Fahrenheit), at night the thermometer drops to –90°C (–130°F). Because the average temperature on Mars is below the freezing point, there is no liquid water today on its surface. But this was not always the case. Photographs of dry riverbeds on the Martian surface taken from orbital spacecraft show that in its distant past Mars was much warmer and wetter than it is today. For this reason, Mars is the most important target for the search for extraterrestrial life, past or present, in our solar system. The Martian day is very similar to that of Earth—24 hours and 37 minutes—and the planet rotates on an axis with a 24° tilt virtually equal to that of Earth, and thus has four seasons of similar relative severity to our own. Because the Martian year is 669 Martian days (or 686 Earth days), however, each of these seasons is nearly twice as long as those on Earth. Mars is a big place; although its diameter is only half that of Earth, the fact that it is not covered with oceans gives the Red Planet a solid surface area equal to that of all of Earth's continents combined. At its closest, Mars comes within 60 million kilometers of our world; at its farthest, about 400 million kilometers. Using present day space propulsion systems, a one-way voyage to Mars would take about six months—much longer than the three-day trip required by the Apollo missions to reach the Moon, but hardly beyond human experience. In the nineteenth century immigrants from Europe frequently took an equal time to sail to Australia. And, as we'll see, the technology required for such a journey is well within our reach.

In fact, as this book goes to press, NASA scientists have announced a startling discovery revealing strong circumstantial evidence of past microbial life within Antarctic rock samples that had previously been ejected from Mars by meteoric impact. The evidence includes complex organic molecules, magnetite, and other typical bacterial mineralogical residues, and ovoid structures consistent with bacterial forms. NASA calls this evidence compelling but not conclusive. If it is the remains of life, it may well be evidence of only the most modest representatives of an ancient Martian biosphere, whose more interesting and complex manifestations are still preserved in fossil beds on Mars. To find them though, it will take more than robotic eyes and remote control. To find them, we'll need human hands and human eyes roving the Red Planet.

WHY MARS?

The question of taking on Mars as an interplanetary goal is not simply one of aerospace accomplishment, but one of reaffirming the pioneering character of our society. Unique among the extraterrestrial bodies of our solar system, Mars is endowed with all the resources needed to support not only life but the actual development of a technological civilization. In contrast to the comparative desert of the Earth's moon, Mars possesses veritable oceans of water frozen into its soil as permafrost, as well as vast quantities of carbon, nitrogen, hydrogen, and oxygen, all in forms readily accessible to those inventive enough to use them. These four elements are not only the basis of food and water, but of plastics, wood, paper, clothing, and—most importantly—rocket fuel. Additionally, Mars has experienced the same sorts of volcanic and hydrologic processes that produced a multitude of mineral ores on Earth. Virtually every element of significant interest to industry is known to exist on the Red Planet. While no liquid water exists on the surface, below ground is a different matter, and there is every reason to believe that geothermal heat sources could be maintaining hot liquid reservoirs beneath the Martian surface today. Such hydrothermal reservoirs may be refuges in which microbial survivors of ancient Martian life continue to persist; they would also represent oases providing abundant water supplies and geothermal power to future human pioneers. With its twenty-four-hour day-night cycle and an atmosphere thick enough to shield its surface against solar flares, Mars is the only extraterrestrial planet that will accommodate large-scale greenhouses lit by natural sunlight. Even at this early date in its exploration, Mars is already known to possess a vital resource that could someday represent a commercial export. Deuterium, the heavy isotope of hydrogen currently valued at $10,000 per kilogram, is five times more common on Mars than it is on Earth.

Mars can be settled. For our generation and many that will follow, Mars is the New World.

GOING NATIVE: THE FAST TRACK TO MARS

Down through history, it has generally been the case that those explorers and settlers who took the trouble to study, learn, and adopt the survival and travel methods of wilderness natives succeeded where others did not. The foreigner sees wilderness where the native sees home—it is no surprise that indigenous peoples possess the best knowledge of how to recognize and use resources present in the wilderness environment.

To the eye of the urban dweller, an Arctic landscape is desolate, resourceless, and impassable. Yet, to the Eskimo it is rich. Thus, during the nineteenth century, the British Navy sent flotillas of steam-powered warships, at great expense, to explore the Canadian Arctic for the Northwest Passage. Loaded with coal and supplies, these expeditions would battle forward against the ice packs for several years at a time, until shortages would force an about-face or even cause the entire crew to perish.

At the same time, however, small teams of explorers working for fur trapping interests were traveling freely over the Arctic by dog sled. Adopting the methods of the natives, they fed themselves and their dog teams on local game and traveled light. At insignificant expense they accomplished far more in the way of exploration than did the naval fleets.

There is a lesson in all of this for space exploration. There are no Martians, yet. But if there are to be, let us ask ourselves some questions. How will they travel? Will they important their rocket fuel from Earth? How about their oxygen? Where will their water come from, their food? How will they survive? There can only be one answer: *When on Mars, do as the Martians will do.*

TO MARS VIA DOGSLED

Many of the concepts advanced for piloted Mars missions have been analogous to the ponderous Royal Navy approach to the Arctic cited above. According to these plans, grand ships are required to haul out to Mars all the supplies and propellant required for the entire mission. Because such ships are too large to be launched in one piece, construction on orbit is required, as is long-term orbital storage of supercold (or *cryogenic*) propellant. Large orbiting facilities are required to enable both of these operations. The cost of such a project soon goes out of sight. One such plan, known as the "90-Day Report," developed in response to President Bush's 1989 call for a Space Exploration Initiative, resulted in a cost estimate of $450 billion. The resulting sticker shock in Congress doomed Bush's program and has deterred most people from seriously considering a humans-to-Mars program ever since.

However, as in the case of Arctic exploration, there is a different way a Mars mission can be approached—a dogsled way, if you will. By making intelligent use of the resources available in the environment to be explored, this approach allows the logistical requirements for launching the mission to be reduced to the point where the endeavor becomes practical.

This is the spirit of Mars Direct, a new approach to Mars exploration that I introduced in 1990 while a senior engineer for the Martin Marietta Astronautics company, working as one of its leaders in development of advanced concepts for interplanetary missions. This plan employs no immense interplanetary spaceships, and thus requires neither orbiting space bases nor storage facilities. Instead, a crew and their habitat are sent directly to Mars by the upper stage of the same booster rocket that lifts them to Earth orbit, in just the same way as the Apollo missions and all unmanned interplanetary probes launched to date have flown.

Flying the mission this way radically simplifies and scales down the required hardware, and eliminates the need for decades of development and hundreds of billions of dollars of expenditure on orbital assembly infrastructure. The key to this plan is the mission's ability to use Mars-native resources to make its return propellant and much of its consumables on the surface of the planet itself.

It is the richness of Mars that makes the Red Planet not only desirable, but attainable.

A piloted Mars mission is not about building enormous interplanetary cruisers—it's about moving a payload capable of supporting a small crew of astronauts from the surface of Earth to the surface of Mars, and then moving that or a similar payload back again to return the crew. Provided we take full advantage of the leverage afforded by the use of local resources to reduce mission logistics to a manageable level, such a task is not at all beyond our technical or fiscal means. Travel light and live off the land—that's the ticket to Mars.

THE GROWTH OF A NEW IDEA

The Mars Direct plan, including its development and mission philosophy, its hardware components and overall architecture, its key operations and logistics requirements, backup plans and abort options, and, finally, its evolutionary potential will be described in this book. In 1990, when I and my key collaborator in its development, David Baker, first put the plan forward, it was viewed as too radical for many in NASA to consider seriously. Some did, however, and over time, through a process of patient explaining and refutation of the alternatives, I managed to gain a significant base of support. Many other people started to pitch in, and with their help the concept moved steadily up the decision-maker ladder. In 1992, I was invited to brief then-NASA associate administrator for exploration Dr. Mike Griffin, who immediately decided to lend his considerable support. Griffin then briefed incoming NASA administrator Dan Goldin, who also became an advocate, going so far as to discuss the plan at several of the "town meetings" NASA held as part of its public outreach during 1992 and 1993.

With the support of Griffin and Goldin, I was able to return to NASA's Johnson Space Center and convince the group in charge of designing human Mars missions to take a good hard look at the plan. They produced a detailed study of a Design Reference Mission based on Mars Direct, but scaled up by nearly a factor of two in expedition size compared to the original concept. They then produced a cost estimate for a Mars exploration program based upon this expanded version of Mars Direct. Their estimate: $50 billion for all the required hardware development and flying three complete missions to Mars. The same costing group had assigned a $450 billion price tag to the traditional cumbersome approach to human Mars exploration embodied in NASA's "90-Day Report." In my opinion, if the JSC Design Reference Mission had been disciplined through the elimination of excess hardware and crew, the cost would have been cut in half—to something in the $20 to $30 billion range.

The Johnson Space Center team also gave Martin Marietta a small amount of money—$47,000 to be exact—to demonstrate what I had claimed was a simple chemical engineering technology for transforming the Martian atmosphere into rocket propellant. We did so, building in the course of three months a full-scale unit that operated at 94 percent efficiency. The demonstration was all the more convincing given that neither I, the project's lead engineer, nor anyone else on the team was actually a chemical engineer by training. If we could build such a machine, then it couldn't be that hard.

WE CAN DO IT

Twenty to thirty billion dollars is not cheap, but it's roughly in the same range as a single major military procurement for a new weapons system; it's in the same range as the money the United States government gave to Mexico in one afternoon in the summer of 1995. Spread over twenty years, with the first ten years developing hardware and the next ten years flying missions, it would represent between 8 percent and 12 percent of the existing NASA budget. For the sake of opening a new world to human civilization, it's a sum that this country can easily afford.

Exploring Mars requires no miraculous new technologies, no orbiting spaceports, no anti-matter propulsion systems or gigantic interplanetary cruisers. We can establish our first outpost on Mars within a decade, using well-demosntrated techniques of brass-tacks engineering backed up by our pioneer forebears' common sense.

How we can do it, and why we should do it, is the dual subject of this book.

ABOUT THIS BOOK

This book presents a condensation, in layman's terms, of many years of technical work devoted to the development of practical plans for the human exploration of Mars. Although, as might be imagined, the details of human Mars mission plans are highly technical in nature, the core issues which determine the fundamental feasibility of such ventures really are not. Rather, they are questions of strategy that can be fully understood by anyone willing to do some clear thinking and who is equipped with some good basic information.

Unfortunately, such information has not been readily available to the public to date. The existing general interest science literature on piloted Mars missions is mostly rather nebulous or naive, while the technical literature is confused, obscure, and frequently distorted by the bias of various technical organizations using the medium of technical publications to argue for their own self-interest. For the educated layman, there really hasn't been a satisfactory book on the subject. In part, *The Case for Mars* is an effort to correct this problem.

I have attempted to walk a fine line between technical detail and simple narrative description in this book. It is simple enough to declare one mission design superior to another, but it is also slightly disingenuous, as it is in the technical details that the reader will find the strongest arguments for or against any mission plan or technology. Some chapters are more technical than others (chapter 4, which describes Mars Direct in detail, and chapter 5, which exposes various arguments against piloted Mars missions for the hobgoblin myths they are, come to mind), but all should be understandable by both novice and expert alike. If for whatever reason you tend to pale before the nitty-gritty of numbers, just read on—you'll get the drift well enough.

I'm an astronautical engineer, but earlier in my career I was a science teacher, and I strive to write and explain technical material in a clear, concise manner. I hold it as a fundamental tenet that (contrary to the witticism popular among some of my scientific colleagues) Clarity is not the enemy of Truth, but her most vital ally. Moreover, I feel very strongly that something as exciting and vital to the human future as the real issues involved in opening a new planet to humanity should not be the property of a technical elite, but must be open to consideration by everybody. Therefore, in writing this book I decided to enlist as a supporting author my long-time friend Richard Wagner, who as former editor of *Ad Astra*, the general interest space exploration magazine published by the National Space Society, has had years of experience bringing scientific arguments to the public at large. With his help, and that of Mitch Horowitz, our capable editor at The Free Press, I believe *The Case for Mars* may well prove successful in finally making the real issues of human Mars exploration comprehensible for the general reader.

Because ultimately it's your understanding that's going to get us to Mars.

THE CASE FOR MARS

1: MARS DIRECT

The planet Mars is a world of breathtaking scenery, with spectacular mountains three times as tall as Mount Everest, canyons three times as deep and five times as long as the Grand Canyon, vast ice fields, and thousands of kilometers of mysterious dry riverbeds. Its unexplored surface may hold unimagined riches and resources for future humanity, as well as answers to some of the deepest philosophical questions that thinking men and women have pondered for millennia. Moreover, Mars may someday provide a home for a dynamic new branch of human civilization, a new frontier, whose settlement and growth will provide an engine of progress for all of humanity for generations to come. But all that Mars holds will forever remain beyond our grasp unless and until men and women walk its rugged landscapes.

Some have said that a human mission to Mars is a venture for the far future, a task for the "next generation." On the contrary, we have in hand all the technologies required for undertaking within a decade an aggressive, continuing program of human Mars exploration. We can reach the Red Planet with relatively small spacecraft launched directly to Mars by boosters embodying the same technology that carried astronauts to the Moon more than forty years ago.

How can this be? Looking at almost any plan for a human mission to Mars, be it from the 1950s or the 1990s, we see enormous spaceships hauling to Mars all the supplies and propellant required for a mission. The size of the spacecraft demands that they be assembled in Earth orbit—they're simply too large to launch from the Earth's surface in one piece. This requires that a virtual parallel universe of gigantic orbiting "dry docks," hangars, cryogenic fuel depots, power stations, checkout points, and construction crew habitation shacks be placed in orbit to enable assembly of the spaceships and storage of the vast quantities of propellant. Based upon such concepts, it has been endlessly repeated that a mission to Mars would have to cost hundreds of billions of dollars and incorporate technologies that won't be available for another thirty years.

Yet landing humans on Mars requires neither miraculous new technologies nor the expenditure of vast sums of money. We don't need to build *Battlestar Galactica*–like futuristic spaceships to go to Mars. Rather, we simply need to use some common sense and employ technologies we have at hand now to travel light and live off the land, just as was done by nearly every successful program of terrestrial exploration undertaken in the past. Living off the land—intelligent use of local resources—is not just the way the West was won; it's the way the Earth was won, and it's also the way Mars can be won. The conventional Mars mission plans are impossibly huge and expensive because they attempt to take all the materials needed for a two- to three-year round-trip Mars mission with them from Earth. But if these consumables can be produced on Mars instead, the story changes, radically.

Starting in the spring of 1990, I led a team of engineers and researchers at Martin Marietta Astronautics in Denver in developing a plan to pioneer Mars in this way. The name of the plan is "Mars Direct," and it represents the quickest, safest, most practical, and least expensive way to undertake the exploration and settlement of Mars.

Mars Direct says what it means. The plan discards unnecessary, expensive, and time-consuming detours: no need for assembly of spaceships in low Earth orbit; no need to refuel in space; no need for spaceship hangars at an enlarged Space Station, and no requirement for drawn-out development of lunar bases as a prelude to Mars exploration. Avoiding these detours brings the first landing on Mars perhaps twenty years earlier than would otherwise happen, and avoids the ballooning administrative costs that tend to afflict extended government programs.

A rough cost estimate for Mars Direct would be about $30 billion to develop all the required hardware, with each individual Mars mission costing about $3 billion once the ships and equipment were in production. While certainly a great sum, spent over a period of ten years it would only represent about 7 percent of the existing combined military and civilian space budgets. Furthermore, this money could drive our economy forward in just the same way as the spending of $100 billion (in today's terms) on science and technology in the Apollo program contributed to the high rates of economic growth of America during the 1960s.

Conventional wisdom might deem Mars Direct attractive because of its simplicity, but it would also deem it infeasible—the mass of the propellant and supplies needed for a human mission to Mars is much too large to be launched directly from Earth to Mars. Conventional wisdom would be right except for one thing: The required propellant and supplies needed for a Mars mission do not have to come from Earth. They can be found on Mars.

From a vantage point of the present, here's how the Mars Direct plan would work:

FEBRUARY 2029

A new, multistage rocket fashioned from currently existing parts rests on the launch pad at Cape Canaveral, its thin metal skin steaming in the morning sunlight. The booster reminds some of the old Saturn V's, the rockets that carried men to the shores of the Sea of Tranquility. The new Ares booster has about the same heavy lift capacity as the Apollo-era *Saturn V*'s, but at its heart are the workhorses of the past several decades, four Space Shuttle main engines and two shuttle solid rocket boosters. The engines ignite. Flame and smoke describe the signature of a new space age as the Ares hurtles skyward. High above Earth's atmosphere, the Ares upper stage separates from the spent booster, fires its single hydrogen-and-oxygen-burning engine, and hurls an unmanned 45-tonne (45-metric-ton) payload to Mars: the Earth return vehicle. (NB: 1 tonne=2204.6 lb.)

The ERV's name says it all. The vehicle is designed to carry a crew of astronauts back from the surface of Mars direct to a splashdown in Earth's waters. On its journey to Mars the ERV carries a small nuclear reactor mounted atop a light truck, an automated chemical processing unit along with a set of compressors, and a few scientific rovers. The ERV's crew cabin stores a life-support system, food, and other necessities to sustain a four-member crew on an eight-month journey back to Earth. Though its two propulsion stages will consume some 96 tonnes of methane/oxygen bipropellant on the return flight, the ERV arrives at Mars with its fuel tanks essentially empty, carrying just 6 tonnes of liquid hydrogen propellant production feedstock.

AUGUST 2029

Traveling across space at an average speed of about 27 kilometers per second, the ERV reaches Mars after a six-month trip. Upon arrival the ERV uses its *aeroshell*—a blunt, mushroom-shaped shield—to plow through the upper reaches of Mars' thin atmosphere. The craft's speed drops, allowing it to brake into orbit. A few days are spent in orbit to allow the flight controllers to perform a final system checkout. Then upon arrival of a clear dawn with low winds and well-defined shadows at the chosen landing site, the craft is targeted back into the atmosphere for final entry. Using its aeroshell again, the ERV decelerates to subsonic speeds until a parachute can pop open and start the spacecraft on a gentle descent toward the surface of Mars. A few hundred meters above the surface, the parachute drops away and small rockets fire up to take the ERV carefully through the last moments before touchdown.

Once settled on the rust-colored soils of Mars, the ERV gets down to the business at hand, making fuel for the return flight home out of thin air—in this case, Martian air. A door pops open on the side of the squat ERV landing stage and a light truck carrying a small nuclear reactor trundles out. Using a small TV camera on board as their eyes, mission controllers in Houston slowly drive the truck a few hundred meters away from the landing site. As the truck wheels along, a power cable snakes off its windlass, keeping the ERV's chemical plant connected to the small reactor. Once the controllers maneuver the truck to an appropriate spot, a winch lifts the reactor from the truck's bed and lowers it into a small crater or other natural depression in the landscape. The reactor kicks in and begins to energize the chemical processing unit with 100 kilowatts of electricity (kWe). Now the chemical plant goes to work, producing rocket propellant by sucking in the Martian air with a set of pumps and reacting it with the hydrogen hauled from Earth aboard the ERV. Martian air is 95 percent carbon dioxide gas (CO_2). The chemical plant combines the carbon dioxide with the hydrogen (H_2), producing methane (CH_4), which the ship will store for later use as rocket fuel, and water (H_2O). This methanation reaction is a simple, straightforward chemical process that has been practiced in industry since the 1890s. As the methanation reaction proceeds, it rids us of a potential problem, that of storing super-cold liquid hydrogen on the Martian surface. The chemical plant continues its work, splitting the water produced by the methanation process into its constituents, hydrogen and oxygen. The oxygen is stored as rocket propellant, while the hydrogen is recycled back into the chemical plant to make more methane and water. Additional oxygen is produced by a third unit which takes Martian carbon dioxide and splits it into oxygen, which is stored, and carbon monoxide, which it vents as waste. At the end of six months of operation, the chemical plant has turned the initial supply of 6 tonnes of liquid hydrogen brought from Earth into 108 tonnes of methane and oxygen—enough for the ERV plus 12 tonnes extra to support the use of combustion powered ground vehicles on the Martian surface. Using Mars' most freely available resource, its air, we have leveraged the portion of our return propellant hauled from Earth eighteen times over.

This chemical synthesis sequence may appear to some to be rather involved, but it's actually all Gaslight Era technology, utterly trivial by comparison with practically every other significant operation required for a successful interplanetary mission of any kind. Moreover, it is this concept of living off the land that makes Mars Direct possible.

If we attempted to haul up to Mars all the propellant required, we indeed would need massive spacecraft requiring multiple launches and on-orbit assembly. The cost of the mission would shoot out of sight. It should come as no surprise that local resources make such a difference in developing a mission to Mars, or anywhere else for that matter. Consider what would have happened if Lewis and Clark had decided to bring all the food, water, and fodder needed for their transcontinental journey. Hundreds of wagons would have been required to carry the supplies. Those supply wagons would have needed hundreds of horses and drivers, who in turn would have required further supplies. A logistics nightmare would have been created that would have sent the costs of the expedition beyond the resources of the America of Jefferson's time. Is it any wonder that Mars mission plans that don't make use of local resources manage to ring up $450 billion price tags?

MARCH 2030

Thirteen months following launch, a fully fueled spacecraft—the ERV—sits on the surface of Mars, awaiting the arrival of a human crew. Engineers at NASA's Johnson Space Center have monitored every step of the chemical production process, and, certifying its successful completion, give the go-ahead for the next step in the Mars Direct mission to proceed. The ERV deploys small robots to examine and photograph the terrain in its immediate vicinity. The crew of the first human expedition, skilled and vitally interested in landing site selection, takes an active role in exploring the ERV's neighborhood via these distant explorers. After several months of robotic exploration, an ideal landing spot is identified. One of the ERV robots ambles across the rough Martian terrain and places a radar transponder at the landing site to help guide the crew to a safe touchdown.

APRIL 2031

The Ares 3 launch vehicle, carrying a spacecraft called the "*Beagle*" after the ship of exploration that carried Charles Darwin on his historic voyage, towers majestically over the flatlands of the Cape, moments away from opening a new era of human history. Just a few weeks ago a similar booster, Ares 2, climbed into the skies over Florida. Identical to the first Ares booster and carrying a similar ERV payload, Ares 2 hurtles toward Mars even as crowds gather to watch the launch of the *Beagle*, the ship that will carry the first four humans to Mars.

The primary component of the *Beagle* is a habitation module that looks a bit like a huge drum. The module stands about 5 meters high and measures about 8 meters in diameter. With two decks each with 2.5 meters (about 8 feet) of headroom and a floor area of 100 square meters (about 1,000 square feet), it is large enough to comfortably accommodate its crew of four. The "hab," as everybody calls it, has a closed-loop life-support system capable of recycling oxygen and water, whole food for three years plus a large supply of dehydrated emergency rations, and a pressurized ground car powered by a methane/oxygen internal combustion engine. (See Figure 1.1.)

The four crew members are true renaissance men and women.

FIGURE 1.1
The Mars Direct hab and Earth return vehicles (ERV) within their aerobrakes.

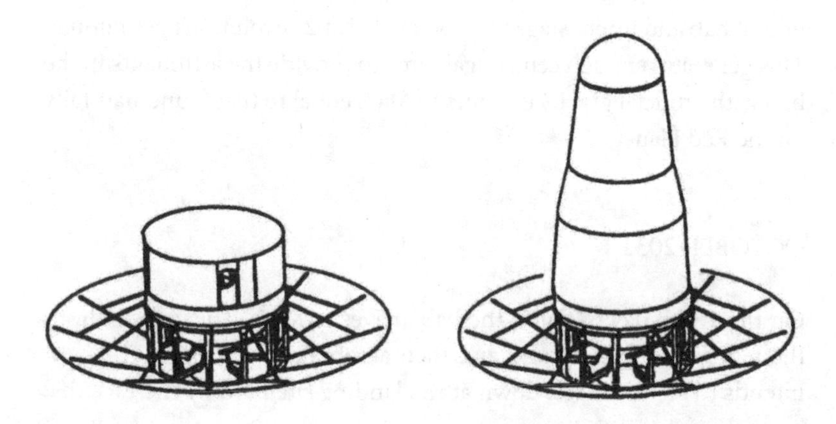

Given the nature of their mission—exploration far from home—all are cross-trained in several disciplines. At heart, though, they are a crew of two field scientists and two mechanics. A biogeochemist and a geologist will complement a pilot who is also a competent flight engineer. The last crew member, a jack-of-all-trades, is primarily a flight engineer, but can also provide common forms of medical treatment and understands the broad means and objectives of the scientific investigations. This person backs up all the specialists in their functions, and provides one more—he or she will be the mission commander.

On board the *Beagle,* four men and women prepare themselves for a journey that will take them to another world and return them home in the span of about two and one half years—about the same amount of time it took explorers centuries before to circumnavigate the globe. Miles distant from their small ship, more than a million people camped around Cape Canaveral gaze in anticipation as the countdown clock approaches zero. The lower-stage engines of the booster erupt, pouring out a sea of flame. A cheer louder than any this country has heard in years sweeps the crowd as the Ares 3 lifts off the pad. The rocket accelerates, propelling the upper stage and its payload through the atmosphere. The upper stage fires its own engines and breaks away, driving the hab to trans-Mars cruise velocity. Four humans are on their way to Mars.

The pilot of the hab directs it to pull away from the burnt-out upper stage of the booster, releasing it on a tether 330 meters long as it goes. A small rocket engine on the hab fires, causing the tethered combination of hab and upper stage to now revolve at 2 revolutions per minute. This generates enough centrifugal force to provide the astronauts in the hab with artificial gravity en route to Mars equal to that found naturally on the Red Planet.

OCTOBER 2031

On the 180th day of flight, the hab arrives at Mars. The vehicle drops the tether and upper stage, and then aerobrakes into orbit. The crew intends to set the *Beagle* down at the landing site hard by the ERV that flew out to Mars in 2029. A radio beacon in the Ares 1 ERV, detailed photos and maps of the landing site, a landing pad radar transponder, and the crew's expert handling of the ship virtually guarantee a precision landing. In the unlikely event that the *Beagle* misses the landing site, the crew has three backup options available. In the first place, they have on board the hab a fueled pressurized rover boasting a one-way range of nearly 1,000 kilometers. So long as they're within that distance of the landing site, the crew can still get to their ERV by driving overland. If some disaster causes the *Beagle* to miss the mark by more than a thousand kilometers, the second backup can be brought into play. This is the ERV launched by Ares 2, which, since it was launched on a slower trajectory than the *Beagle,* is now following the crew to Mars. Even if the crew lands the hab on the wrong side of the planet, this second ERV can be maneuvered to land near them. Finally, as a third-level backup, the crew arrives at Mars with sufficient supplies for three years—if worse came to worst, the four could just tough it out on Mars until additional supplies and another ERV could be sent out in 2033.

The landing, however, is right on target. Though they have studied the landing site in detail, seen it from images captured by rovers and relayed to Earth, nothing can prepare the crew for the sight of the Martian landscape stretching before them. The soils are rust colored, littered with sharp-edged rocks, large and small. In the distance are small hills and dunes. The landscape is akin to the deserts of America's southwest, save for the skies, which are a ruddy, salmon color. There's an immense amount to be done just after touchdown, but they take the moment to gaze out at Mars, to savor the fact that no creature with eyes to see has ever gazed out on this vista in the four-billion-year history of Mars and Earth.

With the *Beagle* safely down at the landing site, the Ares 2 ERV lands some 800 kilometers away, where it begins the process of filling itself with propellant. It will be used as the ERV for the second human expedition, which will arrive at its site in Hab 2 in 2033, along with another ERV that will open up Mars landing site number three. As the missions proceed, a network of exploratory bases will eventually be established, turning large areas of Mars into human territory.

The crew of the *Beagle* will spend five hundred days on the Martian surface. Unlike conventional Mars mission plans based upon orbiting mother ships with small landing parties, Mars Direct places all the crew on the surface of Mars where they can explore and learn how to live in the Martian environment. No one has been left in orbit, vulnerable to the hazards of cosmic rays and zero-gravity living. Instead, the entire crew will have available to them the natural gravity and protection against cosmic rays and solar radiation afforded by the Martian environment, so there is no strong motive for a quick departure. For a crew left in orbit during a conventional mission, there's little to do but soak up cosmic rays, and that tends to create a strong incentive to limit the time allowed for surface exploration, generally to thirty days or so. This leads to spectacularly inefficient missions. After all, if it takes a year and a half for a round trip to Mars, a stay of only thirty days is rather unrewarding. Worse yet, the rush to get back home forces conventional missions to follow trajectories that require far more propellant. But that extra propellant alone won't get a spacecraft back to Earth directly. Because Earth and Mars are constantly changing their positions relative to one another, "quick return" flight plan trajectories have to get a gravitational boost by swinging past Venus—where the Sun's radiation is twice that at Earth.

Even with such a substantial amount of surface time, the crew's days will be filled with projects that will vastly expand our knowledge of the planet and pave the way toward future exploration and, eventually, human facilities and settlements. There will be the geologic characterization of Mars, which will begin to tell us the story of Mars' past climatic history, how and when it lost its warm and wet climate, key clues to reviving Mars and perhaps saving the Earth. Geologic investigations will also include searches for useful mineral and other resources. Above all, astronauts will seek out easily extractable deposits of water ice or, better yet, subsurface bodies of geothermally heated water. Ice or water is key, because once water is found, it will free future Mars missions from the need to import hydrogen from Earth for rocket propellant production, and will enable large-scale greenhouse agriculture to occur once a permanent Mars base is established. Experimentation with agriculture is another item high on the priority list, and an inflatable greenhouse will be brought along for this purpose. The area of exploration that will seize the attention of the people of Earth, though, will be the astronauts' search for Martian life.

Images of Mars taken from orbit show dry riverbeds, indicating that Mars once had flowing liquid water on its surface—in other words, that it was once a place potentially friendly to life. The best geologic evidence indicates that this warm and wet period of Mars' history lasted through the first billion years of its existence as a planet, a period considerably longer than it took life to appear on Earth. Current theories of life hold that the evolution of life from nonliving matter is a lawful, natural process occurring with high probability whenever and wherever conditions are favorable. If this is true, if the theories are indeed correct, then chances are life should have evolved on Mars. It may still lurk somewhere on the planet, or it may be extinct. Either way, the discovery of Martian life, living or fossilized, would virtually prove that life abounds in the universe, and that the billions of stars scintillating in a clear, dark night sky mark the home solar systems of living worlds too numerous to count, harboring species and civilizations too diverse to catalogue. On the other hand, if we find that Mars never produced any life, despite its once clement climate, it would mean that the evolution of life is a process dependent upon freak chance. We could be virtually alone in the universe.

Given the importance of the question, the search for life past or present will be intensive, for there are many different places to look. There are dry riverbeds and dry lake beds that may have been the last redoubts of the retreating Martian biosphere, and thus promising places to look for fossils. Ice sheets covering the planet's poles may hold well-preserved frozen remains of actual organisms, if there were any. There is a high probability that subsurface ground water, geologically heated, may exist on Mars. In such environments living organisms may yet survive. What a find such organisms would be, for they may well be very different from anything that has evolved on Earth. In studying them, we would discover what is incidental to Earth life, and what is fundamental to the very nature of life itself. The results could lead to breakthroughs in medicine, genetic engineering, and all the biological and biochemical sciences.

The search for life and resources will necessarily involve a bit more than ambling a few meters along the Martian landscape and drilling a hole or two. The first explorers to Mars will have to range across the Martian landscape, beyond the horizon of their small base. The pressurized ground rover, which provides a shirtsleeve environment for astronauts, will allow the astronauts to explore far and wide on week-long sorties from their base. The rover burns methane/oxygen fuel, the same as the ERV. Ten percent of the stockpile of the methane/oxygen fuel produced by the ERV chemical plant will be allocated to support ground exploration. With this much fuel to run their car, the astronauts will be able to explore a vast area around their base, racking up over 24,000 kilometers on the vehicle odometer before the end of the first mission. As the rover crew travels, they will leave behind them small remote-controlled robots which will allow the base crew, and those of us on Earth, to continue to explore a multitude of sites via television.

The enormous amount of exploring the astronauts will undertake will necessarily result in a staggering amount of information, all of it new, undoubtedly unique, and certainly more than any one crew member could digest. Each astronaut will confer regularly with panels of the world's top experts in his or her assigned fields, creating a massive flow of information between Earth and Mars. Of course, crew members will also send and receive personal messages, but because there is a time lag in the transmission of radio waves between Mars and Earth, they will have to put up with delays of up to forty minutes before they get their answer. That will be troublesome for people accustomed to telephone conversations, but no problem at all for those who still know how to write a decent letter.

MARCH 2033

At the end of a year and a half on the Martian surface, the astronauts clamber aboard the ERV and blast off to receive a heroes' welcome on Earth some six months later. They leave behind Mars Base 1, with the *Beagle* hab, a rover, a greenhouse, power and chemical plants, a stockpile of methane/oxygen fuel, and nearly all of their scientific instruments. In November 2033, shortly after the first crew reaches Earth, a second crew arrives at Mars in Hab 2 and lands at Mars Base 2. The crew of the second mission will spend most of their time exploring the territory around their own site, but they will probably drive over at some point and revisit the old *Beagle* at Mars Base 1, not just for sentimental reasons, but to continue necessary scientific investigations in that region.

Thus every two years, as shown in Figure 1.2, two Ares boosters will blast off the Cape, one delivering a hab to a previously prepared site, the other an Earth return vehicle to open up a new region of the Red Planet to a visit by the next mission. Two boosters every two years: That's an average launch rate of just one launch per year—12 percent of our heavy-lift launch capability—to support a continuing and expanding program of human Mars exploration. This is certainly affordable and thus sustainable. As an added bonus, the same Ares launch vehicles, habs, and Earth return vehicles (fitted with only one propulsion stage) used in the Mars Direct plan can also be used to build and sustain lunar bases. While Moon bases are most emphatically not needed to support Mars exploration, they are of considerable value in themselves, most notably as sites for superb astronomical observatories. By using common transportation hardware for both lunar and Mars exploration, the Mars Direct approach will save tens of billions of dollars in development costs.

Mars Direct is not without risk. The consequences of extended exposure to Mars' gravity—38 percent that of Earth—are unknown. However, experience with the more severe deconditioning of astronauts in orbiting zero-gravity facilities indicates that most of the ill effects are temporary. Then there is space radiation, which on the six-month transit trajectories necessitated by current or near-term propulsion technology will give the astronauts doses sufficient to cause an additional 0.5 to 1 percent probability of a fatal cancer at some point later in life. This is nothing to scoff at, but those of us who stay home all face a 20 percent risk of fatal cancer anyway.

The Martian environment itself may hold some surprises, yet both the 1970s vintage *Viking* landers, and the more recent Mars Exploration Rovers *Spirit* and *Opportunity*, none of which were designed for ninety days of operation, all functioned without hindrance on the Martian surface for years, unaffected by cold, wind, or dust. The biggest mission

FIGURE 1.2
The Mars Direct mission sequence. The sequence begins with the launch of an unmanned Earth return vehicle (ERV) to Mars, where it will fuel itself with methane and oxygen manufactured on Mars. Thereafter, every two years, two boosters are launched. One sends an ERV to open up a new site, while the other sends a piloted hab to rendezvous with an ERV at a previously prepared site.

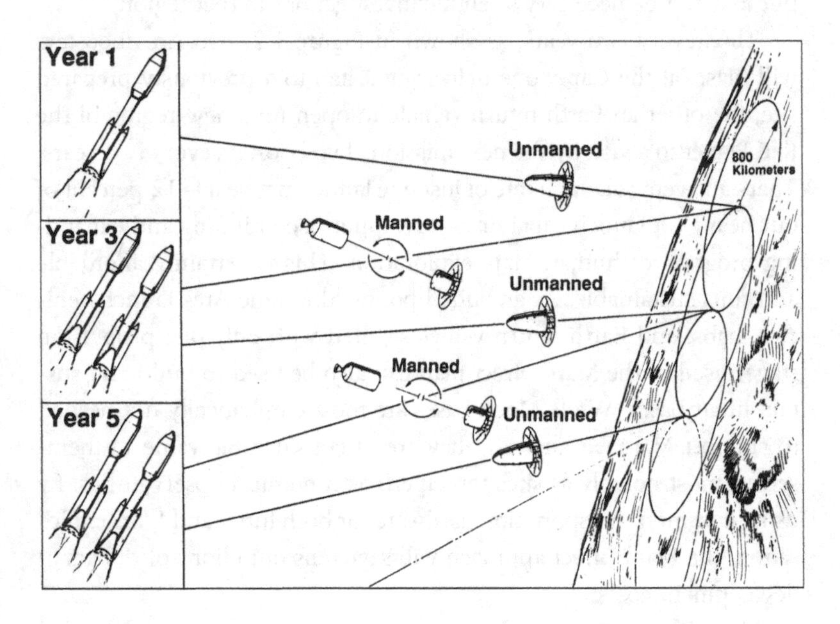

risk arises from possible failures in critical mechanical or electrical systems. Multiple backups for all important systems can minimize the risk, as can the presence of two ace mechanics during the mission. Any way you slice it, though, going to Mars the first time will involve a certain level of risk. This will be true whether we make the attempt with Mars Direct in 2031 or leave it for another generation to try. Nothing great has ever been accomplished without risk. Nothing great has ever been accomplished without courage.

NOVEMBER 2041

Over time, many new exploration bases will be added, but eventually it will have to be determined which of the base regions is the best location to build an actual Mars settlement. Ideally this will be situated above a geothermally heated subsurface reservoir, which will afford the base a copious supply of hot water and electric power. Once that happens, new landings will not go to new sites. Rather, each additional hab will land at the same site. In time, a set of structures resembling a small town will slowly take form. The high cost of transportation between Earth and Mars will create a strong financial incentive to find astronauts willing to extend their surface stay beyond the basic one and a half year tour of duty. As experience is gained in living on Mars, growing food, and producing useful materials of all sorts, astronauts will extend their stay times to four years, six years, and more. As the years go by, the transportation costs to Mars will steadily decrease, driven down by new technologies and competitive bids from contractors offering to deliver cargo to support the base. Photovoltaic panels and windmills manufactured on site and new geothermal wells will add to the power supply, and locally produced inflatable plastic structures will multiply the town's pressurized living space. As more people steadily arrive and stay longer before they leave, the population of the town will grow. In the course of things children will be born, and families raised on Mars—the first true colonists of a new branch of human civilization.

It is possible that someday millions of people will live on Mars, and call it their home. Ultimately, we can employ human technologies to

FIGURE 1.3
Linking Mars Direct habs to establish the beginnings of a Mars base. (Artwork by Carter Emmart)

alter the current frigid, arid climate of Mars and return the planet to the warm, wet climate of its distant past. This feat, the transformation of Mars from a lifeless or near lifeless planet to a living, breathing world supporting multitudes of diverse and novel ecologies and life forms, will be one of the noblest and greatest enterprises of the human spirit. No one will be able to contemplate it and not feel prouder to be human.

That is for the future. Yet we today have a chance to pioneer the way. We can land four men and women on Mars within a decade, and begin the exploration and settlement of the Red Planet. We, and not some distant future generation, can have the eternal honor of opening this new world to humanity. All it takes is present-day technology mixed with some nineteenth-century chemical engineering, a dose of common sense, and a little bit of moxie.

FOCUS SECTION—LIVING OFF THE LAND: AMUNDSEN, FRANKLIN, AND THE NORTHWEST PASSAGE

History has shown time and time again that a small group of people operating on a shoestring budget can succeed brilliantly in carrying out a program of exploration where others with vastly greater backing have repeatedly failed, provided that the small group makes intelligent use of local resources. It is a lesson that explorers of the past have ignored at their peril.

At midnight on June 16, 1903, Roald Amundsen and his crew of six sailed out under the rain-lashed skies of Christiania, Norway, bound for the Canadian Arctic and the Northwest Passage. The Passage hung before Arctic explorers as an elusive prize—nearly three centuries of effort by literally hundreds of expeditions had failed to conquer the fickle ice packs, channels, and waters of the far north.

Amundsen chased the ghost of a boyhood hero, Sir John Franklin, one of the great and ultimately tragic names of Arctic exploration. Franklin had sailed in search of the Passage nearly sixty years earlier. But whereas Amundsen sailed in a thirty-year-old sealing boat bought with money borrowed from his brother and with creditors nipping at his heels, Franklin had set off with the backing of the British Admiralty.

He commanded two ships, the *Erebus* and *Terror*, both displacing well over 300 tonnes, crewed by a complement of 127 men. In the words of historian Pierre Breton, the ships carried ". . . mountains of provisions and fuel and all the accouterments of nineteenth-century naval travel: fine china and cut glass, heavy Victorian silver, testaments and prayer books, copies of Punch, dress uniforms with brass buttons and button polishers to keep them shiny . . ."[1] In a word, Franklin carried all that he needed, save for what he would need to survive.

The *Erebus* and *Terror* set sail on the 19th of May 1845, their commander expecting to discover the Northwest Passage and that feat's attendant glory, but finding only oblivion in the end. Whalers out of Greenland spotted the Franklin expedition's ships tethered to an iceberg on June 25. That was the last any European ever saw of the expedition. Franklin and his ships, his men, all his supplies, sailed into the Arctic wilderness and vanished.

Between 1848 and 1859 more than fifty expeditions set out to discover what befell the expedition. From what could be pieced together in the years that followed—from two brief messages left behind; from the frozen, twisted remains of some of the crew; from bits and pieces of European civilization native Eskimos picked up from the ice or looted from the ships—it became apparent that the expedition ended in disaster because, as one contemporary put it, Franklin had carried his environment to the Arctic.

Trapped in ice near King William Island in the autumn of 1846, Franklin and his men attempted to survive on their provisions of salted meat. The expedition carried meat aplenty, but none was fresh, and salt meat could not protect men from scurvy. Previous explorers had noted the anti-scurvy qualities of fresh meat, but Franklin paid no heed. He was no hunter—the expedition carried shotguns, good for partridge in the British heath perhaps, but not very useful on the Arctic ice—and chose to rely instead on rations of lemon juice. One by one the members of the expedition weakened and died, Franklin apparently on board ship in June 1847. Others, hoping to find a rescue party to the south, abandoned the ships, but literally dropped in their tracks as they dragged heavy iron and oak sledges across the Arctic wastes. All hands died.

Amundsen would follow in Franklin's footsteps, but he would not follow him to his grave. Instead of importing his home environment, he embraced the local environment and adopted a live off the land strategy. He learned about the anti-scurvy qualities of caribou entrails and uncooked blubber. He learned the Eskimo way of Arctic travel, dog sled, which gave him the mobility required to hunt big game. He learned the Eskimo way of building shelters out of ice and chose the deerskin clothing of the Eskimo over the woolens that the British insisted upon.

Amundsen and his crew of six aboard the *Gjoa* were also frozen in, and as a result spent two winters in a small harbor on the southeast corner of King William Island, not far from where Franklin's expedition met disaster, but they did not starve. Making good use of their dog-sled-given mobility they traveled hundreds of kilometers over land to hunt and explore, in the process not only surviving but also making the important geophysical discovery that the Earth's magnetic poles move. The crew of the *Gjoa* thrived in the same environment that destroyed Franklin's expedition. Finally breaking free of the ice in August 1905, the *Gjoa* set sail from King William and within weeks had forced the Northwest Passage. It took another four months of travel for Amundsen to reach an outpost where he could telegraph news of his success to his main backer in Norway, which he did, charges reversed. Six years later Amundsen would use what he had learned on King William to become the first to reach the South Pole.

2: FROM KEPLER TO THE SPACE AGE

> Ships and sails proper for the heavenly air should be fashioned. Then there will also be people, who do not shrink from the dreary vastness of space.
>
> —Johannes Kepler to Galileo Galilei, 1609

We've been to Mars before. On the morning of July 20, 1976, an American spacecraft, *Viking 1*, settled down onto the Chryse Planitia—the Plains of Gold on the Red Planet Mars. At the moment of touchdown, though, with *Viking* resting on the surface of a planet nearly 330 million kilometers distant, no one at NASA's Jet Propulsion Laboratory in Pasadena, California, knew if the unmanned spacecraft had arrived safely or burrowed into the ground. The audience at JPL could do little more than nurse early morning cups of coffee for nearly twenty minutes before learning if *Viking* had safely landed.

Almost immediately after landing, *Viking* set to work. A preprogrammed sequence of commands instructed the lander to take a high resolution picture of the area adjacent to one of its footpads just twenty-five seconds after landing. *Viking* relayed the picture in real time. Imaging data raced toward Earth at the speed of light as the engineers and scientists of the *Viking* program counted down the minutes to the distant radio signals' arrival. And then, with excitement, glee, and undoubtedly some astonishment they watched as, line by line, a photograph of the Martian surface slowly appeared before their eyes.

Granted, the image of a foot pad may seem a bit anticlimactic, but that first photograph delivered an immense amount of important information to the crowd at JPL. It told all that the lander had survived, and that its imaging system was working and working well. The image was clear; small rocks sharply delineated against the Martian soil, the rivets on *Viking's* foot pad as clear as the buttons on an engineer's conservative white shirt. After the first image, the next preprogrammed image *Viking* would capture was a scan of the horizon, a snapshot of the neighborhood. This image will probably reside forever with those who watched as the Martian surface revealed itself. *Viking* gazed out on a barren landscape littered with sharp edged rocks of all sizes. There appeared to be sand dunes in the distance, and small, undulating hills. It was an empty world, familiar, yet utterly alien.

For centuries humans had observed Mars and theorized about the planet. Their study and flights of imagination had opened enormous intellectual vistas for scholars, and revealed to all that the human mind could explore the cosmos and comprehend the complexity of the universe. Now human eyes gazed on a landscape knowing that humanity's grasp of the universe had been extended once again, from the intellectual to the physical. It had been a long journey, one that started not in the late twentieth century, but centuries before, and one that was not without its sacrifices.

OUT OF THE DARKNESS

On the morning of Saturday, February 19, 1600, the great Italian Renaissance humanist Giordano Bruno was taken from his cell and stripped of his clothes. Naked, gagged, bound to a stake, Bruno was led through the streets of Rome, a mocking chorus of chanting Inquisitors following behind. The procession arrived at the Square of Flowers before the Theater of Pompey, the place of execution. Torch in hand, one of Bruno's killers held a portrait of Jesus Christ in front of the condemned and demanded repentance. Bruno angrily turned his face away. The pyre was lit, and one of the most profound minds in human history was burned alive.

Bruno was murdered for having alleged in debate and in writing that the universe was infinite, that the stars were suns like our own, with other planets comprising inhabited worlds like the Earth orbiting around them. Thus, observers on those other worlds would look up and see our Sun with the Earth circling it in their sky, their heavens, and therefore, "We are in heaven."

This insight was a shock to the medieval mind, but why was it necessary to kill to try to stop it? Why was Bruno's younger contemporary Galileo threatened with death and then held under house arrest for decades? Why was astronomy, a science which concerns matters of apparently little practical value, such a torturous subject during the Renaissance? Why, in short, were the stakes so high?

The stakes were so very high because the science of astronomy placed the entire intellectual framework of Western civilization, of knowledge and therefore power, at risk. From the time of Babylon through Bruno's day, the heavens, with their innumerable stars and five wandering planets, were considered divine and unknowable by all save a select few: astrologers and priests in Babylonian times, the Church in Bruno's times. Listen to the second century A.D. librarian of Alexandria Claudius Ptolemy as he defended an astronomy that placed Earth at the center of the universe with the Sun and five known planets traveling along *epicycles*, small circular orbits whose centers move at a constant rate along the path of a greater circle centered around the Earth. Answering objections to the irrational nature of the epicycle scheme (additional epicycles were continually being added to the model to make it match observation), Ptolemy replied, "It is impermissible to consider our human conditions equal to those of the immortal gods and to treat sacred things from the standpoint of others that are entirely dissimilar to them. . . . Thus we must form our judgment about celestial events not on the basis of occurrences on Earth, but rather on the basis of their own inner essence and the immutable course of all heavenly motions." For Ptolemy, the laws of the heavens were completely different from those governing the Earth. The universe was unknowable, unchangeable, and uncontrollable by man. With the divine plan something beyond comprehension, only a ruling priesthood, with its unique access to the mystical and supernatural, could tell the people what was right and what to do.

So it stood for centuries, until the time came when a few thinkers challenged the notion that the universe would forever lie beyond humanity's intellectual grasp. The action started with the work of Nicholas Copernicus, who between 1510 and 1514 redeveloped a long-forgotten *heliocentric* (Sun-centered) theory of the universe first posited by the third century B.C. Greek thinker Aristarchus of Samos. Under the heliocentric system, the planets traveled about the Sun in circular orbits. This concept was revolutionary, heretical even, and could not precisely match the observed planetary motions, yet some scholars of the time saw beauty in the fundamental simplicity of Copernicus' system. Chief among them was Johannes Kepler.

Born in 1571, Kepler grew to be a devout Lutheran, yet also a die-hard Platonist with a passion for seeking the true nature of the universe in the rational laws of geometry. He would write, "Geometry is one and eternal, a reflection out of the mind of God. That mankind shares in it is one of the reasons to call man an image of God."

This quote is the key to the whole affair. If the human mind can understand the universe, it means that the human mind is fundamentally of the same order as the divine mind. If the human mind is of the same order as the divine mind, then everything that appeared rational to God as he constructed the universe, its "geometry," can also be made to appear rational to the human understanding, and so *if we search and think hard enough, we can find a rational explanation and underpinning for everything.* This is the fundamental proposition of *science.* It is this proposition that Bruno died for. It is this proposition that Kepler set out to prove, and by so doing, to lift the darkness off the soul of Western civilization. And that he did, with a significant piece of help from the planet Mars.

In February 1600, the same month as Bruno's execution, Kepler went to work for Tycho Brahe, without question the greatest observational astronomer of his time. Brahe had his own theory of the universe, and entrusted the twenty-eight-year-old Kepler with the task of determining Mars' orbit, all for the glory of Brahe's own theories, of course. When Brahe died in October 1601, the Holy Roman Emperor Rudolph II ordered that Kepler be put in charge of the treasure trove of Brahe's observations and that he succeed Brahe as Imperial Mathematician. Kepler now had the ammunition needed to undertake his assault on Mars in earnest.

Since the time of Aristotle, astronomers had simply assumed that the planets moved in uniform circular orbits because, as Aristotle himself argued, the circle was a perfect form, and only circular motions could come back on themselves and ensure an eternal motion. Try as he might, though, Kepler simply could not get any sort of circular orbit to match Brahe's observations. True, he could have invoked epicycles, but this Kepler refused to do. Ad-hoc systems of epicycles were irrational—and there had to be a rational answer. But if not circular orbits, what could they be? It took eight years of intense intellectual effort for Kepler to discover what Tycho's observations of Mars revealed: Mars traveled in an elliptical orbit, with the Sun as one focus of the ellipse. We now know that Mars' orbit is the most elliptical of all the planets, except Pluto, which was not discovered till the twentieth century, and therefore presented the acid test for any astronomical theory. Indeed, if Mars' orbit had been circular, the Aristarchus/Copernicus theory would probably have passed muster without anyone looking much deeper.

Kepler published the results of his labors in 1609 in a work entitled, in full, *A New Astronomy Based on Causations or a Celestial Physics Derived from Investigations of the Motions of Mars Founded on the Observations of the Noble Tycho Brahe.* Unlike many previous astronomers and philosophers, Kepler declared that this new astronomy was not simply a mathematical construct that reproduced the motions of the heavens. It was, instead, a treatise on the "true reality" of the heavens, an epic work that overthrew two thousand years of dogma and replaced it with an astronomy based on causes. In it he laid out what are now known as Kepler's first two Laws of Planetary Motion; that the planets move in elliptical orbits with the Sun at one focus, and that the radius vector from the Sun to the planet sweeps out equal areas in equal times. These laws are correct and are found today in all textbooks on astrodynamics. Equally important, however, was what strictly speaking can be called Kepler's incorrect hypothesis: that the planets were pulled by a "magnetic" force emanating from the Sun, spreading out from it "in the manner of sunlight." When his opponents accused him of mixing physics with astronomy, Kepler replied, "I believe that both sciences are so closely bound that neither can achieve perfection without the other." In other words, Kepler did not describe a model of the universe whose geometry was merely appealing—he was investigating a universe whose causal relationships could be understood in terms of nature knowable to man. In so doing, Kepler catapulted the status of humanity in the universe. Though no longer residing at the center of the cosmos, humanity, Kepler showed, could comprehend it. Therefore, as Kepler wrote to Galileo in the quote that leads this chapter, not only was the universe within man's intellectual reach, it was, in principle, within physical reach as well.

Ten years of further study followed, until Kepler was able to publish his masterpiece, *The Harmony of the World.* Here he laid out his final great discovery, the Third Law of Planetary Motion; that the square of the periods of revolution of the planets is proportional to the cube of their distances from the Sun. Once you have this law, it is a relatively simple matter to derive mathematically what is now known as Newton's Law of Universal Gravitation. Newton's laws are the basis of what is now known as classical physics, the powerful new body of scientific knowledge that made possible the Industrial Revolution in the eighteenth and nineteenth centuries. With Kepler's study of the planet Mars, the Dark Ages came to an end, and the scientific and industrial revolutions began—humanity's first encounter with Mars had paid off handsomely.

VOYAGES BY TELESCOPE

Kepler had used Mars to prove that the Earth was a planet. By implication, therefore, the planets, those little moving lights in the sky, were really vast worlds like the Earth. But how to explore these incredible new bodies? A tool was soon at hand. Barely a year after Kepler published his *New Astronomy,* Galileo turned a new instrument toward the heavens—a telescope. His discovery of mountains on the Moon and "three little stars" dancing about Jupiter over the course of several weeks of observing gave additional credence to the Keplerian view of the universe. Soon enough, other telescopes were trained on Mars.

The Italian astronomer Francisco Fontana produced in 1636 the first drawing of Mars through the telescope, though viewed today it reveals no recognizable features. In 1659 the Dutch astronomer Christiaan Huygens produced the first drawing that shows a known Martian feature, a roughly triangular dark blotch that appears on the planet's face, today known as Syrtis Major. By carefully observing Syrtis and similar features, early astronomers determined that the Martian day, or sol, was close to Earth's. In 1666, the Italian Giovanni Cassini measured the Martian day at 24 hours, 40 minutes, about two and one-half minutes longer than today's accepted measure of 24 hours, 37 minutes, 22 seconds. Although Cassini was also apparently the first to note one of Mars' polar caps, Huygens in 1672 produced the first sketch of one of the caps. Utilizing observations made between 1777 and 1783, William Herschel, discoverer of Uranus, noted that Mars should have seasons, as its polar axis was tilted about 30° (24° is the modern value) to its orbital plane.

Observations of Mars continued through the decades, especially around "oppositions," those times when Mars (technically, any planet outside Earth's orbit) lies on the opposite side of the Earth from the Sun. At these times, Mars is at its closest to Earth and thus shines most brightly in the sky. By the early nineteenth century, astronomers had collected a basketful of basic Mars statistics: its orbital period, the length of its day, the planet's mass and density, distance from the Sun, and surface gravity. But what truly intrigued observers was the changing face of Mars. Through the years the telescope's eyepiece revealed that Mars' face was mottled with dark patches that came and went with time. Likewise, the bright white spots observers noted at the poles appeared to vary with the Martian seasons, expanding and contracting over the course of a Martian year. And Mars apparently hosted an atmosphere, as some observers spied vague indications of clouds above the Martian surface.

The opposition of 1877 proved especially fruitful for observers and for Martian studies. Asaph Hall of the U.S. Naval Observatory discovered two small moons of Mars and promptly named them Phobos and Deimos—fear and terror, an appropriate entourage for the planet of war. But in hindsight, 1877 is perhaps best remembered for a series of observations that launched a turbulent episode in the history of Mars observations and one of the strangest chapters in the history of astronomy.

Among those who turned a telescopic eye toward Mars in 1877 was the Italian astronomer Giovanni Schiaparelli, director of the Brera Observatory in Milan. Schiaparelli's reports of his observations noted the location of more than sixty features on the Martian surface. But, along with many standard features, he reported sighting linear markings criss-crossing the face of Mars. He named these features after terrestrial rivers—Indus, Ganges—but referred to them in his writings as "canali," the Italian plural for channels or grooves. While not the first to note these strange markings, he was the first to identify an extensive system of "canali." More than a decade later, the enthusiasms of Percival Lowell would catapult Mars and its "canali" to headline status throughout the world.

Born into an illustrious New England family of poets, educators, statesmen, and industrialists (the great poet Amy Lowell was his sister, his brother Abbott was president of Harvard), Lowell while in his late thirties became intrigued with Mars, especially with Schiaparelli's observations. For Lowell there could be only one interpretation—for "canali" Lowell read not channels, but canals. Canals reflect the work of minds in collaboration, of life. For reasons that remain unclear, Lowell decided that Mars demanded his attention and devote his attention to Mars he did, with a passion and pocketbook few could match.

The tool Lowell built for his investigations—the Lowell Observatory in Flagstaff, Arizona—saw first light in April 1894, just a few weeks before Mars reached its biennial opposition with Earth. Lowell and his staff atop Mars Hill spent more than a decade studying and mapping the face of Mars. Lowell and his assistants mapped hundreds of canals. In their number and organization, Percival Lowell saw the history of an alien race trying to survive on an arid, dying world plainly writ.

Lowell captured the popular imagination with his sympathetic picture of an intelligent race of Martians trying to forestall its inevitable doom. The effect of his writings was amplified further by adventure writers such as Edgar Rice Burroughs, who used the Lowellian Mars as the setting for an extraordinary romantic Martian civilization that called its home planet "Barsoom." Burroughs's Mars novels featured swashbuckling heroes rescuing daring and beautiful princesses endangered by monsters, savages, and power-mad Martian tyrants, all set against a rich tapestry of life on Barsoom. In its Barsoomian incarnation, Lowell's Mars enchanted millions of readers.

Over the years though, neither Lowell's eloquence as a writer and speaker nor his energy and enthusiasm could defend his theories against the barbs of the astronomical community. The tide of opinion slowly turned against Lowell, as other observers using more powerful telescopes found no evidence whatsoever of canals. We now know that Lowell was absolutely wrong in his investigations of Mars, but he did leave an important legacy behind: he fired the imaginations of people to make them see a world on Mars. True, that world turned out to be wildly inaccurate, but its envisionment led to a massive uplifting of at least a segment of the popular mind, which three centuries after Kepler was and still is largely addicted to the ancient geocentric view of the Earth as the only world, orbited by tiny lights in the sky. Lowell made Mars habitable in the imagination only, but it is from imagination that reality is created. It was Lowell's works that inspired the pioneers of rocketry, including Robert Goddard and Herman Oberth, to begin their quest to develop the tools that would soon make the solar system accessible, not only to the eye, but to the hand of man. It was the spirit of Lowell that touched the rocky surface of Mars as *Viking* landed.

VIKING'S SEARCH FOR LIFE

Life had brought *Viking* to Mars. Though Lowell's visions had long since died, the idea that Mars might harbor some form of life had itself never died. Streaking by the planet in July 1965, the first spacecraft to visit Mars, the American *Mariner 4*, certainly quashed once and for all the

Lowellian vision of the Red Planet, revealing a barren, cratered surface, more Moon-like than Barsoom-like. Those who hoped for postcards from life's far edge got, instead, funereal images of an aged, dead planet, a *cosmic fossil* in science fiction author Arthur C. Clarke's words. During the summer of 1969, *Mariners* 6 and 7 confirmed their predecessors' findings. Science experiments confirmed *Mariner 4*'s atmospheric findings—the atmospheric pressure of the carbon dioxide-rich atmosphere was low, just 6–8 millibars. (A millibar is 1/1,000th of Earth's sea-level atmospheric pressure, so at 7 millibars, Mars' atmosphere was a bit less than 1 percent as thick as Earth's.) Temperatures measured near the south pole supported the notion that frozen carbon dioxide—dry ice—formed the polar cap. Mars, according to the *Mariner* flybys, was a cold, dead, cratered planet—not a place you want to linger. Then came *Mariner 9*.

Unlike the previous American spacecraft, *Mariner 9* would go into orbit around Mars. Where the early *Mariners* shot by the Red Planet and captured what information they could, *Mariner 9* and a companion spacecraft would map the planet's surface and observe planetary dynamics over a 60-day period. Unfortunately, that companion spacecraft, *Mariner 8*, ended up in the waters of the Atlantic shortly after launch in the spring of 1971. *Mariner 9*, though, lifted off flawlessly on May 30th, bound for Mars. Just days earlier the Soviet Union had launched *Mars 2* and *Mars 3*, combination orbiter/lander spacecraft. No great surprises arose on board the spacecraft as they sped toward their destination. The same couldn't be said for Mars.

On September 22, about two months before the *Mars* probes and *Mariner* were due to arrive, astronomers noticed a bright, white cloud begin to develop over the Noachis region of Mars. The cloud grew quickly, by the hour. Within days the cloud, now recognized as a dust storm, had enveloped the planet. As robotic eyes sped toward Mars, the planet pulled a shroud around itself. Far-encounter photographs of the planet captured by *Mariner 9* on November 12th and 13th showed a blank disk, save for a slight brightening near the south pole, and a few small, dark smudges above the equator. On the 14th, the spacecraft slipped into Mars orbit. *Mariner* gazed down on an essentially featureless planet. The probe's controllers rewrote the mission plans, allowing for some science experiments and photography to be undertaken, but, in essence, told the spacecraft to kick back and ride out the storm.

Mars 2 and *3* didn't have that option. Unlike *Mariner*, the Soviet program did not have adaptive operational capability. On arrival at Mars, the orbiters duly released their landers into the maw of the largest Martian dust storm ever recorded. Parachuting blindly through an atmosphere whipped by 160 km/hr winds, both probes hit the ground too hard for their airbag deceleration systems to save them. *Mars 2* was destroyed on impact; *Mars 3* managed to transmit 20 seconds of data after crashing, and then died.

The Soviet orbiters hardly fared any better than the descent probes. Nearly all data from *Mars 2* was lost because of poor telemetry, and *Mars 3* pulled into a wildly elliptical orbit about Mars, producing only one released photograph.

While the dust storm raged, and the Soviet probes met their respective fates, *Mariner 9* serenely orbited the planet, waiting for the dust to clear, both literally and figuratively. Toward the end of December and into early January 1971, the Martian skies started clearing, and *Mariner* began to return staggeringly vivid images of an unimagined world.

The small smudges *Mariner* imaged during far encounter could now be seen for what they were: enormous mountains whose tops *Mariner* had spied through the dust storm. A century earlier, optical astronomers had noted a bright region in the area of the largest of these massifs, and dubbed the region *Nix Olympica*, the Snows of Olympus. It was an apt name, as Nix Olympica proved to be the largest mountain in the solar system—Olympus Mons—looming some 24 kilometers above the Martian surface and covering an area about the size of the state of Missouri. Another region of Mars well-known to astronomers, the Coprates region, yielded surprises as well. Through the telescope, Coprates appeared as a dark, stubby, bright, cloudlike band. As skies cleared, *Mariner*'s audience of scientists realized they were looking at a dust cloud slowly settling into the bottom of a valley of, again, Olympian proportions. Now known as *Valles Marineris* (in honor of *Mariner 9*), this ragged scar stretches nearly 4,000 kilometers across the planet. Up to 200 kilometers wide and 6 kilometers deep, Valles dwarfs any similar feature on Earth (if need be, you could tuck the Rocky Mountains in one of Valles's side valleys; nobody would see them).

With each orbit of the planet, *Mariner* returned ever more astonishing information. The greatest surprise, though, proved to be images of sinuous channels (yes, canali!) that appeared to have been carved by running water—there were riverbeds on Mars.

Whatever romance the earlier *Mariners* had killed, *Mariner 9* renewed. The probe reinforced many of the earlier *Mariner* findings, but overturned others, including the notion that Mars was simply a knockoff of the Moon. Imagine the Martian globe bisected by a line running at roughly a 50° angle to the planet's equator. Below that line to the south lies the heavily cratered, ancient terrain *Mariners 4, 6,* and 7 discovered and recorded. North of the line, craters are few while evidence of more recent geological activity is plentiful. It just happened that the first three *Mariners* visited the south, offering no clues as to what other regions of the planet might reveal. *Mariner 9's* images (more than 7,000 of them) and data swept away the notion of the Red Planet as a cosmic fossil. Instead, *Mariner 9's* findings told the tale of a planet of fire and ice. In the distant past, Mars' surface had been geologically alive. Volcanoes had roared and resurfaced vast areas of terrain; internal mechanisms of some sort had fractured and split the landscape, lifting the Tharsis region (on which Olympus Mons stood) kilometers above the landscape; and water had flowed across the planet's surface in volumes large enough and for periods long enough to carve the face of the planet. Mars was once warm, wet, and alive with geologic activity. And that begged the question once again: Was Mars now, or perhaps in the past, bustling with biologic activity, with life?

To answer that question, astronomers and biologists found themselves stepping back from the concept of life on Mars to the simpler but still complex concept of, simply, life. What is it? If you can't define what life is, if you can't distinguish between life and nonlife here on Earth, you'll have a devilish time looking for it on a red dot 400 million kilometers distant. So, the search for life on Mars began with a review of the only known sample of life in the universe, terrestrial life. While terrestrial life comes in all forms, shapes, and sizes, its presence invariably causes changes to its local environment. These changes can be small, tiny even, especially if you're dealing with tiny life forms. But, no matter the size, life will still alter its environment simply by the fact of metabolism and respiration, the complex physical and chemical business of keeping something, anything, alive. Seal up an airtight box and the mix of gases (assuming there's no outgassing from the walls) will remain stable. Stick a cat in the same box and the mix will change pretty quickly (as will the state of the cat). So, if you're casting about for signs of life, establish a controlled environment, insert whatever sample you have, and then observe the changes, chemical or physical, inside the box. Chances are, any large changes will be attributable to biological processes. This, in essence, is what the scientists of the Viking project chose to do.

The *Viking* program was fairly straightforward in description—two orbiters, two landers, all to head for Mars in 1973 to search for life—but proved staggeringly difficult in execution. A budget squeeze delayed launch until 1975, which, in retrospect, was a hidden blessing, as the spacecraft simply would not have been ready by 1973 without, in the words of a *Viking* team member, "compromising both capability and reliability."

The four *Viking* spacecraft bristled with instruments for imaging, water-vapor mapping, thermal mapping, seismology, meteorology, and more, but the heart of the mission lay with the landers' biology packages. *Viking* engineers had packaged three biology labs weighing about 9 kilograms total into something that could sit quite comfortably in your bookcase.

The three experiments in the biology package operated on the same basic principle: seal some Martian dirt in a container with a culture medium, incubate it under different conditions, and then measure the gases emitted or absorbed. The experiments differed in the specific approaches they took to incubate samples and in what they sought to detect and measure as evidence for life. The *Viking* landers also carried an X-ray fluorescence instrument capable of assessing the elemental composition of the soil, and a gas chromatograph mass spectrometer (GCMS) capable of detecting and identifying organic compounds in the soil.

The search for life began on *Viking 1*'s eighth Martian day—"sol" 8 in the local time zone, July 28, 1976, here on Earth—as the lander extended its sampler arm, dragged it across the Martian surface, and delivered soil to the biology package. The three experiments received their small allotments of soil and set to work. Over the course of the next three days, incredibly, all three biology experiments reported powerful gas releases, positive signals for life, in some cases virtually immediately after exposure of the culture media to Martian soil.

The *Viking* biology team was, to say the least, stunned. Three experiments, three positive responses, three indications of life . . . maybe. The gas release signals were definite, but their suddenness of both onset and cessation had more of a ring of chemical reaction than biological growth. So caution was called for. The discovery of life anywhere in the solar system would have profound ramifications not just for the world of science but for the entire world community. Once again, as in Kepler's time, humanity would come to know its place in the universe more fully, more truthfully. We would know that while we are not the center of the universe, we are part of a phenomenon that is general throughout the universe. We would know that life owns the universe. This was, most definitely, no small announcement.

No one on the biology team was eager to rush out such an announcement, only to discover that he had jumped the gun. So conservatism prevailed, especially since many on the biology team had strong suspicions that the reactions witnessed were nonbiological in origin. One of the biology team's principal investigators, Norman Horowitz, stated his position quite clearly during a press conference announcing his own experiment's first positive readings. "I want to emphasize," he told an eager group of journalists, "we have not discovered life on Mars—not."

On sol 23, the gas chromatograph mass spectrometer analyzed a sample of Martian soil and found not a trace of organic carbon in the sample. After the reactions recorded by the three biology packages, this came as an enormous surprise and heightened the debate. Scientists had expected the GCMS to find at least some trace of organic compounds of nonbiological origin, such as materials from meteorites. In fact, that was a concern surrounding the GCMS—how to tell biologic organics from nonbiologic. But now, with the GCMS recording absolutely no evidence of organics in Martian surface soils, the search for life on Mars became for some a search for processes that could reconcile the discovery of an evidently lifeless Mars with the biology results.

On September 3, *Viking 2* settled down on to the Utopia Planitia, nearly halfway around the planet, some 6,400 kilometers distant from the *Viking 1* landing site and about 25° farther north. The biology experiments and the GCMS were soon up and running, investigating soils that appeared to be slightly moister than samples from the Chryse site. Again, results from the biology experiments gave positive responses that appeared to be more indicative of chemistry to some, and the GCMS found no trace of organic carbon. Again, the results caused a stir, with some investigators holding out for biology, others chemistry. Again, the results highlighted a basic problem: the *Vikings* could perform four experiments and only four, and three were saying "maybe life," while the other was saying "very doubtful." If the soil samples had been in a terrestrial lab, dozens of additional experiments could have been performed to resolve the argument definitively. On Earth, the samples could even have been incubated in a culture medium and the results observed directly with a microscope. But in *Vikings* limited four-experiment lab on Mars, none of this was possible. In essence, we were left with contradictory results. In the words of writer Leonard David, "*Viking* went to Mars and asked if it had life, and Mars answered by replying 'Could you please rephrase the question?' "

Today, most researchers—but definitely not all—feel that the *Vikings* did not find evidence of life. Instead, they have drawn a picture wherein the Martian soils are rich with peroxides and superoxides. According to this theory, the results from at least two of *Vikings* experiments were the signature of chemical reactions involving these peroxides. The failure of the GCMS to detect any carbon at either site fit neatly with the peroxide/superoxide theory because peroxides destroy organic matter with abandon. But not everybody buys this, with some suggesting that perhaps the GCMS was not sensitive enough to detect vanishingly small amounts of organic material, that is, life. When cultured in *Viking*, it is conceivable that such sparse spores could quickly multiply into a population large enough to give off the positive signals. Similarly, the abrupt end of the signals displayed by the biology packages, easily explained by the chemistry side as the exhaustion of the peroxide supply, could also be explained by an overmultiplying population of organisms in the soil sample poisoning itself with its own wastes. Gilbert Levin, principal investigator for the biology test package called the Labeled Release experiment, to this day believes passionately that his equipment detected evidence for Martian life. A decade after the *Viking* landings, Levin would write ". . . after years of laboratory work trying to duplicate our Mars data by non biological means, we find that the preponderance of scientific analysis makes it more probable than not that living organisms were detected in the LR [labeled release] experiment on Mars. This is not presented as an opinion, but as a position dictated by the objective evaluation of all relevant scientific data."[2] A scant twenty pages earlier in the same volume, another biology team member, Norman Horowitz, writes, "For some Mars will always be inhabited, regardless of the evidence. . . . One does not have to search far to hear the opinion that somewhere on Mars there is a Garden of Eden—a wet, warm place where Martian life is flourishing. This is a daydream."[3]

My own feeling is that Horowitz is too harsh in his assessment of the possibility of Martian life and Levin a bit too enthusiastic. The best bet is that the *Vikings* did not detect life *in the surface soil of Mars.* The reason for this is that there is no liquid water there, and virtually no organics, and so while abstract arguments can be made for the sparse spore hypothesis, it seems almost impossible to construct a rational theory explaining how the life cycle of these putative Martian surface organisms would function. Furthermore, since Mars has very little in the way of an ozone layer in its atmosphere, the surface is bathed in ultraviolet light of sufficient intensity to do a pretty good job of sterilizing the planet's surface of microorganisms. However, regardless of Horowitz's opinion, this does not rule out the possibility of a microbial Garden of Eden *below* the surface. In fact, if terrestrial life has taught us anything, it's that life flourishes not only in Garden of Eden environments, but in hellish ones as well. Indeed, there are families of bacteria known as chemotrophs that derive their energy from various inorganic chemicals as opposed to sunlight (like plants) or organic nutrients (like us). A small group that is adapted to temperatures of 70° to 90° centigrade and lives happily by oxidizing sulfur for their energy requirements would probably feel right at home in some underworld environments that, as more recent discoveries would show, almost certainly exist on Mars. Across our own globe, in the most extreme environments imaginable, scientists have discovered life tenaciously hanging on, making do with scant resources. In the Antarctic, colonies of lichen flourish within surface rocks, protected from the harsh environment by a centimeter or so of porous sandstone. Vast colonies of microorganisms thrive around the mouths of deep-sea vents that spew founts of boiling, mineral-rich water. There are organisms that thrive only in heat, others only in cold; some that grow only in alkaline conditions; others only in acidic environments; some that feed on sulfur; others on iron; others on hydrogen. Not only can life survive extreme environments, it appears it can also survive over unimaginably long time spans. In the late 1980s, a research group in Britain discovered that a group of salt-tolerant microbes called *halobacteria* could become trapped within rock salt and survive for months at a time in their tiny, briny homes. Intrigued, the group set out and collected samples from a natural subterranean salt deposit that dated from the Permian period, more than 230 million years ago. Again, they discovered tiny, fluid-filled cavities within the rock salt, and within a small fraction of these cavities (6 out of 350) they discovered viable halobacteria that could be cultured in the laboratory after a time span of more than 200 million years.[4]

All creatures great and small surviving in extreme environments have one thing in common: Their environment includes a source of water, however meager. The fact that Mars shows a remarkable amount of evidence of both surface and subsurface water in its distant past argues for the possibility of life in the past or perhaps even now in an unexpected "Garden of Eden." Host environments to such life could be thermal hotspots, such as hot springs; subsurface hotspots; subsurface permafrost deposits; subsurface or near surface brines, or perhaps even areas with evaporite deposits, such as the salt formation that was home for millions of years to earthly bugs. Many geologists believe that Mars does have a liquid water table, at least in certain places, perhaps a kilometer or so beneath its surface. Perhaps life which evolved on the surface of the planet in the distant past when it was warm and wet has retreated there. Recently, investigators in the state of Washington discovered a species of bacteria living deep underground, subsisting on the chemical energy derived from the reaction of cold ground water with basalt. There does not seem to be any particular reason for believing that similar organisms could not survive equally well in the subsurface environment that is hypothesized to exist on Mars. The point is that life is tough, even if it may be hard to find on Mars. No one expects to find herds of six-legged Barsoomian thoats thundering across Martian dunes. But life on the level of microorganisms, living in sheltered environments, that's another matter. It could be there now, or may have been there once. To find it will take more than robotic probes with limited mobility, dexterity, and perception.

AFTER *VIKING*

The *Viking* orbiters and landers kept on with their science observations long after the biology experiments came to a close. Orbiter *2*'s last transmission came on July 25, 1978, followed by the demise of Lander 2 on April 11, 1980, nearly two years later. Orbiter *1* sent its last signal on August 17, 1980, while Lander *1* signed off on November 5, 1982.

The Soviet space program attempted two launches in 1988 to explore Mars and its moon Phobos that met with disappointment, continuing a streak of bad luck that has plagued every Soviet or Russian Mars mission. (Out of more than sixteen attempts, none has been successful.) The United States' Mars program has also had to deal with failures. The *Mars Observer* spacecraft carried seven instruments intended to investigate Mars over the course of a Martian year. The mission would rewrite the books on Mars, or so researchers hoped. But just days before the spacecraft was due to enter orbit around the Red Planet, it fell silent. In attempting to reconstruct what may have happened, engineers have surmised that a fuel line ruptured as the spacecraft prepared to fire up its engines to slip into Mars' orbit. Whatever the cause, after a seventeen-year hiatus, America's exploration of Mars appeared headed for the deep freeze.

Fortunately, instead of using the demise of *Mars Observer* as a pretext for thrashing NASA's Mars exploration budget, members of Capitol Hill looked kindly upon continuing the legacy of exploration that the *Vikings* exemplified, though with a twist. With a new focus on "faster, cheaper, better" methods of accomplishing planetary exploration, NASA fashioned a decade-long program of Mars exploration out of the *Mars Observer* failure. Instead of launching a single massive spacecraft to the Red Planet, America's new plan sent a series of small spacecraft to orbit and land on Mars. This program started in late 1996 with the launch of the *Mars Global Surveyor* (MGS) spacecraft and the *Mars Pathfinder* mission. About half the size of *Mars Observer*, the *Surveyor* began mapping the Red Planet from polar orbit in March 1999 and continued to do so successfully through 2006. Among its discoveries have been altimetry data revealing a large basin of depressed and relatively uncratered terrain in Mars' northern hemisphere that is flatter than anything on Earth except for the sea bottoms, indicating the previous existence of a northern ocean.[5] Possibly even more exciting are a pair of photographs that MGS took of the same crater in 2001 and 2005, which show the appearance of a new water erosion gully during the period between them.[6] This could only have been created by a transient outflow of water from the crater side sometime between 2001 and 2005, thus proving the existence of subsurface reservoirs of liquid water on Mars today. *Mars Pathfinder* landed on Mars on July 4, 1997, with the help of parachutes, braking rockets, and airbags. Surviving several 40 to 60 miles per hour bounces along the surface, *Pathfinder* opened up and released a tiny rover dubbed *Sojourner* (after antislavery heroine Sojourner Truth). *Sojourner* then traveled for two months about the Ares Valles runoff channel landing site, collecting geological information and making the notable water-indicative discoveries of rounded cobbles and conglomerate rocks.

While the U.S. robotic Mars exploration program accelerated following its twin successes of 1996–97, budgetary difficulties and bad luck threw the Russian program into chaos. Russia's most recent attempt, entitled *Mars 96*, aimed to place a spacecraft in orbit around Mars, as well as two small science stations and two ground penetrators on the Martian surface, but was thwarted by a launch vehicle failure in the fall of 1996. This caused the indefinite postponement of a second mis-

FIGURE 2.1
Before and after photos of the same crater taken by the Mars Global Survey (MGS) orbiter in 2001 and 2005 show new erosion features created during the interim, indicating the presence of a subsurface aquifer capable of producing a transient outflow. (Photos Malin Space Science Systems/NASA)

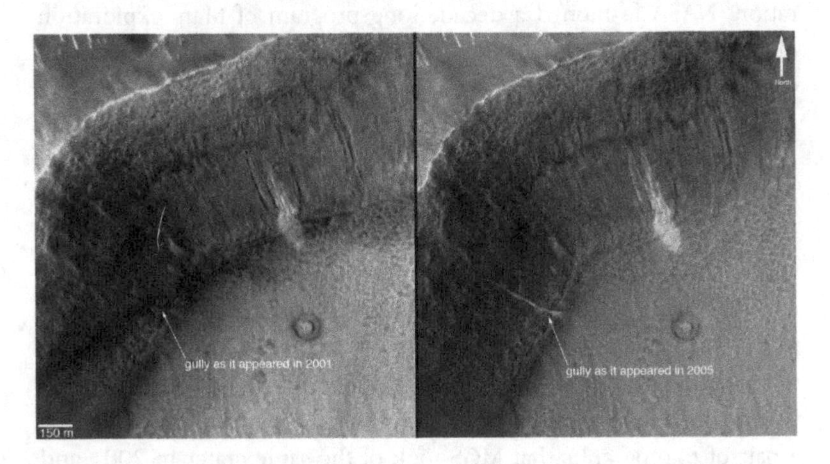

sion, *Mars 98*, which was to deliver an orbiter, rover, and balloon to the planet. The Russian *Marsokhod* rover would have dwarfed the American *Pathfinder* rover and, instead of venturing just 10 meters away from its landing site, could have logged nearly 50 kilometers. Trailing an instrument-laden "snake," the balloon, a product of the French space agency CNES, was designed to soar as high as 4 kilometers into the Martian atmosphere during the day, but settle toward the ground during the Martian night. Designed for a ten-day flight, the balloon would have been able to rack up several thousand kilometers in its windborne wanderings across the Martian surface. As Russia's economy continued to lurch through the next decade, however, hope for this mission has dimmed, and it is now questionable whether it will ever fly.

The U.S. Mars exploration program also had its share of bad luck, losing both its *Mars Climate Orbiter* and *Mars Polar Lander* due to failed orbit capture and landing maneuvers, respectively, during the fall of 1999. NASA pushed on with its decadal plan, however, successfully putting the *Mars Odyssey* spacecraft into orbit around the Red Planet in October 2001. Still operational a decade later, *Mars Odyssey* has used its infrared camera system and gamma ray spectrometer to map the min-

FIGURE 2.2
Mars Exploration Rover Spirit *operating on the Martian surface. (Artwork courtesy NASA/JPL)*

eral content of the Martian surface, discovering, among other things, continent-sized high-latitude regions where the soil is over sixty percent water by weight.[7]

Following up on this success, NASA aimed two medium-sized Mars Exploration Rover (MER) vehicles at the Red Planet in mid-2003. Arriving at Mars six months later, both rovers survived airbag-cushioned landings that set them down at two widely separated spots on Mars. The rover *Spirit* landed January 3, 2004 on the floor of Gusev Crater, a Connecticut-sized feature 15° south of Mars' equator. Gusev intrigued mission planners because of a 900-kilometer long meandering valley that cuts into the crater's rim. The valley appears to have been carved by flowing water in ages past and may have once filled the crater with water. *Spirit* would be looking for present evidence of a watery past within Gusev. As it turned out, *Spirit* would have to do some hard traveling from its landing spot to find that evidence. The second rover,

Opportunity, landed three weeks later nearly halfway around the planet in an area called Meridiani Plenum, one of the smoothest and flattest spots on the planet. Again, mission planners followed evidence of water in selecting the landing site. Instruments aboard Mars Global Surveyor indicated that the area was rich in the mineral gray hematite. This form of iron oxide is found on Earth and usually forms in wet environments. As *Opportunity* began returning its first images, the MER team was both amazed and delighted to realize that the rover had landed, rolled into a small crater, and sat on the surface facing an outcrop of layered rock. Few could imagine a sweeter spot to land.

Both rovers were slated to undertake 90-sol missions. Before their nominal mission time was up, both found hard evidence of liquid water in Mars' past. In March 2004, the Mars Exploration Rover science team

FIGURE 2.3
Spirit *casts its shadow on Mars. (Photographs Malin Space Science Systems/ NASA)*

announced that *Opportunity* had discovered solid evidence for liquid water in both the composition and morphology of rocks the rover inspected. A few days later, the team reported that *Spirit* had found salt deposits, remnants of an ancient shoreline in Gusev crater. With these successes under their respective belts or motherboards, the two rovers really got down to work.

Over the course of more than six years, the rovers continued science investigations, scuttling about the surface of Mars while surviving Martian winters, dust storms, mechanical failures, and transient bouts of amnesia. *Spirit* lived up to its name, overcoming numerous difficulties including a front right wheel that began to fail in June 2004. The finicky wheel prompted the rover's driving team to drive the rover backward, with the cranky wheel dragging along the surface during forays over smooth ground. They have dusted off rocks, ground the surfaces of boulders, examined soils in microscopic detail, witnessed transits of the moons of Phobos and Deimos, captured images of dust devils skipping across the surface of Mars, and even watched for meteors in Martian skies. The library of raw images from the rovers is vast, ranging from exquisite panoramic landscapes to close-ups of tiny grains of sand. They're available for viewing over the internet, assuming you have time to flip through more than a quarter of a million snapshots.[8]

In May 2009, *Spirit*'s wheels broke through a crusty surface layer and sank into loose sand. The sand eventually proved too much of a trap, and NASA declared the tough little rover a stationary science platform after it had logged 7,730 meters of travel (4.8 miles). Even so, researchers examined the soil *Spirit* had disturbed as it tried to free itself, and a detailed study of the exposed soil layers once again revealed evidence of water in Mars' past. Just short of a year later, the rover fell silent, its last transmission occurring in March 2010. Meanwhile, *Opportunity* keeps plugging away. *Spirit*'s robust twin has spent years rolling from one intriguing landmark to the next. As of late 2010, it was more than halfway through a long-range, 15 kilometer trek to Endeavour crater. As long as it survives another winter, and as long as the Martian winds occasionally blow its solar panels clean, it seems *Opportunity* will keep on rolling. While the rovers' accomplishments have proved impressive, their most stunning achievement may have been to get the

FIGURE 2.4
Opportunity's *tracks across Mars, (a) as seen by the rover, and (b) as seen by the* Mars Reconnaissance Orbiter (MRO). *(Photos Malin Space Science Systems/* NASA)

United States Congress to actually agree on something. In March 2009, by unanimous vote, the Congress gave the rovers a pat on their backs, recognizing ". . . the success and significant scientific contributions of NASA's Mars Exploration Rovers."

In 2004, the first European probe reached the Red Planet An ambitious mission, *Mars Express* included both a French- and Italian-made orbiter, and the British lander *Beagle II.* Although the *Beagle* crash-landed, the orbiter proved successful, returning mountains of data including the first detection of trace quantities of methane in the Martian atmosphere.[9] While initially disputed, these measurements have since been conclusively confirmed. In 2009, a team of researchers based at NASA's Goddard Space Flight Center announced that they had not only confirmed the presence of methane in Mars' atmosphere from ground-based observations, but had discovered multiple, "substantial plumes" of the gas that blossomed during the warmth of the spring and summer seasons. Located in the planet's northern hemisphere, the plumes were seen over areas that bear evidence of ancient ground ice or flowing water.[10]

To add to the intrigue, the research team determined that there exists on Mars a mechanism that removes methane from the atmosphere at a rate much higher than can be explained by photochemical destruction by ultraviolet light alone. Rather than taking centuries, something destroys Mars' atmospheric methane in as little as four Earth years, perhaps even as quickly as just a half year. Thus the presence of methane on Mars today demonstrates that there must be a source creating methane on Mars, today. This can only be explained by biology or hydrothermal geology, meaning that there is either life, or, at minimum, a subsurface environment friendly to life. Thus, if human explorers were to drill down and sample such environments, the chances are excellent that they might find life. Examining such organisms in their laboratory, astronauts would then be able to determine whether Martian life followed the same biochemistry exhibited by all life on Earth, or developed along different lines altogether. All Earth life—whether mushrooms, people, crocodiles, or bacteria—exhibits identical biochemistry, incorporating the same amino acids, and using the same RNA/DNA method of transmitting structural information from one

FIGURE 2.5
Mars Reconnaissance Orbiter (MRO). *(Artwork courtesy NASA/JPL)*

generation to the next. But does life have to operate that way? Is life as we know it on Earth the pattern for all life everywhere? Or are we just one peculiar example drawn from a much vaster tapestry of possibilities? These questions are key to our understanding of the nature of life itself. *Mars Express*'s methane discovery tells us that their answers may wait for us on the Red Planet.

NASA's *Mars Reconnaissance Orbiter* (MRO) pulled into orbit in 2006, and began mapping the planet with a camera good enough to actually see the *Spirit* and *Opportunity* rovers from space, and guide them in their travels. With the aid of its photos, we can now select boulder-free landing sites to help ensure the safety of future explorers, both robotic and human. In 2008, the *Phoenix*, so named because it was built using the spare left over from the failed 1999 *Mars Polar Lander* program, redeemed that lost cause by landing successfully on the North Pole of Mars. Since their discovery in 1672 by Christian Huygens, the composition of Mars' bright polar caps have been a subject of scientific dispute.

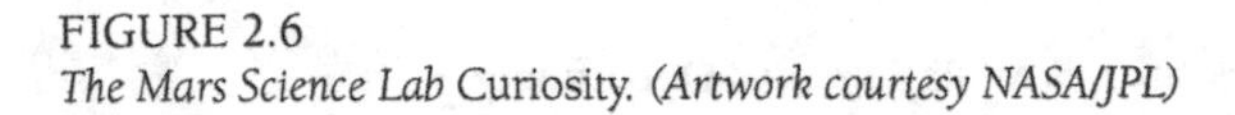

FIGURE 2.6
The Mars Science Lab Curiosity. *(Artwork courtesy NASA/JPL)*

Phoenix found pure-water ice, settling the matter once and for all, at least for the planet's north.

The next NASA mission to go to Mars will be the *Mars Science Laboratory.* Recently renamed *Curiosity* (on the suggestion of Kansas sixth-grader Clara Ma), the probe is scheduled to launch in November 2011 and land in August 2012. Powered by a radioisotope generator that will give it the ability to operate regardless of sunlight or season, *Curiosity* will carry 11 times the payload of scientific instruments wielded by *Spirit* or *Opportunity,* and be able to travel much further and faster. A sense of the advance (and investment) this mission represents may be gathered by the figures in Table 2.1, which compare *Curiosity* to its predecessor MER rovers.

Curiosity will carry a suite of cameras, allowing it to take three-dimensional, true-color photographs and movies, developed by the imaging experts at Malin Space Science Systems, with input from *Avatar* filmmaker James Cameron. Its robotic arm will be equipped with a microscope good enough to allow it to see fossils of microorganisms should they exist in the rocks or soil it encounters. It will also be armed with a Los Alamos lab laser capable of vaporizing rocks from up to 7 meters away, and a French spectrometer to analyze the chemi-

TABLE 2.1
Comparison of Curiosity *with MER's* Spirit *and* Opportunity

Specification	MER	*Curiosity*
Vehicle Length	1.57 m	2.7 m
Vehicle Mass	174 kg	900 kg
Instrument Mass	6.8 kg	80 kg
Power Supply	Solar	Radioisotope
Average Power	24 W	125 W
Average Vehicle Speed	100 meters/day	300 meters/day
Flash Memory	256 MB	2 GB
Computer Speed	35 MIPS	400 MIPS
Landing System	Airbags	Rockets
Cost	$400 million (each)	$2300 million

cal composition of the vapor thus produced. Additional instruments to determine the elemental composition and mineralogy of soil samples include an Canadian alpha-particle X-ray spectrometer, a Russian neutron spectrometer, and an American X-ray diffraction/X-ray fluorescence instrument. It will also carry a very advanced gas analyzer, jointly developed by NASA and the French Space agency, CNRS, that will not only sniff for traces of organic gases, such as methane, in the Martian atmosphere, but be able to distinguish on the basis of isotopic composition whether such gases are of geochemical or biological origin. A Spanish meteorological package will allow it to measure atmospheric humidity, pressure, wind velocity and direction, air and ground temperature, and ultraviolet radiation. Finally, *Curiosity* will be equipped with an American/German instrument called *RAD*, which will measure and characterize the radiation spectrum on the Martian surface for the purpose of preparing the way for human explorers.

Curiosity thus promises to be a very high-payoff mission, but also a high-risk one, as NASA is defying the many-small-probes-instead-of-one-big-one wisdom it learned after the *Mars Observer* failure. Indeed, the program has already experienced one near-death experience, even before launch, when, in 2008, NASA's science chief got cold feet and

FIGURE 2.7
Duplicates of three generations of Mars rovers on display at Jet Propulsion Lab. At center is little Sojourner, *which landed in 1997. At left is one of the MER rovers,* Spirit *and* Opportunity, *which landed in 2004. At right is* Curiosity, *scheduled to land in 2012. (Photo courtesy NASA/JPL)*

attempted to pull the plug on the mission on the basis of a projected 20 percent cost overrun (after the other 80 percent had already been spent). Only a sharp counterreaction by the mission's defenders, including myself, and the gutsy willingness of NASA's tough administrator Mike Griffin to take responsibility saved the day. So, we'll all be holding our breath when *Curiosity* faces her launch and landing trials by fire.

Also on deck for launch in 2011 is the joint Russian-Chinese-Finish mission called *Fobos-Grunt* ("Phobos Ground.") Launched by Russia and controlled from a center in Ukraine, this mission will include China's first interplanetary probe, the Yinghuo-1 orbiter (equipped to study Mars' ionosphere and magnetosphere), two Finnish "Metnet" meteorological Mars landers (take that, you exploration wallflower nations), and a Russian vehicle that will land on Mars' moon Phobos, study it with a Russo-Chinese instrument set, gather a soil sample, and return with it to Earth. If successful, this will thus be the first interplanetary sample-return mission, potentially paving the way for follow-on Russian sample return flights to the asteroids, comets, and the moons of the outer planets.

More robotic missions are planned during the coming decade, including NASA's 2013 *MAVEN* mission to study Mars' ionosphere and atmosphere, and the 2016 *Mars Science Orbiter*, which will search for the methane vents that may lead us to find the underground homes of Martian life. Also scheduled for the 2016 timeframe are the twin *ExoMARS* rovers, developed in collaboration by NASA and the European Space Agency ESA as part of their Mars Joint Exploration Initiative, which will search for the signatures of past or present life on the Martian surface.

Many creative ideas for additional future missions are under discussion, including the *Mars Aerial Platform* mission, or MAP, developed by myself and others at Martin Marietta. MAP is a conceptual design for a low-cost mission that would return tens of thousands of high-resolution photographs of the Martian surface, analyze and map the global circulation of the atmosphere, and examine Mars' surface and subsurface with remote sensing techniques. At the heart of the mission is a high-tech approach to a very low-tech concept—balloons.

This is how MAP would work: A single Delta-class booster would launch the MAP payload on a direct trajectory to Mars. The payload would consist of a spacecraft carrying 8 entry capsules, each capsule packed with a balloon, deployment equipment, and a gondola carrying science instruments. Ten days prior to arrival at Mars, the spacecraft, now spinning like a top, would release the capsules, casting them off in directions that would ensure their entry at widely dispersed locations. As each capsule began its descent through the atmosphere, a parachute would deploy to slow the capsule to the point at which a balloon could be inflated. Each would be made of a commercially available material known as biaxial nylon 6, which is just 12 microns thick, one-third the thickness of a standard plastic trash bag. Though seemingly made of gossamer, these balloons will be surprisingly tough. The material's manufacturing process guarantees that it harbors no pores, which means that balloons made of this nylon simply will not leak and, therefore, can be expected to remain inflated not for days, but years. Following inflation, the parachute, capsule, and inflation equipment drop away, carrying a meteorology package to a soft landing on the Martian surface. Free of extraneous equipment now, each balloon begins the first of perhaps hundreds of days of roaming the highways of Mars, the planet's eternal winds.

The 18-meter-diameter balloons will cruise over the surface of Mars at an altitude of 7 to 8.5 kilometers and, unlike the French balloon that had been planned for Mars 98, will maintain these altitudes day and night. They will be able to do this because, with their new material and compact configuration (enabled by a very lightweight gondola), the balloons will be strong enough that when their gas pressure is increased by the heat of the day they can just hold it in without venting. Since they won't vent gas by day, these "superpressure" balloons have no need to drop ballast at night, and therefore can fly almost forever at constant altitude. Current models of Mars' atmospheric dynamics suggest that the winds will carry the balloons primarily in a west-east direction on the order of 50 to 100 kilometers per hour. At these speeds, each balloon could circumnavigate Mars every ten to twenty days and, assuming a conservative average mean time to failure of a hundred days, we can expect each balloon to circumnavigate Mars at least four times. Each balloon will carry an 8-kilogram instrument package containing atmospheric science instruments, data recording and transmitting equipment, a rechargeable battery, a solar array, and an imaging system, the heart of the package. Comprised of two sets of optics—one for high-resolution images, another for medium-resolution images—the imaging system will significantly advance our understanding of Martian geology, while allowing us to identify landing sites for future missions and possible areas that offer promise for investigations of past or present Martian life. The best *Viking* orbiter images reveal surface features the size of a baseball diamond; *Mars* Global Surveyor images reveal surface details the size of a large car; those from MRO can find the *Opportunity* rover; the cameras MAP carries will be able to detect surface features the size of a cat (which is not to say that they will reveal Martian cats). Every fifteen minutes during daylight hours, the cameras carried by each balloon will take two simultaneous photographs:

one a black-and-white, high-resolution photograph, and the other a color, medium-resolution photograph centered on the area imaged by the high-resolution camera (the latter photograph can be used to determine the location of the high-resolution photograph on a map of the planet). MAP will return a staggering number of photographs. For each hundred days, an eight-balloon fleet sails the Martian winds, MAP will return 32,000 high-resolution photographs, along with an equal number of context photographs at resolutions superior to the best *Viking* images.

MAP will return an avalanche of science data that will transform our understanding of Martian geology and meteorology, atmospherics and geomorphology. Engineers and scientists will have the data at hand to help them design new missions, identify sites for exobiological investigations and, perhaps, prospect for water when exploring the Martian surface. But the greatest return from MAP will be the least tangible: its impact on the intellectual life of humanity at large.

Today, nearly five hundred years since Copernicus and Kepler, Brahe and Galileo, most people still think of Earth as the only world in the universe. The other planets remain mere points of light, their wanderings through the night sky of interest to a select few. They are abstractions, notions taught in schools. The MAP cameras offer the possibility of taking humanity's eyes to another planet in a way that has never been done before. Through the gondola's cameras we will see Mars in its spectacular vastness: its enormous canyons, its towering mountains, its dry lake and river beds, its rocky plains and frozen fields. We will see that Mars is truly another world, no longer a notion but a possible destination. And, just as the New World entranced and enticed mariners here on Earth, so can Mars entice a new generation of voyagers, a generation ready to fashion the ships and sails proper for heavenly air.

THE MARS SAMPLE RETURN MISSION

The Holy Grail of robotic Mars exploration programs is the *Mars Sample Return* (MSR) mission. If only those *Viking* samples had been in one of our labs, we could have subjected them to a battery of tests and examinations that would have left no doubt in interpretation of results. Well then, why not bring a sample back? Currently, NASA's solar system exploration branch has penciled in precisely such a mission for 2020.

There are three ways this might be done. The first, and conceptually the simplest, is the brute-force method. In this case, a launch vehicle in the 30-tonne to orbit class is used to deliver to the surface of Mars a very large payload consisting of a miniature rocket ship, massing perhaps 500 kilograms, completely fueled for an ascent from Mars and flight back to Earth. The lander also has on board a robotic rover that is dispatched to wander about (under human remote control) and collect geologic samples. The samples are then loaded aboard a capsule on the rocket vehicle. When the launch window from Mars back to Earth opens up, about a year and a half after arrival, the rocket blasts off and flies back to Earth. Upon approach to Earth eight months later, the capsule separates from the rest of the rocket vehicle and performs a high-speed reentry, much in the manner of an Apollo manned capsule. Depending upon design, the capsule may be decelerated by a parachute or simply use a crushable material like balsa wood or Styrofoam to cushion the landing shock when it hits the targeted desert landing area.

This brute-force mission is fairly simple conceptually, but the problem is that it is likely to be very expensive, as robotic exploration missions go. The required launch vehicle exceeds the capability of the existing Atlas V-heavy, and the development of both it and the large entry and landing system needed to deliver a fully fueled ascent vehicle to the surface are likely to be very costly. Thus, the brute force approach has always led to cost estimates that have made the mission a nonstarter. In an effort to reduce costs, several other methods have also been studied.

One of the most popular alternatives to the brute-force plan is the Mars Orbital Rendezvous, or MOR plan. In this scheme, two spacecraft are sent to Mars, each launched by a comparatively low-cost ($55 million each) Delta 2 booster. One of the launches delivers to Mars orbit an Earth return vehicle (ERV) and entry capsule, and the other delivers to the Martian surface a fully fueled Mars ascent vehicle (MAV), equipped with a rover and sample can. The rover is deployed to collect samples which are placed in the sample can. When this is completed, the MAV takes off and flies to Mars orbit where it performs an autonomous rendezvous and dock with the ERV The sample can is then transferred from the MAV to the reentry capsule on board the ERV. The two craft then separate, the MAV to be expended and the ERV to wait in Mars orbit until the launch window back to Earth opens up, at which point it fires its engine to send it on a trans-Earth trajectory. The rest of the mission is then performed in the same manner as the brute-force approach.

The main talking point of the MOR plan is that it considerably lowers launch costs relative to the brute-force scheme. Since the MAV only has to fly to Mars orbit, and not all the way back to Earth, and moreover only has to lift the sample can and not the complete reentry system, it can be made much smaller than the ascent vehicle used in the brute-force scheme. Thus the lander required to deliver it can be made smaller, lighter, and cheaper, and a much less muscular launch vehicle can be used to send it to Mars. However, there are major problems associated with the MOR scheme. In the first place, two separate launch vehicles are needed, which doubles the risk of launch failure causing mission failure. Also, two complete spacecraft are needed, each of which has to be designed, built, checked out, and subjected to launch environment testing (when you launch a spacecraft it is subjected to severe vibration and acoustic loads, and these must be simulated in expensive setups before launch), and each must be integrated into a launch vehicle. Basically, doing all this will double mission costs. Furthermore, the interfaces between the two spacecraft must be perfect, not only in the factory, but after launch and years of space flight and thermal cycling both in space and on the Martian surface. Guaranteeing this is a very tough design problem; in fact it probably can't be guaranteed since it can't be tested in advance. Finally, the autonomous rendezvous, dock, and sample transfer in Mars orbit required to do this mission is an undeveloped technology which will be very costly to develop and which *cannot be tested in advance of the mission.* This multiplies the risk associated with this already marginal mission plan still more.

In an effort to make the MOR plan look more attractive, its advocates have taken to innovative accounting techniques, such as assigning the cost of the two required launches to separate missions. In more extreme proposals the rover would be flown out on a prior mission, so that its costs and the costs of its mission operations can be charged to someone else. In this case, the lander carrying the MAV now must satisfy the additional requirement of performing a landing with nearly pinpoint accuracy next to the rover. Once again, this cannot be tested in advance, yet it would necessitate a drastic improvement of the state of the art for targeting unmanned Mars landers, which currently involve landing errors of up to 100 kilometers. For the sake of novelty, apparently, some orbital rendezvous fans have also proposed moving the location of rendezvous from Martian orbit to interplanetary space. This saves propellant on the ERV because now it does not have to capture into or blast out of Mars orbit, but it adds not only a considerable amount of propellant to the MAV but also an untestable requirement that the MAV be able to blast off at exactly the right moment to catch and rendezvous in deep space with an ERV that is zooming past Mars at a speed of 5 km/s. This could be very tough to guarantee, from the point of view of MAV engineering systems alone, putting aside the possibility of bad weather on the appointed take-off date.

So, if the brute-force plan is too costly and the MOR scheme is too risky, what's left?

What's left is a third plan, that I, along with engineers Jim French, Kumar Ramohalli, Robert Ash, Diane Linne, and several others, have been advocating for some years now. This third plan is known as the Mars Sample Return with In-Situ Propellant Production, or MSR-ISPP

In the MSR-ISPP plan a single Delta 2 is used to send a single unfueled Mars ascent vehicle (MAV) to the Martian surface together with a rover. While the rover is collecting samples, the MAV employs a small onboard chemical plant to turn gas pumped in from the Martian atmosphere into rocket propellant (I favor methane/oxygen, though carbon monoxide/oxygen has also been proposed), filling the tanks of the MAV. By the time the launch window back to Earth opens, all the propellant needed for the return flight has been made, and with the samples all collected, the MAV takes off and flies directly back to Earth, just as in the case of the brute force mission. The direct return to Earth is possible with a Delta-launched spacecraft because the Delta and its lander only had to deliver the MAV's dry mass (perhaps 70 kilograms) to the Martian surface, instead of the much larger wet mass needed to perform the brute force mission.

The MSR-ISPP mission is by far the cheapest of the mission plans discussed, because instead of employing a new 30-tonne to orbit launcher with one large spacecraft, or two Deltas with two small spacecraft, it can be flown on a single Delta with one small spacecraft. It is also much lower in risk than the MOR plan, because the advanced technology required, the in-situ propellant production (ISPP) plant, *can be fully tested in advance* in Mars simulation chambers on Earth. In addition, the ISPP unit represents a system of a much lower order of complexity (essentially nineteenth-century chemical engineering) than the avionics required for autonomous Mars orbit rendezvous, let alone deep space rendezvous. As noted earlier (and as I will discuss in more detail later), at Martin Marietta, we built and demonstrated successful operation of a full-scale MSR-ISPP unit making both methane and oxygen for $47,000—an amount of money that would be "in the noise" in an MSR mission budget. Now it's true that the Martin ISPP machine was a working brassboard test device, not mature flight hardware, but what needs to be understood is that the issue of mission risk associated with a new technology is not one of maturity, it is one of testability. Because it is testable, ISPP technology is much lower risk than the in-space rendezvous technologies required for the MOR mission. Furthermore, if it is decided to use two spacecraft on the MSR-ISPP mission, they will be identical spacecraft (and therefore cheaper than the two different spacecraft needed for the MOR mission), and, if either one makes it back, the mission is a success. In contrast, in the MOR mission, if either spacecraft fails, the mission is lost.

As we will see, using in-situ produced propellant is also the only way to make human exploration of Mars affordable. As far as MSR mission planning is concerned, that should be decisive in determining strategy. The MSR mission's value will be greatly increased if it can be used to demonstrate the key technology needed for human flights to Mars. Consider this: The MSR mission will only be able to return a kilogram or so of samples gathered from the surface of Mars within at best a few kilometers of the landing site. Since it is unlikely that there is life today on the Martian surface, the above ground search for Martian biology will largely be a search for fossils. Robotic rovers with their limited range and long communication time delay (up to 40 minutes due to speed limitations of radio signals) in Earth-Mars command sequence data transmission are a very poor tool for conducting such a search. If you doubt that, consider parachuting rovers such as *Spirit* or *Curiosity* into the Rockies. It is likely that the next ice age would arrive before one of them found a dinosaur fossil. Fossil searches require mobility, agility, and the ability to use intuition to immediately follow up very subtle clues. Human investigators—rock hounds—will be needed. The hunt for extant life will require setting up and running drilling rigs, burrowing hundreds of meters into the soil, retrieving samples, and then culturing, imaging, and analyzing them in a lab. Such operations are far beyond the ability of robotic rovers.

If Mars is to be made to give up its secrets, "people who do not shrink from the dreary vastness of space" will have to go there themselves.

3: FINDING A PLAN

MARS THE HARD WAY

On July 20, 1989, President George Bush stood on the steps of the National Air and Space Museum in Washington, D.C. Behind him, within the cool halls of the museum, rested artifacts from America's greatest space explorations, among them a gumdrop-shaped spacecraft named *Columbia*, the Apollo 11 command module. The men who rode the *Columbia* home from lunar orbit—Neil Armstrong, Mike Collins, and Buzz Aldrin, the *Apollo 11* crew—now flanked Bush as the president prepared to announce a bold new venture in space on this the twentieth anniversary of humanity's first landing on the Moon.

Bush spoke of the challenges and allure of space exploration, of committing the nation to a sustained program of human exploration of the solar system and even of the permanent settlement of space. This was heady stuff, even if it did come twenty years after the United States' astronauts first stepped off of Earth's surface and onto another world. He continued, speaking of the need for more than a ten-year plan, of a "long-range, continuing" commitment to space exploration. Then he proclaimed his program: "First, for the coming decade—for the 1990s—Space Station Freedom. . . . And next—for the new century—

back to the Moon. . . . And then—a journey into tomorrow—a journey to another planet—a manned mission to Mars."

Thus was born the program which came to be known as the Space Exploration Initiative, or SEI. It was a good start, but it was all downhill from there.

In response to the speech, a sprawling NASA team representing all the centers in the agency, supported by all the major aerospace contractors, went off to figure out how Bush's program could be realized. The team returned three months later with a document entitled "Report of the 90-Day Study on Human Exploration of the Moon and Mars," which soon became known simply as "The 90-Day Report."[11] Before humans could go to Mars, the report said, the nation would need a space infrastructure buildup of thirty years, and the largest and most costly U.S. government program since World War II.

NASA would build the previously envisioned Space Station, but triple its size with the addition of "dual keels" containing large hangars for the construction of interplanetary spaceships. A plethora of additional orbital facilities would be built too: free-flying orbital cryogenic propellant depots, checkout docks, crew construction shacks, and so forth. This huge and complex array of facilities would be used to construct and service trans-lunar spaceships (which themselves would require three heavy-lift vehicles plus one Space Shuttle launch each for their deployment). Those who recalled the single launch required for each Apollo mission scratched their heads and thought, "It wasn't this hard to get to the Moon the last time. . . ." Over the course of a decade, these lunar spaceships would haul to the Moon all the supplies and equipment necessary to build up a massive lunar base complex. Together with the orbital facilities, the lunar base would then provide the basis for building truly huge—1,000 tonnes plus—"Battlestar Galactica"–class spaceships for, finally, voyages to Mars. These trans-Mars space cruisers would employ propulsion and other technologies totally new and different from the lunar craft, and thus require vast new development expenditures as well as additional infrastructure beyond those needed to support the lunar missions. Initial Mars missions would require about eighteen months in transit (round trip) with a one-month stay in Mars orbit. As for actually landing on Mars, a small craft would descend to the surface and support a small crew of explorers for two weeks or so, thereby enabling "flags and footprints" (and little else) human Mars mission to occur. The trans-Mars spaceships would fly out huge and return to Earth orbit tiny, having dropped bits and pieces—fuel tanks, excursion vehicles, aeroshields—in the course of each mission, thereby imposing a massive expense on each "flags and footprints" exercise that followed. The 90-Day Report did not include a published cost estimate; however, cost estimates for the program were generated that eventually leaked to the press. The bottom line: $450 billion.

It is doubtful that any kind of program could have survived that price tag. Given its long timelines and limited set of advertised accomplishments on the road to colonizing space, which did little to arouse the enthusiasm of the space-interested public, the 90-Day Report proposal certainly could not. Unless that $450 billion number could be radically reduced, the SEI was as good as dead, a fact made clear in the ensuing months and years as Congress proceeded to zero out every SEI appropriation bill that crossed its desks.

In fact, however, there was no real inner logic underlying the 90-Day Report, nor any truly new thoughts. Rather, it was a rehash of fixed ideas that hearkened back to the forty-year-old "Die Marsprojekt," an outline of human missions to Mars that German rocket designer Wernher von Braun and his collaborators had first worked out in the late 1940s, and then updated technically to provide the basis for NASA's failed proposal for a humans-to-Mars Apollo follow-on program in 1969. For von Braun and his collaborators, a manned interplanetary mission was part and parcel of a hardware fabricator's wildest dreams: the huge interplanetary spaceship (or better yet, fleets of huge interplanetary spaceships) assembled and launched from Earth-orbiting space stations. What actually occurred on the surface of Mars became an event of secondary interest. Around this fixed idea—giant space stations assembling gigantic spaceships—the unwieldy 90-Day Report team had proceeded to cast as crucial technologies every existing, planned, or wished-for NASA technology development program. In order to include everybody in the game, they designed the most complex mission architecture they possibly could—exactly the opposite of the correct way to do engineering.

DEFINING A COHERENT SPACE EXPLORATION INITIATIVE

By the end of 1989, it was thus clear to many that the mission architecture described in the 90-Day Report was incoherent. In an effort to develop a systematic critique, I produced the following memo, which I subsequently used as the introduction to each of a large series of papers about the Mars Direct plan (see, for example, note 12). What I had to say then summarizes much of the logic that led to the development of Mars Direct, and I reproduce it here in full, with some bracketed material added for clarity:

> The need currently exists for a coherent architecture for the Space Exploration Initiative (SEI). By a coherent architecture what is meant is a clear and intelligent set of objectives and a simple, robust, and cost-effective plan for accomplishing them. The objectives chosen should offer the maximum payoff, and their accomplishment should enhance our ability to achieve still more ambitious objectives in the future. The plan, in order to be simple, robust, and low cost, should not make inter-dependent missions (i.e. Lunar, Mars, and Earth orbital) that have no real need to be dependent on each other. The plan should, however, employ technology that is versatile enough to play a useful role across a wide range of objectives, so as to reduce costs through commonality of hardware. Finally, and most importantly, technologies must be chosen that maximize the effectiveness of the mission at the planetary destination. It is not enough to go to Mars; it is necessary to be able to do something useful when you get there. Zero capability missions have no value.
>
> While the above principles may appear to be common sense, they were violated in every particular by many recent SEI studies [i.e. the 90-Day Report] and as a result, a picture has been presented of SEI that is so costly and unattractive that congressional funding of the program is very much in doubt. Such architectures have driven costs through the roof by employing totally different launch vehicles for the Moon and Mars; totally different space transfer vehicles and propulsion technologies for the Moon and Mars; totally different excursion vehicles for the Moon and Mars; a completely artificial dependence of the Mars missions on the Lunar missions; and a requirement to base the Lunar missions on a massive orbital assembly, refu-

eling, and refitting infrastructure at Space Station Freedom. Furthermore, both the Lunar and Mars missions studied have been close to zero capability, with no serious attempt made to provide surface mobility, and with Mars explorers spending less than 5 percent of the total Earth–Mars round trip mission transit time on the surface of the Red Planet.

Meeting the demands of coherence drives the design of the SEI architecture in certain very definite directions. To wit:

1. Simplicity and Robustness require that the Lunar and Mars missions not depend upon any LEO [low-Earth-orbit] infrastructure. In addition to being tremendously costly to develop, build, and maintain, such infrastructure is intrinsically unreliable and difficult to repair, and its use adds risk to all planetary missions based on it due to the difficulty in verifying quality control of any space-based construction. The demand for the elimination of LEO infrastructure argues in favor of using both advanced propulsion and/or indigenous propellants, both of which can contribute to reducing mission mass to the point where no on-orbit assembly is required.

2. Low Cost requires that the same launch vehicles, space transfer vehicles and propulsion technology, and to the extent possible, excursion vehicles be used for both the Moon and Mars, as well as other destinations. Low cost also demands the elimination of LEO infrastructure, as the potential cost savings made possible by re-use of space transfer vehicles at such infrastructure are insufficient to balance the cost of the infrastructure. This can be seen by noting that the cost of such infrastructure is currently estimated to be about three orders of magnitude larger than the value of the vehicle hardware elements (engines, avionics) that it would be able to save with each space-based refit. Thus, about a thousand refitted missions would be required before such a facility broke even—a somewhat distant prospect. Low cost also demands that the most cost-effective trajectories be taken at all times (i.e. conjunction class [low energy, long surface stay] trajectories for Mars), and that an initial group of opposition class [high energy, short surface stay] Mars missions using completely different hardware from the main sequence of conjunction class missions not be undertaken.

3. High Effectiveness requires that the astronauts be endowed with three essential elements once they reach their destination. These three essentials are:

a. Time
b. Mobility
c. Power

Time is obviously required if the astronauts are to do any useful exploration, construction, or resource utilization experimentation on the surface of the destination planet. This clearly means that opposition class Mars missions (which involve 1.5 year flight times and 20-day surface stays) are out of the question. It also means that architectures involving Lunar or Mars orbital rendezvous (LOR, MOR) are very undesirable, for the simple reason that if the surface stay time is long, so is the orbit time. The LOR or MOR architectures are therefore left in a predicament of whether to leave someone in the mothership during the extended surface stay, exposed to cosmic rays and the rigors of zero-gravity conditions and accomplishing nothing; or leave the mothership unmanned for an extended period and have the returning crew trust to fate that it will be ship-shape when they return. If it isn't, their predicament may be hopeless. The alternative to LOR and MOR architectures are those that employ direct return to Earth from the planet's surface. This is possible to do on a Lunar mission with all terrestrial-produced propellants, however the mission capability is greatly enhanced if Lunar produced LOX [liquid oxygen] can be used for the return. Direct return from the surface of Mars absolutely demands that indigenous propellants be used.

Mobility is absolutely required if any useful exploratory work is to be accomplished on a body the size of Mars or even the Moon. Mobility is also needed to transport natural resources from distant locations to the base where they can be processed, and is also required to enable crews to visit distant assets, such as optical and radiotelescopic arrays on the Moon. The key to mobility on both the Moon and Mars is the generation of indigenous propellants for use in both high powered ground rovers and rocket propelled flight vehicles. On the Moon the resource of choice is Lunar LOX, which can be burnt with terrestrial fuels such as hydrogen or methane. On Mars, chemical fuel and oxidizer combinations such as methane/oxygen or carbon monoxide/oxygen can be produced for both surface and flight vehicles, and, in addition, rocket thrust for flight propulsion can also be produced by using raw carbon dioxide propellant heated in a nuclear thermal rocket engine.

> Power can be generated in the large amounts required for indigenous propellant production on both the Moon and Mars only by using nuclear reactors. Once indigenous propellants have been produced, they form a very convenient mechanism for storing the nuclear energy, thus providing explorers with mobile power, for example by running a 100 kWe generator off the internal combustion engine on a ground rover. The presence of a power rich environment, both at the base and at remote sites, is essential to allow the astronauts to pursue a wide variety of scientific and resource utilization activities.
>
> We thus see that the requirements for Simplicity, Robustness, Low Cost, and High Effectiveness drive SEI toward an architecture utilizing direct launch to the Moon or Mars with a common launch and space transfer system, and direct return to Earth from the planet's surface utilizing indigenous propellants, which are also used to provide surface mobility and mobile power.[12]

This was the thinking process that led to the development of a radically new type of Mars mission architecture that came to be known as "Mars Direct."

THE BIRTH OF MARS DIRECT

By January 1990, it was clear that the 90-Day Report was going down in flames. A retreat meeting of selected Martin Marietta management was called at the Broadmoor Hotel in Colorado Springs to discuss what to do about the situation. Because we were known within the company as the Mars "idea people," Dr. Ben Clark, a lower level Martin manager who had been one of the four principal investigators on the 1976 Viking mission (he had designed the X-ray fluorescence experiment), and I, only a senior engineer, were invited. We were by far the lowest ranking people there.

Ben and I hit the assembled executives with the idea that Martin ought to assemble a small select team and develop our own "blue sky" approach to Mars—one free from all of NASA's current prejudices. It would be hard enough to develop a sound, cost-effective, near-term approach to a human Mars mission without having a bunch of marketeers come in and tell us that we should design the mission this way or that so as to please some manager or group at NASA Johnson or Marshall Space Centers; our team would have to be independent of such influences. After all, it was precisely such an effort to please everyone that had allowed "The 90-Day Report" to get out of control.

This was a very radical proposal. The standing wisdom among the managerial class in the aerospace industry holds that you should always play back to "the customer" (NASA or the Air Force) what they want to hear, which is to say a reiteration of their own party line. That's clearly the easiest way to make a sale. We were proposing the opposite approach: Come up with some good ideas and then tell the customer what they need to hear, whether they like it or not.

While not the most senior, the dominant figure at the meeting was Al Schallenmuller, the newly appointed vice president of Martin Marietta Civil Space Systems, the subdivision of the company charged with responsibility for the SEI. Schallenmuller had cut his teeth as an engineer working for Kelly Johnson in Lockheed's fabled Skunkworks. He knew that, if approached the right way, big and difficult programs could be accomplished cheap and fast. In 1976, he had been one of the key engineers on the *Viking* program. He could never refrain from talking about the thrill of seeing *Viking*'s first photograph of the Martian surface; Schallenmuller really wanted to get back to Mars. He knew that if nothing better than "The 90-Day Report" was on the table, there would be no program. He supported our proposal.

Thus, in February 1990, Martin Marietta formed a twelve-person Scenario Development Team, chaired by Al Schallenmuller, and charged with developing "broad new strategies" for human space exploration. Most of the members of the team, like Ben, me, and David Baker, a spacecraft systems engineer, were generalists. But there were a few specialists, such as Bill Willcockson, Martin's expert on aerobraking, the use of a planetary atmosphere to decelerate spacecraft (Bill was later to play a key role in the successful aerobraking of the Magellan spacecraft at Venus); Al Thompson, a leader in the artificial gravity area; and Steve Price, Martin's specialist in the design of planetary ground rovers.

The two personalities with the most strongly held ideas on Mars mission design were Ben and I, but we were in only limited agreement. We agreed that low-energy, conjunction-class missions were the way to go, that a lunar base was unnecessary to support Mars missions, and that the use of an orbital infrastructure to support on-orbit assembly was a clear minus in mission design. After that, we parted ways. Ben felt that with sufficient use of robotics, on-orbit assembly could be accommodated by using onboard manipulators to allow the spacecraft to build itself up from a series of pieces of hardware delivered to it on-orbit. Because he was willing to assemble his spacecraft on-orbit, Ben did not have the same drive as I did to reduce mission mass. So, while he had for years maintained a significant interest in the possibilities of manufacturing propellants on Mars, he saw no need to introduce such strategies into his mission plan. Ben also did not see the need to maximize time on the surface of Mars. His crew spent a year and a half at the Red Planet, but occupied nearly all that time in orbit, with only a comparatively brief thirty-day sortie to the surface in a small landing craft. Ben used chemical propulsion, obtainable off-the-shelf from existing vendors. The result of this thinking was a relatively conventional mission (if one calls the then dominant 90-Day Report thinking conventional) involving constructing a 700-metric-tonne orbiting spacecraft, but avoiding the 90-Day Report's costly detours in development and construction of lunar and orbital infrastructure. Ben initially called his mission plan "Concept Six," but later changed its name to "The Straight Arrow Approach."

I did not agree with Ben's thinking. I found his robotic self-assembly scheme not credible. Moreover, with a requirement to launch 700 tonnes to low Earth orbit (LEO) for each flight, there would not be very many future missions launched to Mars, and the thirty-day surface stays would not allow sufficient time for much real exploration to occur. As far as I was concerned, we were not going to Mars to set a new altitude record; we were going there to explore and develop a planet. The possibility of a sustained presence on Mars required a large number of repeat missions, and the only way this could be done was if the mass, and therefore the cost of the mission, could be brought way down. The best way this could be done would be by manufacturing an entire mission's return propellant on the surface of Mars. In fact, in 1989 I had done studies showing that if such a strategy were combined with the use of nuclear propulsion for the out-bound leg of the mission, a single booster in the class of the Apollo-era Saturn V could launch an entire human Mars mission. Launched with a single booster, the whole system could be integrated on the ground at Cape Canaveral, and the issue of on-orbit assembly of interplanetary spacecraft would be moot. Moreover, by using locally produced propellant, the whole mission could be landed on Mars, with no liabilities left in Mars orbit, thereby enabling the kind of long surface stays I felt to be absolutely necessary if the program were to do anything useful. Direct launch with a single throw of a heavy-lift booster, use of nuclear propulsion on the outbound trajectory, and direct return from the planet's surface using in-situ produced propellant—this was the way to go.

Enter David Baker. Baker was a very sharp engineer working at the time as the systems and design lead on Martin's Space Transfer Vehicle (STV, a ferry for lunar missions) program. On STV, Baker was being driven out of his mind by arbitrary requirements levied by NASA. For example, the STV would have to be able to land on the Moon if any two engines failed. (The Apollo lunar lander had only one engine.) For reasons of thrust symmetry, this meant you had to have five engines, when one would do the job. Now you had too much thrust, which meant the engines had to be made to throttle back to 10 percent power, which they weren't designed for, thus requiring a new costly engine development program. Furthermore, NASA required that the engines be reused. That meant dragging five heavy engines all the way to the Moon and back—adding enormously to the launch mass and cost of the mission—and then checking the engines out and refitting them in a multibillion dollar orbital facility. All this to do a job that could easily be done better with a single, off-the-shelf Pratt and Whitney RL-10 engine costing $2 million. On STV, Baker contributed what he could as a team player, but he confided to me at the time, "None of this makes any sense."

Baker had participated in earlier Mars mission studies whose direction had been dictated by 90-Day Report type thinking, but it was clear that the logic behind them (or lack thereof) left him feeling distinctly uncomfortable. I hit him with my ideas; some he already agreed with, others, such as the central role of using locally produced return propellants in human Mars missions, I gradually was able to convince him to accept. On other points, however, he would not budge. In particular, he wouldn't buy the idea of using nuclear propulsion as the basis for the first Mars missions. Its development would cost too much, he argued, and public acceptance would be a problem. 1 didn't accept those arguments; on a sustained human Mars program the cost of developing a nuclear rocket engine would be paid back in reduced launch costs after only two or three missions, and if the public wanted a sustained Mars program, they would come to accept nuclear propulsion on that basis. But, Baker said, if you insist on using nuclear propulsion starting on the very first mission, you'll delay the whole program, perhaps fatally.

That point struck home. I felt very strongly that a humans-to-Mars program had to be done on a rapid schedule. Fast schedules reduce program cost: cost equals people multiplied by time. Moreover, every year any major program has to go before Congress for continued funding where it faces risk of termination, often caused by deals or interpersonal frictions that have nothing to do with the program itself. Every time a program goes before Congress for funds it is forced to play another game of Russian roulette. You can only expect to be lucky so many times.

In 1961, John F. Kennedy called for reaching the Moon by 1970. By 1968, administrations had changed, and even as the Apollo astronauts were landing on the Moon, President Richard Nixon was ripping the program to shreds. If Kennedy had called upon the nation to reach the Moon in twenty years as opposed to a decade, 1969 would have seen NASA in the final stages of the Mercury program, with the Moon still a long way off. The program would have been canceled, and reaching the Moon would still be considered an impossible dream today. If you want to get humans to Mars, you can't do it in thirty years; you can't do it in twenty years—ten years is the most you can hope for.

Nuclear propulsion, I conceded, might have to wait but the Mars mission couldn't. By all means, use nuclear propulsion whenever it should materialize; it will increase mission payload capability and cut launch costs (by about a factor of two). But don't delay the mission until you've got it. Go as soon as you can with what's at hand. Improvements can come later. As Baker and I started spending a lot of time in conversation, debating many issues of vehicle and mission design, both technical and philosophical, we increasingly began to converge. We resolved to collaborate.

In many respects we were an unlikely pair for collaboration. I'm on the short side; Baker is extremely tall. I'm mercuric; he's phlegmatic. I'm an optimist; he's a pessimist. I'm a romantic; he's an existentialist. My favorite movie is *Casablanca*; his *Brazil*. My thought process moves in leaps and jumps; his in a steady tread. My credo agrees with that of Hegel: "Nothing great has ever been accomplished without passion." When on one occasion I told that to Baker, he winced and walked out of the room. For Baker, passion and engineering do not mix. For Baker, apparently, it is sufficient to do excellent work and to live well. I want to change the world.

Nevertheless, collaborate we did, and for some time in 1990, extremely effectively We had complementary strengths. I had a much better academic education in broad areas of math, science, and engineering; he had a lot more engineering experience and "knew the ropes." I supplied creative drive, he supplied discipline. We never became close friends, but as a team, we worked.

As alluded to above, in 1989 I had shown in a number of papers that, if nuclear propulsion were available and in-situ produced propellants were used for Mars ascent and Earth return, a human Mars mission could be launched with a single booster in the Saturn V class. Baker had designed such a heavy lift vehicle for NASA. He called the vehicle "Shuttle Z" after Code Z, the NASA organization in charge of developing human space exploration plans at the time. Shuttle Z was basically a growth variant of NASA's Shuttle C design, which replaced the orbiter on the Space Shuttle launch stack with an expendable cargo pod. Shuttle C could deliver about 70 tonnes to low Earth orbit (LEO). By adding a powerful hydrogen/oxygen upper stage inside an enlarged side-mounted cargo pod, Baker created Shuttle Z, and increased the LEO capability of the vehicle to about 130 tonnes, just 10 tonnes short of the Saturn V's capability. Because all the key components of Shuttle Z were drawn from the Space Shuttle inventory, it would be possible to develop the vehicle quickly and inexpensively, a key requirement for a decade-long program.

So, we had our booster, but we didn't have nuclear propulsion for either the outbound or homebound leg of the mission. Throwing our hardware to Mars without nuclear propulsion would now require two launches. In itself, this was not a show stopper, but it did make our mission architecture inelegant, at best. In our design, an Earth return vehicle (ERV) sat atop a habitat, which, in turn, sat atop a partially filled Shuttle Z upper stage, which sat on another nearly filled upper stage. This stack was assembled in orbit via a rendezvous and docking maneuver, with the first three elements (ERV, habitat, and one partially filled upper stage) delivered by one Shuttle Z launch and the fourth (the second nearly filled upper stage) by another.

For a number of reasons this design was not very attractive. To begin with, the long stack was awkward, and whichever booster launched first would leave its payload in LEO for several months, during which time a significant amount of propellant in the upper stage would boil off. On arrival at Mars, the ERV/habitat payload would ride behind an aeroshell—a blunt, mushroom-shaped shield—and decelerate as it plowed through the Martian atmosphere. However, the combined ERV/habitat payload was so heavy that it was questionable whether even a folding aeroshell big enough for the job could be made to fit inside the Shuttle Z payload fairing. But there was an even bigger problem once at Mars.

When nuclear propulsion had been in the cards, I had designed a propulsion system that simply compressed and stored Martian carbon dioxide and then heated it with a nuclear reactor to produce a high-temperature vapor rocket exhaust. (About 95 percent of the Martian atmosphere is carbon dioxide, which liquefies under Martian temperatures when compressed to about 100 pounds per square inch.) Mechanically, such a propellant manufacturing system is very simple; basically all you need is a pump. In such a plan, it was reasonable to suggest that the astronauts could acquire their return propellant after they landed on Mars. Without nuclear propulsion, however, any propellant produced on Mars would have to be manufactured by some form of chemical synthesis. This would be considerably more complex than simply compressing and storing carbon dioxide. Undoubtedly, NASA would quite reasonably insist that all the propellant required for a return to Earth be produced before a crew headed off for Mars; otherwise the crew could find themselves stranded on the planet if fuel production failed.

In 1989, Jim French, an independent engineering consultant, had published an article in the *Journal of the British Interplanetary Society* with some of these considerations in mind. French suggested landing a propellant manufacturing plant on Mars before a crew arrived. The plant would produce and stockpile fuel for the return leg of the crew's journey. But this left the problem of landing the crew's spaceship within a hose length of the propellant depot. This proved so problematical that French ended up conceding that the use of in-situ propellant would not be practical until after a manned base was well established on Mars, along with local infrastructure to provide backup against all sorts of contingencies.

So, there matters stood: By dropping nuclear propulsion we gained the possibility of an accelerated schedule, but a host of problems appeared to come along with the bargain. The most intractable was the question of transferring prearrival-produced chemical propellant from storage to the ERV. Depend on a prelanded mobile robotic fuel truck? Too risky. Wrestling with this problem, I hit upon a novel architectural idea, which now seems so obvious. *Don't send the crew out with their own return vehicle—send the return vehicle, with the propellant plant integrated into it, out first.* At one blow, this idea solved practically everything. The habitat and the ERV were by themselves each light enough that either could be launched *directly to Mars* by a single Shuttle Z. We would still require two launches, but now one Shuttle Z could launch the ERV and another could send the crew and their hab. Combined, the ERV/hab payload would have required a huge aerobrake that would have presented serious challenges to our aerobrake designers, but, flown singly, a manageable aerobrake for each could be made to fit inside the Shuttle Z fairing. To ensure our Mars crew would not be stranded, the ERV would fly one launch opportunity, or twenty-six months, prior to the launch of the astronauts. Thus all the propellant would be made before the crew ever left Earth, and, since the propellant plant was flown to Mars integrated with the ERV there was no question about landing "within a hose length." The plumbing that would deliver the Mars-manufactured propellant from the chemical synthesis unit into the ERV's fuel tanks would be hardwired, installed on Earth.

Best of all, at no point would the mission require on-orbit assembly or orbital rendezvous of any type. The only rendezvous required would be on the Martian surface, and that's easy. During Apollo, we landed an Apollo crew within 200 meters of a *Surveyor* spacecraft that had landed on the Moon several years before, and we have much better avionics today. In an orbital rendezvous, if you miss by 10 meters, you've missed. But in a surface rendezvous, you can be off by 10 kilometers and it doesn't matter—you can just walk or drive over. Moreover, as part of the habitat's cargo, we included a pressurized ground rover with a one-way driving range of 1,000 kilometers; it would take *really bad* piloting to land further from the ERV than that. And say what you will about the NASA bureaucracy, the NASA astronaut corps contains pilots that are among the world's best. Without question, the surface rendezvous would succeed.

Though unorthodox and apparently daring, sending the crew to Mars separate from the Earth return vehicle is actually much safer than landing the crew on Mars together with the vehicle that will deliver them back to Mars orbit. The reason is simple: if the ERV is sent first, the crew will know before they even leave Earth that they have a fully functional Mars ascent and Earth return system waiting for them on the Martian surface, one that has already survived the trauma of landing. In contrast, a crew that lands with their ascent system can only guess in what shape their Mars ascent vehicle will be after they've hit the surface. Moreover, in our plan the crew would launch to Mars in convoy with another ERV which would land at a site within rover long-distance driving range. This second ERV would start propellant production for a second human mission to Mars, but in the event of an emergency, it could serve as a backup vehicle for the crew of the first mission. In addition, the two ERVs on the Martian surface together with the crew's own habitat module gives the crew a total of three redundant units capable of housing them and providing life support. As far as safety is concerned on Mars missions, it doesn't get any better than that.

The more closely we examined it, the better this new mission architecture looked. So, we proceeded to elaborate the required subsystems and vehicle designs. I focused on the chemistry of Mars propellant synthesis. The dominant school working in this area in 1990 was investigating a novel process for splitting carbon dioxide (CO_2) into carbon monoxide (CO) and oxygen (O_2), which could then be burned together as rocket propellant. The only raw material required—CO_2—was free as air on Mars.

There were many disadvantages however. The process was relatively unproved, and, at the scale of production needed to support a manned Mars mission, it would require the use of tens of thousands of small, brittle ceramic tubes with high-temperature (~1,000° Centigrade) seals at each end of each tube to make a reactor. In addition, the carbon monoxide/oxygen bipropellant it produced was a poor quality rocket propellant with a specific impulse of only about 270 seconds. ("Specific impulse," or "Isp" is the number of seconds a pound of rocket propellant can be made to produce a pound of thrust. The higher that number is, the better. German V-2 rocket engines used in World War II had an Isp of about 230 seconds; current-day Pratt and Whitney RL-10 engines burning hydrogen and oxygen have an Isp of 450 seconds; and a nuclear rocket engine using hydrogen propellant could have an Isp of 900 seconds.) The low performance of the CO/O_2 propellant mixture would mean that very large and heavy propellant tanks would need to be carried to Mars to hold the required propellant for the return flight. Also, the carbon monoxide/oxygen propellant combination burns with a very high flame temperature, and no engine is yet available that can operate with it. The development of such an engine would be a significant cost and schedule risk.

An alternative was to make a methane/oxygen (CH_4/O_2) propellant mixture. The advantage here was that methane/oxygen is the highest-performing (Isp = 380 seconds) chemical combination that can be readily stored for long stretches on the surface of Mars. While no flight-rated CH_4/O_2 engines were available, the combination had been run successfully in RL-10 engines on the test stand, and the engines' makers, Pratt and Whitney, had published assessments showing that the modification of the RL-10 to CH_4/O_2 operation would be a relatively straightforward and low-cost program. But there's a problem: you need hydrogen (the H in CH_4) to make methane, and it's not easy to come by on Mars. So where on Mars do you get the hydrogen? In 1976, Professor Robert Ash, now of Old Dominion University, and some JPL collaborators published a paper laying out some extremely simple, robust, and well established (Gaslight Era, to be precise) chemical engineering processes that would produce a methane/oxygen bipropellant on Mars, provided a source of water (H_2O) could be found. But that was the sticking point—water. Mining water out of Martian permafrost is not really a credible option for an initial automated mission, and condensing it out of the extremely arid Martian atmosphere would be extremely difficult. So Ash had moved on to examining CO/O_2 production. But in looking at Ash's proposal, I realized that the only problem his group really had was excessive purism—an insistence that *all* of the propellant come from Mars. But in fact, the hydrogen required to support their chemical process was just 5 percent by weight of the total propellant produced. So why not just bring that relatively small amount of hydrogen from Earth? I checked with Martin's cryogenic (super-cold) fluid storage experts, and they were unanimous in their opinion that storing the 6 tonnes or so of hydrogen needed during an eight-month outbound trip from Earth to Mars was nothing they couldn't handle, provided we started out with about 15 percent extra to cover evaporation ("boiloff") losses in transit (once on Mars any hydrogen boiloff could be sent straight into the methane reactor and thus would not be lost). Conceptually, this solved the problem of how to produce a viable rocket propellant combination on Mars.

Meanwhile, with the help of Sid Early, a Martin launch vehicle trajectory analyst, Baker redesigned the Shuttle Z into the Ares, a booster optimized not to lift payloads to low Earth orbit, but to throw them directly into interplanetary space. (See Figure 3.1.) He also came up with a scheme for using the burnt-out upper stage of the Ares as a counterweight on the end of a rotating tether to create artificial gravity in the crew habitat during the outbound trip to Mars. Tether-generated artificial gravity schemes were nothing new, but our plan was much more robust than others as the object at the far end of the tether from the crew was not mission critical. In more conventional missions, because of the huge mass of the outbound spaceship, tether-generated artificial gravity had to be created by dividing the ship and putting vital components, such as chemical propulsion stages needed for Earth return, at the far end of the tether. If such a tether should snag when the time came to reel it in, the mission would be lost. In contrast, the tether on our mission would never have to be reeled in. Rather, when the hab reached Mars, it would simply be released or cut by the firing of an explosive bolt. This was a major reduction in mission risk, and a strong illustration of one of the derivative advantages of our mission architecture.

So it went. Baker proposed using two space station modules as the basis for the hab, since those presumably would be off-the-shelf items by the date of our mission. Space station modules are long and thin, like an airplane fuselage, designed that way because they have to fit

FIGURE 3.1
Booster evolution from Shuttle C to Shuttle Z to the Ares.

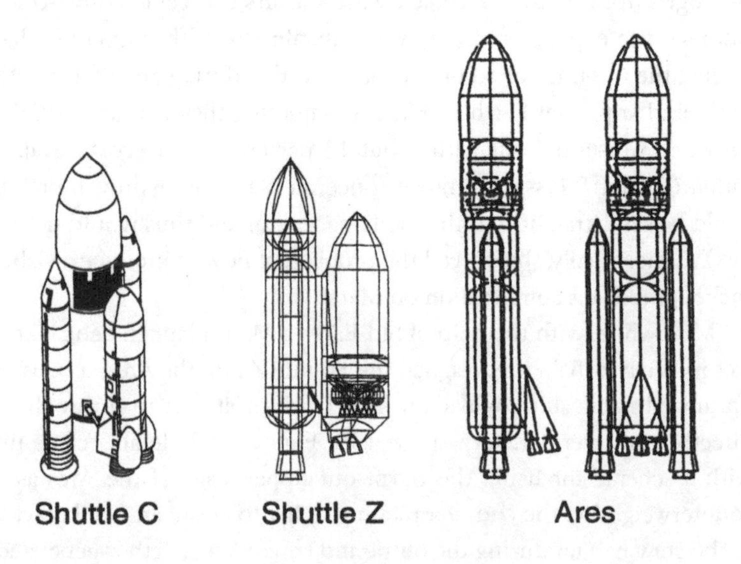

inside the 5-meter-diameter space shuttle payload bay. I pointed out that the real development work associated with the space station module revolved around its life-support and other interior systems, not the manufacture of its hull. A fatter tuna-can shape that took better advantage of the 10-meter-diameter fairing available on the Ares offered much better possibilities for a "human friendly" design for a long duration habitat than a pair of space station modules, I argued, and would be considerably lighter to boot. After playing around with various sketch designs of habitat interiors, Baker satisfied himself that this indeed was the case and so we chose the tuna-can. The tuna-can hab fit beautifully and symmetrically upon the dead center of one of Bill Willcockson's folding aerobrake designs. The hab would nestle comfortably within a folded aerobrake that fit inside the Ares' fairing. Since we wanted a set of vehicles that could be used to accomplish lunar missions as well as Mars (as a sort of side benefit, not as an intermediate step), we decided to split the ERV's propulsion into two stages. The upper stage by itself would have just the right propellant load to allow a direct return to Earth from the surface of the Moon, while the two stages together could drive the ERV home from Mars. Because the upper stage by itself was much smaller than the lower stage, an Ares could be used to deliver a fully fueled ERV to the lunar surface (making rocket propellant on the Moon is possible but unlikely for an initial mission, as it requires breaking down rocks). So designed, the Ares, the hab, the two-stage ERV and the aerobrake module provided a compact (and therefore low-cost) set of elements that could be mixed and matched to accomplish both the lunar and Mars objectives of the Space Exploration Initiative. With the help of CAD (computer assisted design) engineer Bob Spencer and company artist Robert Murray (that's right, artist: a good engineering artist can make extraordinary contributions to a design effort by forcing you to think and explain how this fits into that, and how someone could get from here to there), all the sketch designs were fleshed out into three-dimensional engineering drawings.

On grounds of minimalism, Baker preferred a crew of three; I wanted five. We worked through the mission logistics and it turned out we had enough payload delivery capability to support a crew of four.

So four it was. (The choice was that simple. For reasons explained in a later chapter, however, I have since become convinced that four is just the right number to bring on an initial human Mars expedition.)

One day toward the end of our design work, I walked into Baker's office and sat down on his desk. "We need a name for the plan," I said, "something that captures its essence. We're going to Mars directly, both programmatically in terms of avoiding orbital and lunar development, and physically, in terms of doing the mission by direct throw of the booster outbound, and direct return from the surface inbound. So I was thinking of something like the 'Direct Plan,' or the 'Direct Mars' approach." Baker looked at me and said, "Okay . . . how about . . . 'Mars Direct?' " He only had to say it once. The plan had a name.

Upon completion, we presented the plan to the scenario development team and to a management group for scrutiny. Ben Clark wrote out several pages of tough questions and criticisms about the plan that we had to, and did, answer successfully in writing. Al Schallenmuller, the Martin Marietta Civil Space vice president, was very excited about the plan. Everything needed to accomplish our mission was near-term and relatively simple. Based on his Skunkworks experience, he agreed with my assessment that the Mars Direct plan held the potential for humans to Mars within ten years. He decided to fly us down to Marshall Space Flight Center, in Huntsville, Alabama, to present the plan to NASA.

Neither Baker nor I expected the briefing at Marshall to be well received. Marshall is one of the most conservative of the NASA centers, and it seemed unlikely that any Marshall audience would be favorably inclined toward an idea as radical as Mars Direct. Regionalism would be a real barrier too, acting to greatly reinforce the "not invented here" factor. As I predicted to Baker at the time, only half in jest, the Marshall response was likely to be, "My daddy didn't fly Mars missions that way, and his daddy didn't fly Mars missions that way, and we don't need no damnyankees coming down here to tell us how to fly Mars missions."

I couldn't have been more wrong. Baker and I pitched the mission plan jointly, tag team style. The response was electric. The Marshall SEI planning group was indeed conservative, and, precisely for that reason, became very excited about Mars Direct. For months they had been deluged with grandiose plans for the on-orbit assembly of huge interplanetary spaceships, all of which they regarded as complete hogwash. When we explained how a manned Mars mission could be accomplished with two launches of a Saturn V equivalent booster, the eyes of Apollo program veterans sitting in the room began to light up: "Hey, this is something that we could actually do!" Gene Austin, the head of the Marshall SEI organization, took Baker and me into his office to talk about the plan for two hours (unheard of), first exploring the concept, and then giving us advice as to how to present the plan at Johnson Space Center and elsewhere.

The Marshall briefing took place on April 20, 1990. In the following weeks we visited every major NASA center involved in SEI with similar presentations, and everywhere we went we lit fires. On Memorial Day, I was given the opportunity to present the closing plenary talk at the national conference of the National Space Society in Anaheim. This was the first public presentation of Mars Direct. I received a *standing ovation*. A week later, Baker and I presented the plan to the "Case for Mars" conference (the triennial gathering of the "Mars Underground," more about which to come) in Boulder, and basically took the place over. The next day, under the byline of its veteran science reporter David Chandler, the *Boston Globe* ran a front-page story, "New Mars Plan Proposed," that was carried by wire and printed in hundreds of other papers as well. Mars Direct was out of the box.

As the summer wore on, Baker and I, singly or together, continued to make presentations at open conferences and NASA briefings. We published a detailed description of the mission as a feature magazine article in *Aerospace America*, the industry monthly, as well. Everywhere we went we made converts, but a counterattack was in the works. Powerful forces within NASA linked to the Space Station program were not happy at all about Mars Direct. Since we didn't use the Station or even its (to be created) heritage of on-orbit assembly techniques, we were, in their view, "de-justifying" their program. People in NASA who had been friendly to Mars Direct were told to keep their distance. This slowed us down. Some (but not all) factions in the advanced propulsion community were also hostile. They felt that Mars Direct was "de-justifying" their programs as well, and argued for mission requirements that only their systems could meet. Refuting the need for these requirements slowed us down some more. What had begun as an intellectual blitzkrieg began turning into trench warfare.

Such an extended fight was incompatible with Baker's temperament. As the difficulty in changing an intellectual paradigm became more and more apparent, reinforcing his natural pessimism, and as the NASA bureaucracy's obtuseness in sticking with their $450 billion mega-fantasy approach continued to result in congressional rejection of SEI funding requests, Baker grew demoralized. In February of 1991, he quit Martin to go back to school at the University of Colorado for his master's degree and to start his own consulting firm.

Ever the optimist, I persisted, touring the country giving dozens of talks and papers, writing numerous magazine articles. The Bush administration pulled together a blue-ribbon "Synthesis Group," chaired by former Apollo astronaut General Thomas Stafford, to try to find a new architecture for the Space Exploration Initiative to replace the failed 90-Day Report. I briefed them, and followed up the briefing by continuing to work on key individuals in the committee. When the Synthesis Group report[13] came out in May 1991, it was a disappointment; they had ignored Mars Direct and instead opted for doing Mars exploration with a slightly updated version of Wernher von Braun's nuclear propulsion mega-spacecraft plan of 1969. However, while my plan did not make it into the report, many of my key axioms did. On-orbit assembly was now seen as a clear minus, not a plus. Time spent on Mars was now viewed as a plus, not a minus—accomplishing something useful on Mars, not just getting there and back, was finally seen as important. So, while an opposition-class (high-energy/short-staytime) mission was retained as a mental vestige for the first Mars mission, all subsequent ones would be conjunction (low-energy/long-staytime). My methane/oxygen Martian propellant production process was identified explicitly as something that needed to be developed, if only for use on downstream missions. All this represented progress. Then, in the fall of 1991 more light appeared on the horizon when Mike Griffin, representing the best elements within the Synthesis Group, was appointed Associate Administrator for Exploration within NASA, in charge of the SEI. Griffin was reported to be someone with intellect, not your closed-off bureaucrat type at all. "If only I could get to him," I thought. Griffin was inaccessible so I started by working on his friends, some of whom were also friends of mine. Finally, in June 1992, I got a chance to brief Griffin himself, in his office. It went well. Griffin had read some of my articles, but had some questions. In person I was able to resolve them. Griffin called up Bill Ballhaus, the president of Martin Marietta Civil Space (Schallenmuller was no longer directly involved by this time) and "asked" him (a "request" from a NASA associate administrator is treated as more than a "request" in the aerospace industry) to allocate funds for me to develop a more detailed briefing on Mars Direct to present to his exploration planning group at NASA Johnson Space Center (JSC). He would see that they would take it seriously.

All this came to pass, and more, because what I didn't know at the time was that Griffin liked Mars Direct so much that he proceeded to brief it to incoming NASA administrator Dan Goldin, who also became a supporter. The bottom line was that when I showed up at JSC in October 1992 for a series of detailed briefings on Mars Direct, people were definitely prepared to listen.

The exploration program group at JSC listened, and they liked what they heard, but they still had concerns. They felt that my estimates of mission mass were on the light side, and they wanted a crew of six, so a more powerful booster than the Ares would be needed. Dave Weaver, the lead mission architect in the group, was also leery about making the whole mission architecture critically dependent upon Mars-based production of propellant. True, it was made before the crew that would need it left Earth, so no one would be stranded, but if the propellant production failed, the program would still be a failure. Weaver and I went into his office and got out the chalk and worked out a compromise mission architecture that answered his concerns.[14] I call this plan "Mars Semi-Direct" (Figure 3.2). Instead of two launches per mission there would be three, one delivering a self-fueling Mars ascent vehicle to the surface together with a lot of equipment and supplies, one delivering an Earth return crew cabin together with a methane/oxygen chemical propulsion stage to a high orbit about Mars, and a final launch delivering a hab with the crew to the Martian surface. Now, instead of having to make enough propellant to send an Earth return vehicle directly from the Martian surface to Earth, all that would be needed would be enough to send the Mars ascent vehicle from the surface to rendezvous with the crew cabin in orbit, after which the orbiting chemical stage would push the crew the rest of the way home. The Mars ascent vehicle is light enough that if no extra cargo is sent with it, a fully fueled version could be delivered to the Martian surface with a single heavy lift booster launch. Thus if in-situ propellant manufacturing should fail, the program could still be saved with the aid of a fourth booster launch. I didn't like this architecture as much as the classic Mars Direct, because limiting the application of Martian propellant production also limits its benefits. Instead of Mars Direct's two launches and two spacecraft per mission, three of each would be required by the Mars Semi-Direct plan, and the extra launches and vehicles would make it more expensive. Furthermore, a mission-critical Mars orbit rendezvous had been introduced on the return leg. But it was clearly an extraordinary advance over previous NASA thinking, with all payloads being delivered to Mars with direct throw of the booster, no on-orbit assembly of mega-spacecraft, and long duration surface stays and use of in-situ resources starting on the very first mission. It was a compromise, but a viable one, a plan I could support. Mike Duke and Humbolt "Hum" Mandell, two relatively senior personalities at JSC, also became early and strong advocates of the Mars Semi-Direct plan, and support within JSC solidified rapidly thereafter.

In 1993, Weaver pulled together a large cross-NASA team to undertake an elaborate design study of the Mars Semi-Direct plan. I participated in this study as an advisor. Once again, with the large team in play, centrifugal tendencies were evident. Representatives of various programs tried to skew things to ensure a leading role for their systems. In dealing with this large team, Weaver was basically in the position of someone trying to herd cats. Nevertheless, the team developed a workable, if bloated, plan based on Mars Semi-Direct. This expanded version of the Mars Semi-Direct plan was then subjected to a cost analysis by the same JSC costing group that had developed the $450 billion estimate for the 90-Day Report. The analysis incorporated the development of all required technology, including the large booster (i.e., no sharing of the cost of that system's development with a lunar exploration program was assumed), and flying three complete human missions to Mars. The bottom line: $55 billion, or one-eighth the cost of the traditional plan. In July 1994, word of this work reached *Newsweek* magazine and made the cover. "A manned mission to Mars?" *Newsweek* asked. "The technology is already in place. And at $50 billion—one-tenth of previous estimates—it's a bargain."

Among those who have studied the problem, there is now a

FIGURE 3.2
Mission Sequence for the Mars Semi-Direct plan. Every two years, three boosters are launched. One to deliver a crew to Mars in the hab, the others to deliver unmanned payloads consisting of a self-fueling Mars ascent vehicle (MAV) and an Earth return vehicle (ERV). When it's time to return home, the crew transfers to the MAV and rides it to a rendezvous with the orbiting ERV, which then carries the crew to Earth. The Year 1 hab is flown to Mars without a crew, creating a reserve hab for the first piloted flight, which arrives at Mars in the Year 3 hab.

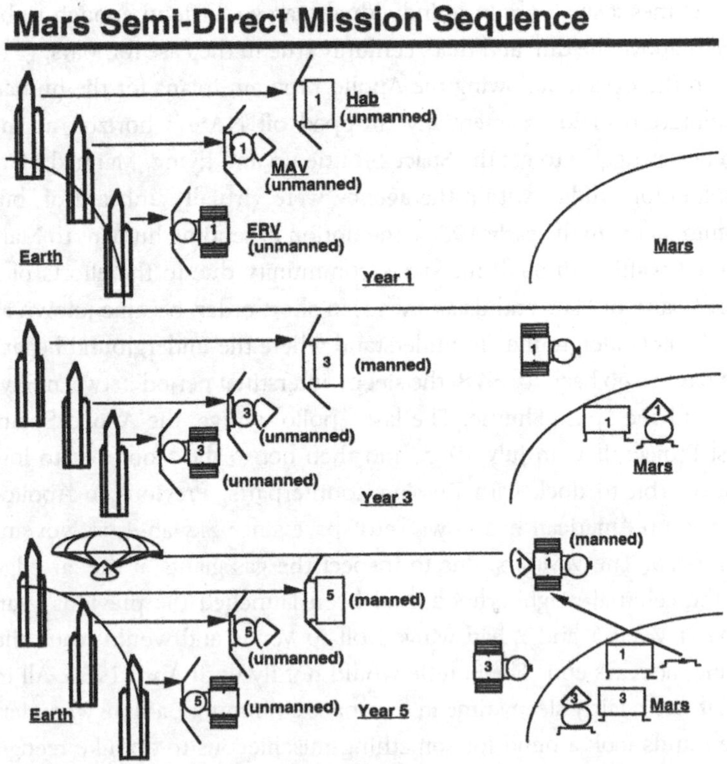

Mars Semi-Direct Mission Sequence

consensus that an affordable, technologically doable, politically supportable plan exists that can get humans to Mars—one with the Mars Direct concept as its basis. This is not a program for some distant future generation, but for us. It is a mission that can be designed by the engineers of today and flown by people who are in the astronaut corps, today.

In the following chapters we'll take a closer look at the Mars Direct plan, and see how it works, step by step and piece by piece. And we'll show what it holds not only for sending humans to Mars, but for exploration, human settlement—and even transformation of the Red Planet itself.

FOCUS SECTION—THE MARS UNDERGROUND

Sometimes a small group of individuals can shout loud enough to be heard above the din, and that's certainly true in the case for Mars.

In the decade following the Apollo program, plans for the human exploration of Mars essentially dropped off NASA's horizon as the agency struggled to get the Space Shuttle up and flying. Manned Mars exploration studies within the agency were virtually unheard of, but commencing in the early 1980s, the notion of sending humans to Mars started wafting through the space community due to the efforts of a small band of Mars enthusiasts who, in short order, became known as the Mars Underground. To understand where the underground began, we have to go back to 1978, the sleepy interstitial period between Skylab and the Space Shuttle. The last Apollo voyage, the Apollo-Soyuz Test Project, flew in July 1975, and then not to the Moon but to low Earth orbit to dock with Russian counterparts. Previous to Apollo-Soyuz, no American had flown into space since Skylab 4, in November 1973. The *Voyagers,* due to inspect the gas giants at the far edge of the celestial neighborhood, had been launched the previous year. *Pioneer-Venus* 1 and 2 had winged off to Venus and would reach the planet at year's end. The Shuttle would not fly until April 1981. All in all, it was a fairly sleepy time in the space community, a time when fertile minds look around for something mischievous to do, like reengineering a planet. So it was that Chris McKay, then a graduate student in astrogeophysics at the University of Colorado, started up a seminar on terraforming Mars.

The seminar arose out of hallway discussions and graduate student lounge beer and bull sessions prompted by the *Vikings*' dismal but intriguing findings. Mars appeared lifeless, according to *Viking*, but it also appeared that Mars needn't remain that way—a bit of wisely applied planetary engineering, *terraforming*, could bring Mars back to the future as, once again, a warm, wet planet. Joining McKay were Carol Stoker, a fellow astrogeophysics graduate student; Penelope Boston, an undergraduate biology major and longtime friend of McKay's; Tom Meyer, president of his own engineering firm and a friend of Stoker's from years past; computer scientist Steve Welch; and a gaggle of others, perhaps twenty-five in all. Charles Barth, the director of the Laboratory for Atmospheric and Space Physics at the University of Colorado, acted as mentor and counselor to the group, helping them transform informal conversations into a formal seminar on "The Habitability of Mars."

Over the course of the first semester, the seminar's participants, with some gentle nudges from Barth, recognized that terraforming Mars was a tall order, even for graduate students. They also realized that they were theory-rich and data-poor. While entertaining and intriguing perhaps, discussions of terraforming Mars without more data would really lead nowhere. They needed more information about Mars—its present atmosphere, its past atmosphere, volatiles, resources, a multitude of items—data that human missions could collect. So, the group began to focus on near-term human missions to Mars, and eventually wrote up their findings as "The Preliminary Report of the Mars Study Group." Barth shepherded the report to NASA headquarters and word soon got out that a band of graduate students and others out in Boulder were enthusiastically—and intelligently—investigating human missions to Mars as well as a new science known as *terraforming*, more about which to come. Some of the seminar's members scraped money together and piled into cars for cross-country drives to various space science conferences and meetings where they would occasionally bump into others of their own flavor, individuals who were entranced by the Boulder group's enthusiasm, vision, and intelligence.

During the spring of 1980, McKay and Boston crossed paths with Leonard David at an American Astronautical Society meeting in Washington, D.C. David had spent the past few years arranging student forums on space exploration and had heard about the Boulder crew. The three hit it off fairly quickly and what started as a chat about Mars exploration ended with David suggesting that some effort ought to be made to hold a conference on human Mars exploration. This was something of a novel idea, as twenty-something graduate students usually didn't organize and host planetary exploration conferences, but adopting something of a "Why not?" attitude (they really didn't have anything to lose), a cluster of Mars devotees, began some low-key planning. McKay, Boston, Welch, Meyer, Stoker, and Roger Wilson, another University of Colorado student, started working on a list of possible topics to be addressed and possible speakers. Via some graduate student guerrilla methods, they ran off a hundred or so copies of a conference announcement and bundled them off for distribution. Much to everyone's surprise, calls started coming in, both from those who wanted to attend and from researchers interested in delivering papers. Deriving the forum's name from a seminal article entitled "The Case for Humans on Mars" that *Viking* scientist Ben Clark had written in 1978, in late April, 1981, the Boulder group hosted the first "Case for Mars" conference.

It was a small conference—just one hundred or so people eventually attended—but to the organizers, these were legions. Before the conference, the Colorado group felt more or less alone in the wilderness. There were not many, they thought, who had the interest and know-how to undertake a serious study of humans-to-Mars missions. But now here they were, in the midst of their conference with presentations on resource utilization, life support on the Martian surface, propulsion options. It was heartening, thrilling, indeed liberating to know that there were others who shared their passion. Leonard David had arrived from Washington with a bundle of red buttons. Imprinted on the buttons, below a Case for Mars logo designed by artist Carter Emmart (depicting a DaVinci-style human figure inside the ancient astrological symbol for Mars), were the words "Mars Underground." A short note accompanied each button, stating that the wearer was now a member of the Mars Underground, an ad hoc collective of Mars aficionados ("tightly knit but loosely woven"), and that the button should be worn discreetly, under the lapel, or perhaps inside the coat. Over the course of four days, numerous workshops, and a slew of presentations, the Mars Underground formulated plans for the human exploration of Mars: the whys and wherefores of the program, precursor missions to the human flights, mission profiles, and rosters of surface activities for explorers. Not a bad result from a conference conceived and organized by a bunch of graduate students.

The conferences continued for fifteen years, one every three years, each building on what preceded and reflecting the character of the times. The second conference, in 1984, resulted in a complete end-to-end design for a Mars mission that Underground members used as the basis for a two-hour presentation on Mars exploration they delivered at NASA headquarters and other NASA centers. The 1984 conference was also notable in that it drew to the group people of greater political influence, such as former NASA administrator Thomas Paine. In 1985, President Reagan appointed Paine to head the blue-ribbon National Commission on Space, who then proceeded to guide it to a recommendation that the United States make the establishment of a human outpost on Mars the thirty-year goal of the space program. The White House responded to the report by setting up the "Code Z" organization and "Pathfinder" programs at NASA headquarters, to respectively plan mission strategies and develop the key technologies required for human expansion to the Moon and Mars. It was these organizations that formed the insider network which provided the policy input for Bush's call for a Space Exploration Initiative in July 1989.

The third Case for Mars conference accelerated the trend, with Carl Sagan giving the keynote address to an audience of over a thousand people, including a substantial representation of the international press. I had first heard about the Mars Underground after Case for Mars II, and along with more than four hundred other out-of-town technical types, went to Case for Mars III to participate in some of the nearly two hundred presentations and sixteen workshops. The two-volume set of papers arising out of Case for Mars III outlines strategies for Mars exploration that touch upon both the technical requirements as well as public policy and political requirements for transforming the vision into a reality. By the fourth conference in 1990 (and held, as always, in Boulder) what had a decade earlier been nearly verboten to speak of in NASA—humans to Mars—had become the current president's stated long-term goal in space. Carol Stoker, who was in charge of the conference scheduling, had attended a private Mars Direct pitch at NASA's Ames Research Center in California, and liked the plan. She gave the bully pulpit at the opening plenary session to David Baker and me to present Mars Direct to the assembled Mars Underground. The next day, news that a low-cost manned Mars mission plan was now on the table appeared in the *Boston Globe* and dozens of syndicated papers.

Mapping the trajectory of a spacecraft is a relatively straightforward business, bounded only by the laws of physics. Mapping the trajectory of an idea through a political system, on the other hand, can be a dicey business. There were many reasons why George Bush stood on the Air and Space Museum's steps in 1989 and declared Mars to be a necessary destination for human exploration, but I have no doubt that the Case for Mars conferences and the small group of individuals who form the core of the Mars Underground were instrumental in positioning a human journey to Mars as an attainable, realistic goal for the United States space program. The conferences provided the cauldron for a brew of ideas, all of which served to heighten the profile of human missions to Mars and all of which energized the community of Mars researchers and enthusiasts. For an organization where membership is defined by enthusiasm and effort rather than a spot on a membership roll or a card carried in a wallet, the Mars Underground and the Case for Mars conferences have to be credited with having influence far beyond their modest size.

It is in honor of their efforts that I have chosen the title of this book.

4: GETTING THERE

FAST MISSIONS AND GOOD MISSIONS

In planning a long journey, you first choose a route and a mode of transportation. So it is with Mars.

Many believe a voyage to Mars is impossible because the Red Planet is so far away from Earth. Until such time as radically more advanced types of space propulsion become available, they argue, the trip will simply take too long. Let's take a look at this objection.

Mars is indeed far away. At its closest approach, when it stands directly on the opposite side of the Earth from the Sun (a condition that ancient astrologers, with their geocentric world view, described as an "opposition," of which more anon) it never gets nearer than 56 million kilometers, or 38 million miles. At its farthest, when it stands behind the Sun as seen from the Earth (what the ancient astrologers called a "conjunction"), it lies about 400 million kilometers distant. (See Figure 4.1.) Now, no propulsion system is even on the drawing boards that can push directly away from the Sun and perform the transit between Earth and Mars in a straight line when the two are in opposition. This is because a spacecraft leaving Earth possesses the velocity of the Earth, some 30 kilometers per second (km/s), and thus, unless massive amounts of propellant are expended to alter course, the spacecraft will continue to circle the Sun in the same direction as the Earth. In fact, as the German mathematician Walter Hohmann discovered in 1925, if you want to go easy on the gas, the best time to travel from Earth to Mars occurs when the two planets are in conjunction; at their maximum distance from each other on opposite sides of the Sun. (See Figure 4.2.) This is the easiest way to go, because if you take this path you can travel along an ellipse which is tangent to the Earth's orbit at one end, and tangent to Mars' orbit at the other, thus minimizing the course change which is required for the spacecraft to depart or rendezvous with each. You can deviate from such a flight plan if you wish, but the more you do, the harder your propulsion job, and the costlier your mission. But even if you do decide to pour on some extra gas to cut corners and avoid the full Hohmann transfer, roughly speaking you'll likely need to traverse at least 400 million kilometers along some curving arc to get from Earth to Mars. Four hundred million kilometers.

FIGURE 4.1
Opposition and conjunction. At opposition, Mars stands directly on the opposite side of the Earth from the Sun. At conjunction, Mars stands behind the Sun as seen from the Earth.

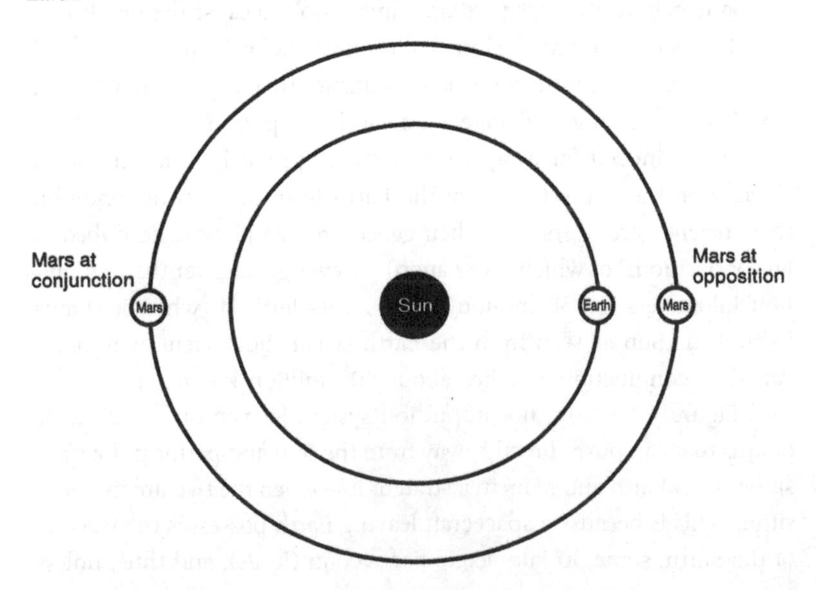

FIGURE 4.2
Trajectory Options to Mars: (A) Hohmann transfer orbit; (B) Fast conjunction mission; (C) Opposition mission.

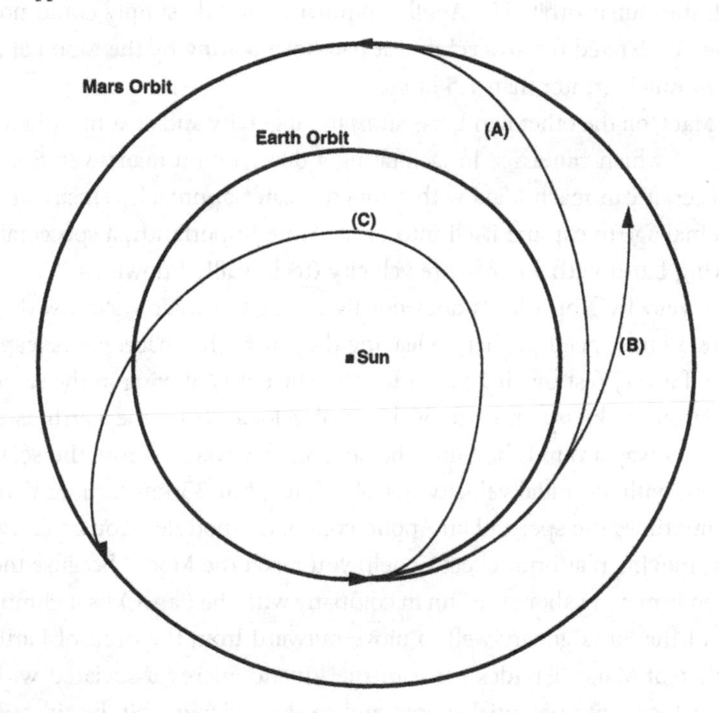

That's a lot. In contrast, the Earth's Moon is "only" 400,000 kilometers away. So, to get to Mars, you'll have to travel a thousand times farther than the Apollo astronauts did when they voyaged to the Moon. It took three days for the Apollo spacecraft to make a one-way lunar transit. Will it take 3,000 days, eight years, to reach Mars?

Fortunately, no. Apollo astronauts traveled between the Earth and Moon with an average speed of about 1.5 km/s. This speed limit was set not by the limits of the propulsion technology of the time—the third stage of the Saturn V could have rocketed the Apollo spacecraft toward the Moon at double or even triple this velocity—but by the nature of the mission geometry. The Apollo astronauts could have been fired off to the Moon at 4.5 km/s and reached it in a single day, but there would have been a huge price to pay: They would have been unable to stop. Because of weak Lunar gravity, a spacecraft's propulsion system has to do nearly all the work required to capture a trans-lunar spacecraft into lunar orbit. The Apollo command module simply could not have decelerated the spacecraft if it had been tearing by the Moon at a speed much greater than 1.5 km/s.

Mars, on the other hand, has substantial gravity and an atmosphere, both of which can assist in facilitating a deceleration maneuver. So, a spacecraft can reach Mars with a much greater approach velocity and still manage to capture itself into orbit. More importantly, a spacecraft leaving Earth with a departure velocity (technically known as "hyperbolic velocity") of 3 km/s does not fly across the solar system with a mere 3 km/s speed. Rather, in leaving the Earth, the spacecraft is leaping off a very fast moving platform, and since it is moving in the same direction, picks up an extra 30 km/s of velocity from the Earth as it races its way around the Sun. The spacecraft voyages across the solar system with an initial velocity not of 3 km/s, but 33 km/s, more than twenty times the speed of an Apollo command module. (You can't use this "moving platform" effect to help you reach the Moon, because the Moon is moving about the Sun in company with the Earth.) As it climbs out of the Sun's gravity well to move outward from the orbit of Earth to that of Mars, it trades some of the kinetic energy associated with this velocity into potential energy, and so slows down a bit, but it's still moving very fast. Fortunately, Mars will be cruising along its orbit with a velocity of 24 km/s in roughly the same direction as the spacecraft. When the spacecraft reaches Mars' orbit, its velocity relative to Mars will be only 3 km/s (since it's traveling along at about 21 km/s), and that's slow enough to allow orbit capture. By the time the spacecraft reaches Mars, it's traveled a thousand times farther than the Apollo astronauts, but, on average, about twenty times faster. One thousand times farther divided by twenty times faster gives us a travel time that is a factor of fifty greater than the three-day transits for the Apollo astronauts—150 days, Earth to Mars. This then is a rough estimate of the travel time for a one-way transit to Mars using Apollo-era or present-day technology for propulsion. It's not a bad estimate either; while a Hohmann transfer actually takes 258 days, a 150-day transit is certainly achievable at the cost of some extra propellant.

But getting to Mars is only half the problem—you also have to get back. Earth and Mars are constantly moving around the Sun, and, since they travel at different speeds, they are constantly changing their positions relative to one another. Because only certain Earth-Mars relative positions are appropriate for launching a return flight, the trajectory you pick not only determines how long you will have to travel, it also determines when you can leave. Determining flight plans for complete round-trip missions thus starts to get pretty complicated, but when all is said and done, you are left with basically two options for a piloted Mars mission round-trip flight plan. These two options are known as conjunction- and opposition-class missions. Typical parameters for each of these mission types are given in Table 4.1.

One example of a conjunction mission would be a "minimum energy" mission performed with two Hohmann transfers between Earth and Mars. Such a mission would be the cheapest way to travel, but would require 258-day transits each way. This is fine for cargo, but if you are flying a crew it is desirable to speed things up a bit. It turns out that it doesn't take too much extra propellant to cut the conjunction transfer time to about 180 days, so that's what we propose to do

TABLE 4.1
Flight Times and Stay Times of Mars Missions

	Conjunction	Opposition
Outbound transit time	180 days	180 days
Inbound transit time	180 days	430 days
Mars stay time	550 days	30 days
Total mission time	910 days	640 days
Mission propulsive ΔV	6.0 km/s	7.8 km/s
Venus flyby needed?	No	Yes
Average mission radiation dose	52 rem	58 rem
Zero gravity exposure	360 days	610 days
Mission cost	Lowest	Highest
Mission accomplishment	Highest	Lowest
Mission risk	Lowest	Highest

in the Mars Direct mission. However, if you adopt such a flight plan, you'll need to stay 550 days on the Martian surface before your return launch window opens up, and thus your total round-trip mission time will be about 910 days.

The opposition mission has one of its legs, for example the outbound trip, accomplished in the same manner as the conjunction. The return leg, however, is then radically different. On the way home, a great deal of extra propellant is used to send the spacecraft from Mars, not back to Earth, but all the way to the inner solar system. There it swings by Venus, getting a gravity assist that slingshots it toward Earth. This procedure opens up a departure window soon after arrival at Mars. Even though the return trajectory takes considerably longer than a Hohmann transfer, an opposition mission cuts total mission time away from Earth by nearly ten months, from about 900 days down to 600 days or so.

Because it minimizes the total mission duration, NASA's "90-Day Report" designers favored the opposition mission. Others have followed suit, and assumed that opposition is the only way to go to Mars. But does such a preference make sense? The opposition mission has significantly larger propulsion requirements, needing 7.8 km/s of "delta-V" or ΔV of total velocity change to speed up or slow down the spacecraft, compared to 6.0 km/s for the conjunction mission. (ΔV is the velocity change required to move a spacecraft from one orbit to another.) In fact, if space-storable propellants are used for the rocket burn that drives the spacecraft from its Mars parking orbit onto its trans-Earth trajectory, the opposition mission shown will have a launch mass about double that of the conjunction mission. It gets worse, however. The ΔV requirements given in Table 4.1 are only for the mission's acceleration maneuvers departing LEO and a highly elliptical Mars' parking orbit; it is assumed that the spacecraft can aerobrake into Earth or Mars orbit. But the opposition mission spacecraft may be so massive that aerobraking proves impractical if not impossible. If that's the case, rocket propulsion would also have to be used for deceleration, adding more ΔV to the mission, which will multiply the mission's mass and cost still more. The situation rapidly gets out of control and leads to the conclusion that the opposition mission is virtually impossible unless nuclear thermal rocket (NTR) propulsion, which can have twice the exhaust velocity of a chemical rocket, or something even better is available. (For this reason the opposition mission has also been favored by some people advocating development of such systems.)

But why do we want to minimize mission time? The reasons classically given are that it is essential to minimize crew exposure to the twin menaces of zero-gravity and various types of space radiation. Because it spends nearly all of its time in interplanetary space, however, the opposition mission actually maximizes the zero-gravity exposure time of the crew. Furthermore, because radiation doses experienced per unit time in interplanetary space are estimated to be about a factor of four greater than those on the surface of Mars, where the atmosphere and surface material provide substantial shielding (even if no measures, such as throwing sand bags on the roof of the surface habitat, are taken), the radiation dose experienced by the crew on an opposition mission is likely to be slightly greater than that on a conjunction mission.

Despite all the hand wringing over the danger of radiation on the way to Mars, it needs to be understood that neither of the doses shown in Table 4.1 is especially threatening. To place them in perspective, we should note that every 60 rem of radiation received over an extended period, such as a several-year round-trip Mars mission, adds 1 percent of extra risk of a fatal cancer at some point later in life to a thirty-five-year-old woman, while 80 rem adds 1 percent of extra risk of fatal cancer to a thirty-five-year-old man. Radiation is not a major risk driver of a piloted Mars mission.

So, the advantages of the opposition mission are illusory, but its disadvantages are not. An opposition mission increases propulsion requirements, thereby increasing mission launch mass, and, consequently, mission cost. Hammering the hardware together for this enormous mass requires on-orbit assembly, which, when compared to integrating all hardware on Earth, boasts nearly zero quality control. Further, it maximizes the extent and complexity of such assembly, thereby maximizing the risk of misassembly. But it doesn't stop there. The opposition mission requires more propellant than any other, and that translates into longer engine burns over the course of the mission, which maximizes the risk of engine failure. The opposition mission also maximizes the transit flight time experienced on a single leg of the mission, which maximizes the reliability requirement for the ship's life-support system (the conjunction ship's life-support system must only be guaranteed for 180 consecutive days; the opposition must go for 430). The life-support system of the opposition-class mission also must handle exterior temperature excursions caused by traversing the solar system not just from Mars to Earth, but from Mars all the way in to Venus, where solar heating is twice that at Earth. (This is why some mission designers call the opposition mission's Venus flyby maneuver a "Venus fryby.") Finally, at the end of the mission, when the opposition-class spacecraft reaches the Earth, it hits the Earth's atmosphere a lot harder than would be the case in a conjunction-class spacecraft. This maximizes the reentry deceleration forces on the spacecraft and the crew, and increases the risk that the spacecraft will, in the event of a misaligned reentry, either burn up or skip out of the atmosphere to leave the crew stranded in interplanetary space.

One enormous and absurd fault, though, looms over all of these flaws—the opposition mission accomplishes next to nothing. After six months of traveling nearly 400 million kilometers to reach Mars, the spacecraft and crew spend just thirty days on Mars. With just a month in Mars orbit, the crew could hope, at best, to get two weeks on the surface before departure. In fact, if the weather is bad when the crew arrives, they might not get to land at all. The entire mission would have been a bust (recall *Mariner 9*'s need to wait four months for a dust storm to clear after arrival). I liken the opposition mission plan to that of a family that decides to fly to Hawaii for Christmas vacation, spending ten days in transit flying from one airport to another, with half a day at the beach, weather permitting. Put simply, the opposition mission plan is just plain silly. It maximizes both cost and risk, and minimizes mission science return. It is favored only by those who wish to present a piloted Mars mission as a pipe dream, or who wish to increase the mission's technical difficulty so as to justify funding some new propulsion system that they are pushing. Among those who are interested in actually getting humans to Mars, the opposition mission has been dropped from all further consideration.

Now, between the various types of conjunction mission plans there is more room for rational choice. The minimum energy plan is the cheapest mission, but a fast-flight plan leads to a better mission, with a larger fraction of the total mission time spent on Mars doing useful exploration and less time wasted in transit. Flying to Mars on a fast conjunction profile drastically reduces the crew's zero-gravity exposure time; cuts back on radiation exposure; and minimizes the in-space life-support system's reliability requirements. However, because the minimum energy mission is not being pushed as fast, its ship can be made heavier, with more backups to various mission critical propulsion, control, and life-support systems. So, while a minimum energy spaceship needs to be more reliable than a fast conjunction craft, the mass budget is there to make it more reliable. (The opposition spacecraft, which must be made the *most* reliable, will have the *smallest* mass budget of all to support subsystem redundancy and reliability.)

There's clearly a tradeoff here, and it's necessary to make an intelligent compromise between spacecraft speed and system redundancy. But there's another consideration. For certain departure velocities it is possible to go to Mars on a trajectory that will take you straight back to Earth if you decide not to (or for some reason can't) go ahead with orbit capture at Mars. These are known as *free-return trajectories.* If the ship's propulsion system fails completely during the outbound transit, or if the mission needs to be aborted for any other reason, following such trajectories can allow the crew to get home safely, just as was done in the nearly disastrous Apollo 13 mission, which employed a free-return trajectory to get to the Moon. The safety advantage of setting off to Mars on such trajectories is so obvious that considering outbound nonfree-return trajectories that might save thirty days, at best, is hardly worthwhile. In Table 4.2 we list the options for free-return trajectories to Mars. The near minimum energy 3.34 km/s departure (option A) takes 250 days to get to Mars, and three years to get back to Earth (involving two 1.5 year period orbits), making it fine for cargo but a mediocre choice for a piloted flight. The 5.08 km/s departure (option B) cuts the transit to Mars down to 180 days and the free-return flight time to two years. This is clearly the best choice for piloted flight, because flying to Mars on the higher-energy free-return trajectories (options C and D) not only requires expending a lot more propellant in exchange for a modest reduction in outbound transit time, but because their orbits will loop well out beyond Mars actually would cause the crew to take longer to get home if carrying out the free return should become necessary. In addition, the high-energy options will arrive at Mars traveling too fast for safe aerobraking.

Free-return capability is not a factor in choosing trajectories that go from Mars to Earth. However, transit-time reduction reaches a condition of diminishing returns when departure velocity starts to exceed 4 km/s. Trying to go much faster than this will simply force us to give up ship payload, and thus critical system redundancy, without reducing transit time appreciably.

Thus, we find that the best trajectories between Earth and Mars for a piloted Mars mission are those that leave Earth with a departure velocity of 5.08 km/s (and no more) and leave Mars with a departure velocity of about 4 km/s. For unpiloted cargo, either the Hohmann transfer or option A's close-to-minimum-energy 3.34 km/s departure free-return trajectory is clearly best. So what's the punch line? Simply this: These optimal trajectories can easily be accomplished using state-of-the-art chemical propulsion.

Note: The ΔV needed by a mission and its departure velocity are related but are not the same. For those who are interested, the mathematical relationships relating them to each other and to rocket specific impulse and mission mass are explained in a technical note at the end of this chapter.

TABLE 4.2
Free Return Trajectories between Earth and Mars

Departure Velocity	Orbit Period	Time to Earth Return	Transit to Mars	Mars Aeroentry
A 3.34 km/s	1.5 years	3 years	250 days	Easy
B 5.08 km/s	2.0 years	2 years	180 days	Acceptable
C 6.93 km/s	3.0 years	3 years	140 days	Dangerous
D 7.93 km/s	4.0 years	4 years	130 days	Impossible

WHO GETS TO GO

So, now that we've determined our trajectory, we need to select our crew—who, and how many.

"The more the merrier" summarizes much of the human-factors literature concerning crew size for a long duration Mars mission. However, since the number of crew members drives the mass of all habitats, transportation stages, and launch vehicles, it is essential to keep crew size to a minimum for reasons of cost and technical feasibility. Furthermore, no matter how many backup plans and abort options the mission design includes, we must understand that in sending a crew to Mars, we are, one way or another, sending a group of humans into harm's way Thus, from a moral point of view, the fewer people we have on board the initial missions, the better. Finally, no matter how desirable a large social group may appear to be for company on a long trip, any examination of the history of human exploration on Earth will show that it is entirely possible for long-duration expeditions to be carried out successfully by *one person*, two people, or any other number.

The question then is how many people are really needed on a piloted Mars mission. Put another way, whom do we really need? If the mission were to fail, far and away the most likely cause would be failure of one or more of the mission-critical mechanical and electrical systems employed (propulsion, control, life support). The most important member of the crew then, the one on whom all others depend for their lives, is the *mechanic*. Call this person a flight engineer if you wish (he or she is an engineer in the sense of an old-time railroad locomotive or steamship engineer), but the mission needs an ace mechanic who can sniff out problems before they occur and fix anything that can be fixed. This job is so critical that, despite the need for small crews, I recommend carrying two people capable of handling it.

The next most important job on the mission is that of *field scientist*. Remember, the exploration of Mars is the raison d'être of a human mission to the Red Planet. After those needed to get the crew to Mars and back, the next most important personnel are those essential to competently carry out the mission's exploration goals. Since no science return would effectively be a form of mission failure, once again I recommend carrying two people for the job. One of the field scientists should be a geologist, oriented toward exploring the resources and understanding the geologic history of Mars. The other should be a biologist, directed toward exploring those aspects of Mars upon which hinge the question of past or present life. The biologist would also conduct experiments to determine the chemical and biological toxicity of Martian substances to terrestrial plants and animals, and the suitability of local soils to support greenhouse agriculture.

And that's it. With two mechanics and two field scientists, we have the ability to split the crew into two groups in which no one is alone (one out in the field with a ground rover, say, while the other remains at the base camp) and someone expert at fixing malfunctioning equipment and someone capable of doing scientific work are present at all times. There is no need for people whose dedicated function is "mission commander," "pilot," or "doctor." True, the mission will need someone who is in command, and a second in command for that matter, because in dangerous circumstances it is necessary to have someone who can make quick decisions for all without electioneering or debate. But there is no room for someone whose sole function is to manage others to get the job done. Similarly, there can be no one on board whose job description is "pilot." The spacecraft will be capable of fully automated landing, and piloting skills would at most be useful as a contingency backup to the automated flight system during a few minutes of the two-and-one-half-year mission. If such a manual flight control backup is desired, then one or more members of the crew could be given cross training as a pilot (it's much easier to train a geologist to be a pilot than a pilot to be a geologist). Finally, no ship's doctor as such. The great Norwegian explorer Roald Amundsen always refused to carry doctors on his expeditions, noting that they were injurious to morale and that the large majority of medical emergencies that occur on expeditions can be handled just as well by experienced explorers. And, truth be known, behind their public relations facade, nearly all astronauts hate space doctors. You would too in their shoes—just think about trying to get a hard job done while somebody constantly jabs you with needles, wires, and thermometers. In place of a physician, all crew members will be trained in first aid, and expert systems on board and medical consultation from Earth will be available to diagnose readily treatable conditions (ear infections and the like). Such diagnoses could be assisted by having a crew member be someone who had either practiced general medicine at some point earlier in his or her career or who had been crossed-trained to a medical assistant's level of knowledge, and equipping this individual with a country doctor's black bag and a stock of broad-spectrum antibiotics. The biologist would be a natural candidate for such a cross-trained role. However, the idea of having a dedicated top-notch doctor on board who spends his or her time reading medical texts and honing skills by practicing surgery with virtual reality gear, or worse, being a pest by subjecting the rest of the crew to an in-depth medical study, is cumbersome and unnecessary.

To summarize in *Star Trek* terminology, what a piloted Mars mission needs are two "Scottys" and two "Spocks." No "Kirks," "Sulus," or "McCoys" are needed, and more importantly, neither are the berths and rations to accommodate them.

We can do the mission with a crew of four.

DIRECT LAUNCH

All interplanetary missions flown to date have been flown by "direct launch"—a launch vehicle lifts a spacecraft to LEO, and then uses its upper stage to throw the spacecraft on a trajectory to its planetary destination. It's the way the *Mariner* and *Viking* missions reached Mars, and it's also the way the Apollo lunar missions reached the Moon. No missions beyond LEO have ever been flown by lifting the payload to an orbiting spaceport and transferring it to a freshly fueled interplanetary cruiser just back from Saturn. No missions beyond LEO have ever been flown on an interplanetary spaceship constructed in space. The association in many people's minds of humans-to-Mars missions with such futuristic spaceship/spaceport scenarios has caused human Mars exploration to be relegated out of today's world and into the world of "The Future." But, if a manned Mars mission can be done by *direct launch*, then *we* can do it. Get rid of the spaceships and spaceports, and a human mission to Mars moves from the parallel universe of The Future into our universe. If we can do it by direct launch, then 90 percent of everything we need to send humans to Mars is available now

We've chosen the trajectory and the crew size. Now, can a realistic heavy-lift launch vehicle deliver, in no more than two tandem launches per mission, everything that's needed to conduct a four-person Mars mission in accord with the flight plan we've chosen? Let's see.

There is nothing magical about a heavy-lift vehicle—the United States built and operated one forty-five years ago. The Saturn V booster that sent the Apollo astronauts to the Moon went into operation in 1967 after a five-year development program, and operated without a single failure for an eight-year period until 1975 when the last working unit launched the Apollo-Soyuz mission. The Saturn V could lift 140 tonnes to LEO. If we wanted equivalent capability today, one foolproof way of doing it would be to reengineer the dies and start producing Saturn Vs. There are other ways to get the job done, though. For example, using Space Shuttle hardware it's possible to produce a heavy-lift booster in the same class by attaching a pod of four Space Shuttle main engines (SSME) to the bottom of a Space Shuttle external tank (ET), attaching two Space Shuttle solid rocket boosters (SRBs) to either side of the ET, and positioning a hydrogen/oxygen upper stage on top of the ET. This is the Ares booster design David Baker developed for Mars Direct. Depending on the thrust of the upper-stage motor used, an Ares could deliver between 121 tonnes (with a 250,000 lb upper-stage thrust) and 135 tonnes (with a 500,000 lb thrust SSME on the upper stage) to LEO. The Russians had a heavy-lift vehicle during the 1980s and 1990s called the Energia, which could also be revived. The demonstrated model can only lift 100 tonnes to LEO, but an upgraded design, the Energia-B, boasts a 200 tonne capability. During the Space Exploration Initiative's short life, NASA developed dozens of heavy-lift booster designs of all sorts with capacities between 80 and 250 tonnes. In short, if the United States wants a heavy-lift booster, we can certainly get one.

While on paper a booster can be designed to any size desired, reality is different. Some super boosters have been designed with 1,000-tonnes-to-LEO capability. Sounds great, but they would probably blow away Orlando when they took off (or at least the Kennedy

Space Center). So, let's be exceptionally conservative and assume that the United States—today—can build a heavy-lift booster with a capability no greater than the one we fielded in the 1960s. Let's baseline our booster at 140-tonnes-to-LEO capability, exactly the same as a Saturn V. Would such a launch system be good enough to launch the Mars Direct mission by direct throw?

Part of the answer to this question is given in Table 4.3, which shows the amount of payload that a single launch of our 140-tonne-to-LEO booster can deliver to the Martian surface after a preliminary aerocapture at Mars. The table shows variants for both the cargo and piloted outbound trajectory, and for the assumption that the third stage of this vehicle is either a state-of-the-art hydrogen/oxygen chemical stage with a specific impulse of 450 seconds, or a near-term nuclear thermal rocket (NTR) with a specific impulse of 900 seconds.

The payload delivery capabilities shown in Table 4.3 assume that aerobraking is employed to capture the spacecraft into Mars orbit. This is clearly the optimal way to perform Mars orbital capture (MOC) in the Mars Direct missions because all the payload is destined for the Martian surface, and so it must carry an aeroshield in any case. Using aerocapture in the Mars Direct mission thus eliminates a significant propulsive ΔV essentially for free. If rocket propulsion had to be employed for this maneuver instead of aerobraking, the payloads delivered would be about 25 percent less. In mission plans such as that in the NASA 90-Day Report, aerocapture faced many technical difficulties. Aerobraking the enormous Battlestar Galactica spacecraft the plan called for

TABLE 4.3
Payload Delivery to Martian Surface from 140 Tonne to LEO HLV

Mission	Trans-Mars Injection Stage	Trans-Mars Throw Capability	Payload Delivered to Surface
Cargo	H_2/O_2	46.2 tonnes	28.6 tonnes
Piloted	H_2/O_2	40.6 tonnes	25.2 tonnes
Cargo	Nuclear thermal rocket	74.6 tonnes	46.3 tonnes
Piloted	Nuclear thermal rocket	69.8 tonnes	43.3 tonnes

would require huge aeroshields that could only be built in orbit, and that, as I've noted, is really not a credible proposition. Furthermore, the opposition-class trajectories employed by those missions hit Mars really hard, thereby increasing the heating and mechanical loads on the aerobrakes during atmospheric entry. Mars Direct uses lower energy conjunction-class trajectories that have lower entry velocities and thus lower heating rates and experience much lower aerodynamic deceleration forces. More decisively, the spacecraft that need to be decelerated in the Mars Direct plan are relatively small, so that aeroshields big enough to protect them can easily be made to fit inside the launch vehicle's payload fairing. This can be done in one of two ways: either by using flexible fabric umbrella-shaped aerobrakes that fold around the bottom of the payloads, as in the original Mars Direct design, or by replacing the fairing of the launch vehicle with a rigid, bullet-shaped shell that fits over the payload from on top. Both are feasible, and when used with Mars Direct-sized payloads, either can be launched "all-up" without any need for on-orbit assembly In addition, the guidance, navigation, and control requirements on Mars Direct aerocapture are less than in those plans where a subsequent Mars orbit rendezvous is anticipated, because it does not really matter exactly what orbit the vehicle captures into (since the orbit will be erased after the vehicle lands), so long as its orbital inclination is within the broad tolerances that will allow access to the designated landing site.

To deliver payloads, we can also employ an approach known as direct entry. As with aerocapture, aerodynamic drag against a planet's atmosphere, not rocket propulsion, decelerates the payload. There is a difference, though. With aerocapture, the spacecraft dips into the planet's atmosphere just enough to slow down and then reemerge from the atmosphere to place itself in orbit. In the case of direct entry, the entering spacecraft plunges deep into the atmosphere until all of its velocity is shed and then proceeds directly to landing. Most people consider aerocapture the better flight plan for a piloted Mars mission, because if the weather is bad, it allows the crew to assume a station on orbit until conditions improve for a landing. With direct entry, the vehicle is completely committed for a landing immediately after Mars arrival. Nevertheless, direct entry was successfully used on the *Mars Pathfinder, Spirit, Opportunity,* and *Phoenix* missions. This track record has created a data base that may encourage mission designers to employ direct entry on the piloted Mars mission as well.

The bottom line in all of this, however, is the payload delivered to the surface. If chemical propulsion is used, then the unmanned cargo flight launched by a single 140-tonne-to-LEO booster can deliver 28.6 tonnes to the Martian surface, while the faster piloted flight can deliver 25.2 tonnes. Can a manned Mars mission be designed within these mass limits? If it can't, we could always design a bigger booster or go ahead and develop the NTR stage. But let's see if we can make it with nothing better than a Saturn V and chemical propulsion. If we can, then more advanced technologies or propulsion capabilities and their associated benefits are icing on the cake.

SUPPLIES FOR THE CREW

Is our mass delivery capability sufficient? Well, let's take a look at the mission's supply needs. In Table 4.4 we see the consumables required for each member of the crew per day for each leg of the mission and the totals required to support a crew of four in each of the two habitation systems, the hab (which houses the crew during the outward voyage and during the surface stay), and the Earth return vehicle (ERV) cabin. The numbers given under the need/man-day column are NASA standards (quite liberal in the washing water category, as you may notice), except that I have replaced 0.13 kg/day of dehydrated food with 1.0 kg/day of whole (wet) food. Such a mixed diet is much better for crew morale on a long mission than dehydrated rations only, and actually costs the mission very little in the way of added mass, since the water content of the whole food serves to make up for the losses in the potable water recycling system. The life-support system assumed for the crew is a fairly low-efficiency physical/chemical one that recycles 80 percent of the oxygen and drinking water and 90 percent of the washing water (which can be of lower quality). Such a system is much simpler and less power hungry than futuristic ones based upon a closed ecology wherein 100 percent of food, oxygen, and water will supposedly be recycled.

Read between the lines of Table 4.4 and you'll immediately note the huge advantages Martian resources give us. In addition to manufacturing fuel, the ERVs produce copious amounts of water and oxygen. Without the ERV chemical processing plant, we would need to ship out an additional 7 tonnes of consumables with the hab. This would increase the required consumables from 7 tonnes to 14 tonnes, which since we only have the capability of delivering a 25-tonne hab, would be very difficult to accommodate. The 9 tonnes of water each ERV produces provide an excess over NASA nominal water requirements and that should be a real plus for the morale of a hardworking crew on a desert planet. For these reasons, in Table 4.4 there is no requirement to transport oxygen or water to support the hab surface stay. We also see that each hab flies out to Mars with enough food for an 800-day mission, which gives it more than enough provisions to handle a two-year free-return abort. In the latter case, the crew in the hab will have to exploit the 5 tonnes of methane/oxygen propellant in the lander stage to provide extra water and oxygen (unneeded as propellant in the event of a free return, which is concluded by aerocapture into Earth orbit), and reduce their use of wash water to 40 percent NASA nomi-

TABLE 4.4
Consumable Requirements for Mars Direct Mission with Crew of Four

Item	Need/ man-day	Fraction recycled (kg)	Wasted/ man-day	ERV Reqs 200 days in (kg)	Hab Reqs 200 days out (kg)	Hab Reqs 600 day Surface	Hab Reqs Total kg
Oxygen	1.0	0.8	0.2	160	160	0	160
Dry food	0.5	0.0	0.5	400	400	1200	1600
Whole food	1.0	0.0	1.0	800	800	2400	3200
Potable water	4.0	0.8	0.0	0	0	0	0
Wash water	26.0	0.9	2.6	2080	2080	0	2080
Total	32.5	0.87	4.3	3440	3440	3600	7040

nal levels. This will be uncomfortable and bad for morale, but it could be endured and survived, which is the only issue in the event of such an abort. Also, in Table 4.4, there is no wastage of potable water shown because potable water lost due to inefficient recycling is made up by water added to the system from the use of whole food.

Given these consumable requirements, the mass allocations for the ERV cabin and the hab can be assigned, and are presented in Table 4.5.

The ERV payload shown above will, after landing, convert its 6.3 tonnes of hydrogen feedstock into 94 tonnes of methane/oxygen propellant and 9 tonnes of water. Of the 94 tonnes of propellant produced, 82 tonnes will be used by the ERV for rocket propulsion to return the crew to Earth, while 12 tonnes will be available to support the use of

TABLE 4.5
Mass Allocations for Mars Direct Mission Plan

Earth Return Vehicle	Tonnes	Hab	Tonnes
ERV cabin structure	3.0	Hab structure	5.0
Life-support system	1.0	Life-support system	3.0
Consumables	3.4	Consumables	7.0
Electrical power (5 kWe solar)	1.0	Electric power (5 kWe solar)	1.0
Reaction control system	0.5	Reaction control system	0.5
Communications and information management	0.1	Communications and information management	0.2
Furniture and interior	0.5	Lab equipment	0.5
EVA suits (4)	0.4	Crew	0.4
Spares and margin (16 percent)	1.6	EVA suits (4)	0.4
ERV cabin total	11.5	Furniture and interior	1.0
Aeroshell	1.8	Open rovers (2)	0.8
Light truck	0.5	Pressurized rover	1.4
Hydrogen feedstock	6.3	Field science equipment	0.5
ERV propulsion stages	4.5	Spares and margin (16 percent)	3.5
Propellant production plant	0.5		
Power reactor (80 kWe)	3.5		
ERV total	28.6	Hab total	25.2

ground vehicles using internal combustion engines. If we count only the water and the 12 tonnes of rover propellant, and add them to those other parts of the ERV payload that are useful while on the Martian surface (such as the ERV cabin with its power and life-support system, the power reactor, the EVA suits, light truck, etc.), we find that each ERV flight delivers 36.5 tonnes of useful *surface* payload. The first Mars mission crew will have with them on the Martian surface two ERVs (the precursor, which made its propellant in advance of the crew launch, plus the backup, which flew out in tandem with the crew) plus one hab (which has 24.7 tonnes of useful surface payload). This adds up to 97.7 tonnes of useful surface payload available to the crew, roughly four times that of the traditional opposition-class mission featured in the NASA 90-Day Report (which had more than double this mission's initial launch mass). The surface payload available to the crew includes four pressurized volumes capable of supporting life: the hab, the two ERV cabins, and the pressurized rover. The crew thus has many safe havens available in case the primary life-support system in the hab should malfunction. In addition, they have 12 EVA (extra-vehicular activity) suits, five motorized vehicles (the pressurized rover, the two open rovers, and two light trucks), five primary power supplies (two 80 kWe nuclear reactors plus three 5 kWe solar power systems in the hab and the two ERVs), five backup power supplies (the engines on each of the motorized vehicles can be used to turn a generator), a *thousand kilograms* of combined field and lab scientific equipment, 14 tonnes of consumables from Earth plus 18 tonnes of Mars-produced water and 24 tonnes of rover propellant, plus two chemical plant systems, either of which is capable of producing oxygen from the Martian atmosphere at a rate roughly *fifty times* that required by the crew for life support. The plan must therefore be considered *extremely* robust. And in case *that's* not good enough for you, the redundancy can be multiplied further by taking advantage of the first launch window in which no crew is sent to Mars to send a complete hab, loaded with supplies but containing no crew, to accompany the precursor ERV to the first landing site (thus making the program's launch schedule two heavy-lift booster flights every other year, including the first). In that case, the crew would have available to them six habitable volumes including two complete habs, plus two complete ERV cabins, plus . . . but I think you get the point. No program of exploration on Earth has ever been conducted with anything approaching this level of backup redundancy. And we've done it all with 1960s technology Saturn V's, chemical propulsion, and no on-orbit infrastructure, assembly, or docking operations, or orbital rendezvous of any type at any point in the mission.

This ability to pile nearly unlimited but useful redundancy into the Mars surface camp compared to what can be provided to the crew in transit is another reason why Mars mission planners should try to maximize the crew's time on the Martian surface and minimize the time spent in transit. *Mission assets can be concentrated cumulatively on the Martian surface.* If this is done, then *the Martian surface becomes the second safest place in the solar system.*

BACKUP OR ABORT?

In the past, many Mars mission plans were constructed around the following scenario: Days before arrival at Mars, perhaps on arrival, the crew of a Mars expedition realizes they have to abort the mission. Our concern now is not why they have to abort, but how. How do they reach a safe haven? Well, obviously they have to return to Earth, and, though they had planned for a lengthy conjunction-class surface stay, they have fortunately brought along enough fuel for an opposition-class quick return to Earth. They can power away from Mars and head for Earth via a Venus flyby. They don't have to wait for a Hohmann transfer window to open, and who would in the case of an emergency? But let's think about this. There are costs involved in planning a mission around this abort option, and they are not trivial. First off, such missions require the extra payload needed both for a long surface stay and for a long trans-Earth cruise, as well as the extra propellant to send all this stuff onto a very high-energy opposition-class trajectory. It is hard to imagine a more costly approach to mission design. Furthermore, if the abort is not exercised, all of the extra mass delivery entailed by such a strategy is for naught. Moreover, the opposition return trajectory subjects the crew to 1.5 years of continuous deep space radiation doses (probably in zero gravity as well), high solar radiation during a close-in pass through the inner solar system, and very high g-loads at Earth return. All in all, such an abort return may be problematical to survive, and obviously, even if the crew does survive, the mission is a complete loss from the standpoint of exploration.

In the end, mission plans of this sort would do little to increase effectiveness, yet greatly increase mission mass and cost. Fortunately, we can solve the problem of what to do in an emergency by questioning one very basic assumption: does Earth have to be the only safe haven? The answer is a resounding no. Rather than design the mission around Earth-oriented abort options, the right strategy is to base the plan upon creating a safe haven in advance on the Martian surface, and aborting to it as our primary option. Such a haven can be reached much more quickly by an outbound crew than Earth can, and is thus much more likely to represent a real source of help in the event of trouble. The primary abort option is thus the same as the primary mission mode, imposes no mass penalty, and its invocation still allows the mission to be carried out. There are secondary aborts that do not involve carrying out the mission, but *the mission is not designed around them.* Put another way, rather than design the mission around *abort options*, you need to design it around a hierarchy of *backup plans.* This is the way the Mars Direct mission handles things.

Let's start the mission in LEO and see what kinds of aborts and backup plans are available to the crew as the mission proceeds. The first major event of the mission is the engine burn that will send the spacecraft onto trans-Mars injection (TMI). The total ΔV required to perform this maneuver is 4.3 km/s to put the spacecraft onto a fast conjunction, two-year free-return trajectory, which will get the crew to Mars in 180 days or so. However, a ΔV of 3.7 km/s is good enough to send the crew on a 250-day minimum energy trajectory to Mars, so if the engine burn is successful to at least that point, the crew will be sent on its way to carry out the mission. If the propulsion system on the TMI stage should fail to provide a ΔV of 3.3 km/s—the ΔV required to escape from Earth—the spacecraft will be left in an elliptical orbit about the Earth. In this case, the crew will use the hab's own propulsion system to dip the perigee (lowest point) of their orbit very slightly into the uppermost part of the Earth's atmosphere. After a number of orbits the drag induced by this maneuver will lower the apogee (highest point) of the orbit to altitudes that can be reached by the Space Shuttle Orion (such slow-aerobraking apogee-reduction maneuvers were successfully undertaken by the *Magellan* spacecraft at Venus in 1994, by Mars Global Surveyor at Mars in 1997, and every Mars orbiter since), after which a small propulsion burn by the hab will raise its perigee out of the atmosphere, circularizing and stabilizing the orbit. Once this is done, the crew can be retrieved (although there's not any rush; they have enough supplies for almost three years on board). If the TMI stage propulsion system should fail between 3.3 and 3.7 km/s ΔV the crew can get back to Earth orbit by retrofiring the hab's propulsion system; between its midcourse correction, Mars orbit propulsion maneuver system, and landing propellant, the hab has a total ΔV capability of 0.7 km/s—more than enough to negate the maximum excess ΔV of 0.4 km/s that could strand the crew between Mars and Earth. All of this, however, is hypothetical. A properly designed TMI stage would use multiple engines, each with a reliability for a burn of this length on the order of 0.99. The probability that two such engines would both fail is about 1 in 10,000, a negligible portion of total mission risk.

Once the TMI and midcourse burns have been successfully completed, the hab is targeted for an aerocapture pass at Mars. During the first 95 percent of the outbound flight, several options, including free-return aborts and powered flyby maneuvers, can be undertaken. However, once the lander has been targeted for an aerocapture trajectory (typically several days prior to aeroentry), the options of a free-return or powered flyby trajectory abort back to Earth become increasingly tenuous. At some point, on the order of several hours to one day prior to aerocapture, the ability to perform any trajectory abort is lost completely. But you have to make up your mind sometime, and the fact that the free return is available for the first 175 days of your 180-day trip is nothing to sneeze at.

Since an orbital rendezvous is not required in the Mars Direct mission plan, the accuracy of the capture orbit is unimportant, so long as the inclination of the orbit is such that it permits access to the surface site (that is, the inclination of the capture orbit must be greater than or equal to the latitude of the desired landing site). With this in mind, so long as the crew simply captures into an orbit around Mars, they'll be able to descend to the surface outpost that was delivered in the previous opportunity. This relaxation of aerocapture accuracy requirements translates into a relaxation of aerocapture guidance, navigation, and control requirements, significantly enhancing the attractiveness of aerobrake technology for the orbit capture maneuver in Mars Direct. If the maneuver is unsuccessful to the point that the hab will not capture into orbit, the crew could use the propulsive capability of the lander (up to 700 m/s) to augment the performance of the aerobrake. The crew might now be unable to descend to the surface in the hab, but they would be captured into orbit around Mars. They would then have two potential options. First, the crew could remain in orbit and rendezvous with either of the ERVs (the precursor or the one that followed them to Mars, either of which could be flown up to them by remote control) after 600 days in orbit. They could then transfer to the ERV and return to Earth. Alternatively, they could wait in Mars orbit for only 90 days or so until the ERV that is following them to Mars arrives, and then rendezvous with it prior to its landing. They could then have the option of transferring some propellant from the ERV to their hab, thereby enabling a hab landing (but sacrificing the ERV). Or, they could transfer to the ERV and land in it, leaving the hab behind in orbit. This could be done immediately after rendezvous, if there already was a hab on Mars to support surface operations (left behind by the crew of another mission). If not, the landing could be delayed, allowing the crew to spend most of the Mars stay time on orbit (where they could use the large supplies of consumables and ample quarters aboard the hab), and then perform a short-duration surface mission using the quarters of the two ERVs as their surface base.

However, since a safe haven and the possibility of completing the mission successfully exist on the surface of Mars, clearly the best option is to go there. For this reason, the aerocapture maneuver will choose to err on the side of entering the atmosphere too deeply, rather than risk skipping out into interplanetary space. Since the Mars Direct plan does not require capturing the ship into a loosely bound, highly elliptical orbit (as traditional missions do—such orbits require less propellant to get out of), the craft can be targeted toward a more tightly bound, slightly elliptical or circular orbit about Mars, which would make a skip-out almost impossible. If the ship enters the atmosphere too deeply to capture into a stable orbit, the crew can simply land the hab. After all, the plan is to ride it to the surface anyway.

Eliminating the need for a Mars orbit rendezvous prior to descent represents a major addition to mission safety, because it jettisons the need to cut it fine with an aerocapture maneuver that might risk skip-out. But, of course, Mars Direct exchanges an orbital rendezvous for a surface rendezvous. What about that? Well, let's consider. The surface rendezvous plan used in the Mars Direct mission also has several levels of backup to assure mission success. First of all, the ERV has been on-site for the two years prior to the crew's arrival, affording the opportunity to deploy robotic roving vehicles to fully characterize the rendezvous site well in advance of the crew's arrival, as well as to place a transponder on the best possible landing site identified in the vicinity. The ERV also mounts a radio beacon much like an instrument landing system transmitter at an airport that gives the crew exact position and velocity data during approach and terminal landing. We should remember that both *Viking* landers touched down within 30 kilometers of their targeted sites without active guidance, while the Apollo manned lunar landers were able to land within 200 meters of a target *Surveyor* spacecraft. With the aid of a feedback targeting control system and a guiding radio beacon, the landing should be within a few meters of the targeted point. However, if the landing should prove inaccurate by tens or even hundreds of kilometers, surface rendezvous can still be achieved through the use of the ground rover carried in the hab, which has a one-way surface range of up to 1,000 kilometers. Because the crew has landed in a full habitat, and not a short-duration small landing vehicle, they have the ability to support themselves for a long period of time if they should land in an isolated location, and therefore third and fourth levels of backup are also available. As the third level, if the landing rendezvous fails by distances of planetary dimensions, the second ERV following (by several months) the piloted hab to Mars can be redirected to the hab's landing site. As a fourth level of backup, the entire crew has landed on Mars in a hab carrying enough supplies for a two-year surface stay. If all else fails, they can simply tough it out and wait for the next launch window when more supplies and another ERV can be sent out to them.

Because it uses in-situ propellant for ascent, the Mars Direct plan does not have the capability for abort-to-orbit during descent. Once you start for the surface of Mars, there's no turning around. However, it is extremely doubtful that any lander, fully fueled for ascent though it might be, could really ascend successfully to orbit by taking off from the back of an aeroshield buffeting its way at hypersonic speeds through the Martian atmosphere. (Such a maneuver would require flying the ascent vehicle *through a hypersonic shock wave* to the rear of the aeroshell, and then turning it around in midair to shift its engines from a deceleration to acceleration orientation!) In exchange for giving up the illusory option of abort-to-orbit during landing (which the crew of a traditional mission landing in a fully fueled ascent vehicle would like to believe they have, but don't) the Mars Direct crew gains an element of real safety. That is, they know, not only before atmospheric entry but even before they have left Earth, that they have a fully fueled Earth return vehicle that has already survived the trauma of landing waiting for them on the Martian surface. Furthermore, in carrying out their own landing, they will be traveling in a large and sturdy hab containing multiple pressurized compartments and an up-and-running long-duration life-support system, and one that is almost empty of rocket propellant at the time of touchdown. In contrast, the crew who descends in a fueled Mars ascent vehicle will have to make their landing in a small vehicle with minimal long-duration life support—one that is filled to the gills with high explosive rocket propellant.

As discussed in the previous section, because the Mars Direct mission concentrates its assets on the surface, and not in orbit, all required systems to sustain the crew during the long 600-day surface stay are multiply backed up, with the degree of redundancy increasing as the sequence of missions proceeds and hab after hab is added to the set of available surface facilities. When it comes to return to Earth, the crew has two complete Earth return vehicles with them on the surface, either of which can get them home without any other assistance and both of which can be checked out manually prior to departure. This is a radical improvement over the situation in a traditional mission plan, where a crew must ride the only available Mars ascent vehicle to a mission-critical Mars orbit rendezvous with a mothership that may have been in orbit for up to a year and a half with nobody minding the store and little in the way of onboard resources to effect repairs. The Mars Direct crew can personally check out their ERV before they commit themselves to ride on it, and have all the resources of the Mars base camp behind them if they need to effect repairs or adjustments. And if both ERVs are unsatisfactory, they can just sit tight at the Mars base because another hab loaded with supplies and another ERV will arrive within a few months of their scheduled departure time. In that case, they would have to extend their Mars stay by two years over the original plan, but it sure beats dying.

ADVANCED TECHNOLOGY OPTIONS

The transportation system used by the Mars Direct plan as described so far in this book can be executed with all existing technology: Saturn V's or equivalent heavy-lift boosters, chemical propulsion, and so forth. But certainly if some more advanced technologies should materialize, the plan can and should be prepared to take advantage of them. While many forms of advanced space transportation systems have been proposed—nuclear and solar electric (ion drive) propulsion, solar and magnetic sails, fusion and even antimatter rockets, to name some of the most prominent examples—only a few of these systems have the potential of materializing within the time frame of interest to initial manned Mars missions. These are nuclear thermal rockets (NTRs) and the closely related solar thermal rockets (STRs), which could replace chemical rockets for space transportation, and single-stage-to-orbit vehicles (SSTOs), which could replace expendable multistaged heavy-lift boosters for launch from Earth. That is not to say that nuclear electric ion drives, magnetic sails, fusion rockets, and other advanced systems are infeasible. Quite the contrary, they are thoroughly feasible and will probably dominate interplanetary commerce a century from now. For that reason we shall discuss them further in some of the later chapters of this book that deal with the more futuristic aspects of Mars colonization. However, just as Columbus would not have traveled very far if he had held his expedition on the dock until an iron steamship or a Boeing 747 was available for trans-Atlantic transport, so the first generation of Mars explorers will have to settle their hopes upon a more primitive set of technologies than will be available to travelers of a later era. Columbus crossed the Atlantic with vessels designed for Mediterranean and Atlantic coastal traffic. It was only after European outposts were created in the Americas that the technology driver came into being to propel naval architecture from Columbus's primitive craft to three-masted caravels, to clipper ships, to ocean liners, and to airliners. Similarly, establishing human settlements on Mars will drive the creation of more advanced forms of space propulsion. For that reason, up till now we have based our discussion of Mars missions entirely upon the current primitive state of space technology. That's the conservative approach. But there are technologies that could potentially be put into play in the relatively near future that could significantly improve mission performance or cut costs. Let's take a look at them.

Thermal rockets, either nuclear or solar, are the most likely candidates for a space propulsion system capable of replacing chemical rockets. The idea behind such systems is very simple. A heat source, either a nuclear reactor or a parabolic mirror focusing sunlight, heats a fluid to very high temperatures, turning it into ultrahot gas that is then expelled out a rocket nozzle to produce thrust. In other words, a thermal rocket is just a flying steam kettle. The performance of such a system is limited primarily by the maximum temperature the engine materials can tolerate, and this is generally believed to be in the neighborhood of 2,500° centigrade. The highest exhaust velocity, and thus specific impulse, obtainable by such a rocket will be supplied by the propellant gas having the lowest possible molecular weight. Therefore, hydrogen is the propellant of choice for thermal rockets. An NTR or STR using hydrogen propellant can achieve a specific impulse of 900 seconds (9 km/s exhaust velocity), twice that of the best hydrogen/oxygen chemical rocket engines.

And thermal rockets are not just a theory. In the 1960s, the United States had a program called NERVA (Nuclear Engine for Rocket Vehicle

Applications) that built and ground tested about a dozen NTR engines in sizes ranging from 10,000 lb thrust all the way up to 250,000 lb thrust. These engines really worked, and really delivered specific impulses of over 800 seconds, far beyond the wildest dreams of any chemical rocket engineer. Wernher von Braun planned to use NTRs as the propulsion system for the manned Mars mission NASA hoped to follow Apollo with by the early 1980s. But when the Nixon administration pulled the plug on NASA's post-Apollo Mars plans, the NERVA program went down the drain as well. The engines were never flight tested and the ground test facilities were left to rust. Many of the veterans of the NERVA program are still around, although most have passed retirement age. Even as I write, their priceless expertise with such systems is evaporating. Still, feasibility has been demonstrated.

During the period when the Space Exploration Initiative (SEI) was still alive, a faction in NASA led in spirit (but not in command authority) by Dr. Stan Borowski of NASA's Lewis (now Glenn) Research Center in Cleveland made an effort to revive America's NTR research and development program. This effort, which I vigorously supported, faced many hurdles in the political environment, not the least of which was the fact that the SEI's enormous projected price tag had disposed Congress not to spend a penny on anything associated with it. There were other problems too. In the 1960s, the antinuclear movement had not yet materialized as a serious political force and it was routine practice to test NTR engines in the open, with their potentially radioactive exhaust spewing straight up into the air of the Nevada test site. This procedure would not pass muster today. Instead, modern NTRs would have to be tested inside closed facilities containing scrubbers that would remove all radioactive products from the exhaust gases before releasing them into the environment. Depending on the size of the NTR engine, such a facility could be very large and costly, possibly on the order of a billion dollars, and the required environmental permits needed to build one could hold the program up for years. There was an existing facility, called the *LOFT*, already certified at the Idaho National Engineering Lab that, with minor modifications, could have been used to support testing a *small* NTR engine of perhaps 15,000 lb thrust. This would have saved a great deal of time and money. Such a small NTR would have been *large enough* to push the relatively small Mars Direct spacecraft out of LEO onto a trans-Mars trajectory, and would also have been *small enough* to be useful for a multitude of non-SEI applications, including sending unmanned probes to the outer solar system and military satellites to geosynchronous orbit. These other missions had real budgets, whereas SEI did not.

For this reason, I and several others argued long and loud for this option. However, in the early 1990s, when this debate was underway, NASA had not yet accepted Mars Direct, and the 15,000-lb thrust NTR would have been much too small to send "Battlestar Galactica" to Mars. So, because of NASA planners' unwieldy mission design, engines between 75,000 lb and 250,000 lb thrust were baselined. Furthermore, many of the people who had rallied around Borowski were representatives of institutions that hoped to receive large infusions of cash for the job of building the new giant test facility, and they pressured him accordingly. In addition, Borowski's bosses in the NTR program were NASA managers who generally favored the idea of making the NTR a big, long-duration program, and therefore were opposed to any shortcut, small-scale, quicker, cheaper approach. So, in the end, the big engine school won out. NASA dawdled away the SEI opportunity, drawing up plans for a $6 billion NTR program dedicated solely to SEI, with big facilities and a twelve-year development timeline. When SEI was canceled, the NTR went with it. Once the program was terminated, the rats jumped ship, leaving Borowski to argue for starting a small NTR program. Since then, everything has been on hold.

If the United States wanted to, I believe that we could launch a small NTR program that could produce a flight-ready, 15,000-lb thrust, 850 seconds Isp engine within four years for between $500 million and $1 billion. These estimates are based upon detailed discussions and studies done in conjunction with NERVA veterans and other experts working in industry and at several national labs. The cost is not trivial, but it is on the order of the cost of a single Space Shuttle launch and would create a whole new spectrum of space capabilities for the nation. Because such an engine has all sorts of potential applications, its development would be a wise thing to undertake whether we are planning on sending humans to Mars or not.

There's no denying however that pulling off a space nuclear program is a tall order nowadays. So, on the principle that half a loaf is better than none, a group of engineers associated with the U.S. Air Force's Phillips Lab in Albuquerque, New Mexico, have been pushing for the development of a solar thermal rocket system. The STR is an old concept that was first proposed by German V-2 veteran Krafft Ehricke in the 1950s, but it has never been flown. Concentrated sunlight supplies the power for an STR, thereby eliminating the nuclear baggage, but because of the diffuse nature of solar power, it's hard to make an STR with a thrust of more than 100 lbs or so. Furthermore, for obvious reasons, the system is completely ineffective in the outer solar system. Because the thrust is so limited, the STR could not be used to push a Mars Direct spacecraft all the way from LEO to trans-Mars injection. But it could be used in a long series (taking several weeks) of maneuvers known as "perigee kicks," in which the engine is fired for about thirty minutes each time the spacecraft passes through the lowest portion of its orbit. This would raise a Mars Direct spacecraft from LEO to a highly elliptical orbit just short of Earth escape. From this orbit the spacecraft would fire off to Mars with a short chemical burn, while the STR stage would either be expended or cycled back to LEO to lift another Mars spacecraft. Since the STR ΔV required to lift the spacecraft to near escape is about 3.1 km/s, while the total trans-Mars burn has a ΔV of 3.7 km/s (for cargo) to 4.3 km/s (for crew), the STR is accomplishing 72 to 83 percent of the trans-Mars propulsion job. It thus offers benefits comparable to, but somewhat less than, NTR.

What would be the benefits of these systems for Mars Direct? As we have seen, they would not be used to achieve fast flights to Mars. Short of the introduction of very futuristic propulsion systems (fusion engines, antimatter, etc.) that do not use ballistic trajectories, the right trajectory to take humans to Mars is the two-year free-return option, which takes about 180 days to reach Mars regardless of what propulsion system is used. But what the STR or NTR would allow us to do is take a lot more payload for the same launch mass. As we have seen, the use of NTR allows the delivery of 60 to 70 percent more payload to Mars than hydrogen/oxygen chemical propulsion when used for the trans-Mars-injection burn of the mission on our chosen trajectories. The STR would allow about 40 to 50 percent more payload to be sent than chemical propulsion allows. Therefore, if we use the same 140-tonne-to-LEO booster that we baselined for our chemically propelled missions, the use of these propulsion systems would allow expansion of the crew size to about six (three mechanics, three field scientists—no doctors!) with wider mass margins available for all mission components.

Alternatively, the superior throw capabilities of these systems could be used to reduce the size of the required launch booster, while keeping all payload allocations the same. Instead of a 140-tonne-to-LEO booster, the mission could be launched with a booster with a capability in the 85-tonne (for NTR) to 100-tonne (for STR) to LEO range. The former figure is about the capability for a "Shuttle C" (basically, a Shuttle launch stack but with an empty payload fairing replacing the Orbiter space plane, a launch vehicle that NASA estimates it could develop rapidly for a lot less than a Saturn V-class vehicle). The latter figure (100 tonnes) is the launch capability of a Russian Energia booster, although the comparatively narrow Energia payload fairing would have to be widened to accommodate the bulky hydrogen propellant that an STR or NTR driven mission would require.

But maybe the mission can be done without a heavy-lift vehicle at all. During the 1990s, the United States initiated a very ambitious program to develop a fully reusable single-stage-to-orbit (SSTO) vehicle. This program was inspired by space visionaries Gary Hudson and Max Hunter, and then given a big boost by the successful demonstration of a subscale, suborbital, reusable rocket (the McDonnell Douglas DC-X) in a "quick and dirty" program sponsored by Col. Pete Worden's team in the Ballistic Missile Defense Organization. (Bill Gaubatz, the DC-X program manager, pulled it off for $60 million, which is a useful fact to throw in the face of the next person who tells you that something you want done will have to cost $10 billion and take forever.) The project, then taken over by NASA and renamed "X-33," faced many technical hurdles, because assuming hydrogen/oxygen rocket propulsion (as all the X-33 designs did), the SSTO needs to have a dry mass equal to only 10 percent of its mass when filled with propellant. This is very difficult to do from a structural standpoint, as the hydrogen fuel is very bulky and the vehicle must be armored with a thermal protection system that can withstand reentry (expendable vehicles can be built a lot more flimsily). In order to make the SSTO work, many advances beyond the current state of the art will have to be implemented in lightweight structural materials, engines, and thermal protection systems. There is no guarantee that the required technical achievements will be accomplished, and in fact, the X-33 program ultimately ran out of steam and was cancelled when its prime contractor, Lockheed Martin, failed to meet its objectives on time and within the allowed budget. Still, a major national effort in this direction could be launched again, and American ingenuity has rarely failed when adequately funded and then pitted with true determination against a problem of this kind. Let's say that such a program were to be implemented, and that it succeeds. What then for Mars Direct?

Well, in order for the SSTO to be really useful for a Mars Direct mission, we would want to have a version whose engines can be made to burn both hydrogen/oxygen and methane/oxygen. (A straight methane/oxygen SSTO might also do. According to SSTO leader Max Hunter, a methane/oxygen system is just as promising for SSTO applications as hydrogen/oxygen. The greater density of the methane fuel allows for more compact and therefore lighter tanks, thereby compensating for its lower specific impulse compared to hydrogen.) This is not impossible; Pratt and Whitney RL-10 engines, which are designed for use with hydrogen/oxygen, have been run successfully with methane/oxygen on the test stand. Furthermore, some Russian rocket engine technology reportedly allows hydrogen/oxygen engines to be run alternatively with kerosene/oxygen, which is more of a stretch than a hydrogen/methane/oxygen tripropellant system (because methane is a lot more like hydrogen than kerosene is).

Okay, let's say that's what we've got. The SSTO has a dry mass of 60 tonnes, carries 600 tonnes of propellant (86 tonnes hydrogen and 514 tonnes oxygen), and can deliver a payload of 10 tonnes to low Earth orbit. So we fly one of these things to orbit with 10 tonnes of payload needed for the Mars mission and leave it there. Then, through a series of more than twenty additional SSTO flights, we deliver another 200 tonnes of propellant to the orbiting SSTO along with another

30 tonnes of cargo. (This cargo includes 20 tonnes of liquid hydrogen that will not be burned as propellant on the outbound trip, but used instead as hydrogen feedstock for the in situ propellant production process on Mars. It can still be stored along with the propellant hydrogen in the vehicle's fuel tanks, though.) So, now we have an orbiting SSTO loaded with 40 tonnes of cargo and enough propellant to send it off on a Minimum Energy trajectory to Mars. We'll call this spacecraft "ERV/SSTO 1." Off it goes, to aerocapture and land on Mars with the full cargo carried by the regular Mars Direct ERV (any SSTO designed for Earth reentry has more than enough thermal protection to deal with a Mars reentry). As in the standard Mars Direct plan, it would now deploy its reactor and start its propellant plant to turn its 20 tonnes of hydrogen cargo into 332 tonnes of methane/oxygen bipropellant (320 tonnes for the trip back, 12 tonnes for the surface rovers) plus 9 tonnes of water. (It must produce far more methane/oxygen than in the standard Mars Direct plan because it is a single-stage vehicle, whereas the Mars Direct ERV is a two-stage vehicle, and it carries a relatively massive structure required for reusable operations. Both of these exert a price when it comes to propellant requirements.) While this is going on, another SSTO has lifted to LEO carrying 10 tonnes of cargo. A series of 24 further flights by another SSTO loads the first with another 20 tonnes of cargo, an additional 220 tonnes of propellant, and on the final flight, the crew. This second SSTO, the "Hab/SSTO 1," now has a crew, 30 tonnes of cargo, and enough propellant to send it to Mars on a 180-day fast-conjunction trajectory. Presumably, the loading of the second SSTO is timed so that it ends shortly before the opening of the Earth-Mars launch window. That being the case, and the refueling of the first SSTO having been completed on the Martian surface, the crew can then set off for Mars. Arriving at the Red Planet 180 days later, they rendezvous on the surface with ERV/SSTO 1. Shortly after the crew arrives, a second unmanned cargo SSTO arrives at the site, ERV/SSTO 2, and starts making propellant for the next manned mission (and otherwise provides the crew of Hab/SSTO 1 with a backup) in analogous manner to the standard Mars Direct mission sequence. The crew remains on the surface for 600 days, and then abandons their Hab/SSTO 1 on the surface, flying ERV/SSTO 1 back to Earth. Shortly

after they depart Mars, another SSTO (Hab/SSTO 2), containing a crew of four astronauts will arrive at the base to continue exploration, trailed in convoy by another unmanned Earth-return type SSTO, ERV/SSTO 3. The crew of Hab/SSTO 2 will return to Earth in ERV/SSTO 2, and so forth, the sequence of missions continuing in this way indefinitely, with each mission adding another Hab/SSTO to the base. All SSTOs that do not remain on Mars return to Earth for reuse, so nothing is expended, making such a plan potentially highly economical.

Note that each piloted Mars mission conducted in this way requires a total of 49 SSTO flights. This would be totally ludicrous if the SSTOs were to operate in any way similar to existing launch vehicles, with flight rates on the order of one per month. However, if as SSTO proponents advertise, the SSTOs could be made to operate more in the manner of airplanes, with quick turnarounds and flights on the order of several per week, or faster, it's conceivable that this could be a viable plan. It's also a very high-tech approach, however. In addition to the demand that the SSTO achieve as yet unapproached performance and operability, the scenario also requires that both liquid oxygen and liquid hydrogen be transferred from one orbiting SSTO to another in zero gravity. Now both liquid oxygen and liquid hydrogen are cryogenic (ultracold) fluids, and no zero-gravity transfer of cryogens from one tank to another has ever been done. It is an operation fraught with problems. The cryogens would freeze a flexible bladder if you tried to use such a device to move the fluids from one tank to another, and pumps won't work because in zero gravity there is no way to get a fluid to go to the point of suction (the pump gets to bite out one mouthful of fluid and after that just sits there with an empty space in front of it). It might be possible to settle a tank by slowly accelerating the vehicles with rocket thrusters, or rotating them on a spinning platform, and capillary and other devices that rely upon surface tension to control fluid movements have also been proposed. In addition, at least for oxygen, there is the possibility of controlling fluid motions with magnets. (Liquid oxygen is paramagnetic—you can pick it up with a magnet.) In short, while the situation is not hopeless, a lot of work needs to be done before this plan can be regarded as credible.

So for now, my bets are on old-fashioned Mars Direct, complete with expendable heavy-lift boosters, chemical propulsion, horse-drawn rovers (well, not quite), and the rest of the primitive paraphernalia of our current Dark Age of space exploration. There may be better ways of getting to Mars, and when they come to hand we will use them. But chances are they won't materialize until and unless we use what we've got now to get to Mars and get the ball rolling. What was it the old salts use to say about who and what conquered the seven seas? Iron men and wooden ships, not wooden men and iron ships. So it will be with Mars. We can do it with what we have now.

ΔV AND HYPERBOLIC VELOCITY

In this chapter I've talked a lot about ΔVs and hyperbolic velocity. The two concepts are not the same, but they are related.

Velocity change, or ΔV ("delta-V"), measured in units of speed, such as kilometers per second (km/s), is the fundamental currency of rocketry. If you have spacecraft with a given dry mass M (i.e., empty of propellant), and a certain amount of propellant, P, and a rocket engine with an exhaust velocity C, the following equation, known as "the rocket equation," shows how big a ΔV the system can generate:

$$(M + P)/M = e^{\Delta V/C} \quad (1)$$

So the quantity (M + P)/M, known as the vehicle's "mass ratio," increases exponentially in proportion to ΔV/C. If ΔV/C = 1, then the mass ratio equals $e^1 = 2.72$. If ΔV/C = 2, the mass ratio equals $e^2 = 7.4$. If ΔV/C = 3, the mass ratio equals 20.1. If ΔV/C = 4, the mass ratio equals 54.6. The exponential is a very strong function; a small increase in ΔV or decrease in C can cause a very big jump in the mass ratio. In fact, the situation is worse than this, because the dry mass M has to include not only the payload you are trying to push, but also includes the mass of the tanks required to hold the propellant and the engines big enough to push the spacecraft with its propellant, and both of these parasitic weights also increase in proportion to P. So as ΔV/C goes up, the mass of the spacecraft goes up faster than the exponential, so much so that depending on the lightness of the structural materials and the density of the propellants employed, somewhere between ΔV/C = 2 and ΔV/C = 3 the mass of a single-stage spacecraft will go to infinity! This is the reason why rocket engineers will kill to get ΔV down and C up.

By the way, if you're interested, you can get a rocket's exhaust velocity in meters per second by multiplying its specific impulse, or Isp, by 9.8. If you want C in units of kilometers per second, multiply the Isp by 0.0098.

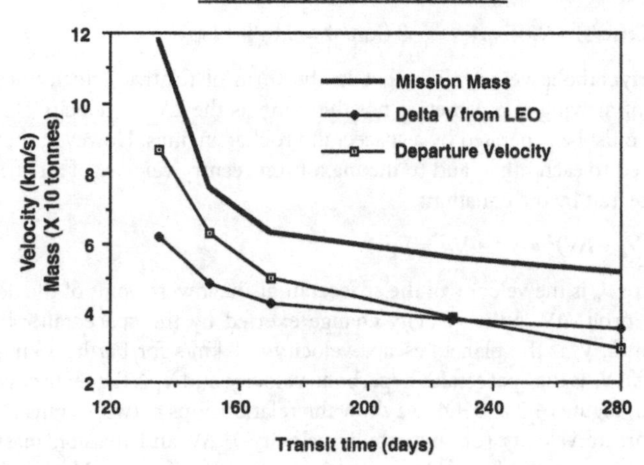

FIGURE 4.3
Relationship between average transit time, departure velocity, ΔV, and spacecraft mass for a 20-tonne spacecraft leaving low Earth orbit (LEO) for Mars. Spacecraft propulsion is hydrogen/oxygen with a specific impulse of 450 seconds. Note that mission mass rises steeply for transits less than 170 days.

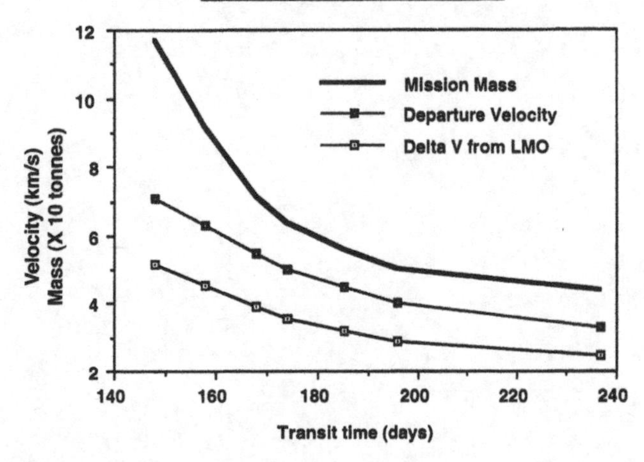

FIGURE 4.4
Relationship between transit time, departure velocity, ΔV, and spacecraft mass for a 20-tonne spacecraft leaving low Mars orbit (LMO) for Earth. Spacecraft propulsion is methane/oxygen with a specific impulse of 380 seconds. Note that mission mass does not begin to rise sharply until you attempt to reduce transit times below 170 days.

$$C\ (m/s) = 9.8(Isp) \qquad C\ (km/s) = 0.0098(Isp) \tag{2}$$

Hyperbolic velocity, whether in the form of relative velocity departing or arriving at a planet, is not the same as the ΔV or velocity change, that must be generated by a spacecraft's rocket engines. However, they are related to each other, and to the maximum reentry velocity of an arriving spacecraft by the equation:

$$(V_0 + \Delta V)^2 = V_e^2 + V_h^2 = V_r^2 \tag{3}$$

where V_0 is the velocity of the spacecraft at the lowest point of the departure orbit, ΔV is the velocity change exerted by the spacecraft's rocket engines, V_e is the planet's escape velocity (11 km/s for Earth, 5 km/s for Mars), V_h is the spacecraft's hyperbolic velocity, and V_r is the reentry velocity. In Figures 4.3 and 4.4, we show the relationships between transit time, departure velocity (or "hyperbolic velocity"), ΔV and mission mass for 20-tonne spacecraft leaving low orbit about either Earth or Mars to make an interplanetary transit.

5: KILLING THE DRAGONS, AVOIDING THE SIRENS

In olden days, before the Earth was well explored, mapmakers used to decorate unknown regions of their maps with various imaginative creations, not the least of which were menacing dragons that could swallow a ship whole, and delightful but equally dangerous sirens who could lure sailors to wreck on rocky shores by the temptation of their sweet songs. The dragons may have been imaginary, but even imaginary dragons can and did prey upon the minds of would-be voyagers, and by so doing stifled human exploration for centuries. And sirens never did need to be real in order to be heard, and heard they were, drawing many a hopeful venture fatally off course.

Well, things haven't changed that much. Today, those who hope to raise a mission to Mars find their charts filled with dragons too. Reports of horrible beasts with names such as radiation, zero-g, human factors, dust storms, and back contamination intrude into the discussion of mission plans, and do their worst to terrorize would-be crews (unsuccessfully), would-be mission planners (somewhat successfully), and would-be mission sponsors (very successfully). A siren is there too, named Diana, the moon goddess, and her songs can be heard calling the Martian mariners to divert their ships once more to a barren destination. If we're going to get to Mars, we're going to have to clear the maps. The dragons, cyclops, and other monsters of the mind must be killed, and the siren exposed for the fraud she is.

RADIATION HAZARDS

One of the leading dragons barring the path to Mars goes by the name *radiation.* Radiation is deadly, we are told, and only by using ultrafast spacecraft that can speed through the supposedly radiation-infested seas of space in impossibly short times can we be sure of a safe voyage. Or alternatively, we are told that only by using huge spacecraft with masses approaching those of asteroids can we shield the crew well enough to assure their health. We are further warned that cosmic radiation is something totally new in character, and only after we have spent decades studying its long-term effects on humans in interplanetary space can a trip to Mars be risked.

But, in fact, almost all the assertions quoted in the above paragraph are sheer nonsense. The only one of them that is even close to being true is the first, that "radiation is deadly," which it certainly is, but only if taken in excessive quantities.

Human beings have evolved in an environment featuring a significant amount of natural background radiation. In the United States today, people who live near sea level receive an annual radiation dose of about 150 millirem. (A millirem is a thousandth of a rem, the basic unit used to measure radiation doses in the United States. Europeans use Sieverts. One Sievert equals 100 rem.) Those who can afford to live in Vail or Aspen, on the other hand, take an annual dose of more than 300 millirem in consequence of their willingness to forgo a significant fraction of the cosmic ray shielding offered to them by the Earth's atmosphere. Because we have evolved in a radiation field, humans actually need radiation to stay healthy. It may be counter to popular belief and the orientation of various governmental regulatory agencies, but numerous studies of individuals subjected to an unnaturally radiation-free environment have shown significant health deterioration relative to controls exposed to natural levels of ionizing radiation. This phenomenon, known as *hormesis*,[15,16] is caused by the fact that the human body needs a certain amount of pummeling by natural radiation in order to keep its self-repair mechanisms stimulated. It is unclear what the optimum radiation exposure level for human health is, but it is not zero.

That said, it is certainly true that very large amounts of radiation delivered over very short amounts of time, such as the exposure to a huge dose within seconds via the gamma-ray flash from an atomic bomb blast, or within minutes by exposure to unshielded release products from a disabled nuclear reactor, can and will kill. The effects of such prompt doses of radiation are well-known from studies of the victims of the Hiroshima and Nagasaki bombings. These studies have revealed that prompt doses of less than 75 rem result in no apparent health effects. If the doses are between 75 and 200 rem, radiation sickness (whose symptoms are vomiting, fatigue, and loss of appetite) will appear in from 5 percent to 50 percent of exposed individuals, with the percentages increasing from the low to high end of this range as the dose increases from 75 to 200 rem. At this level of exposure almost everyone recovers within a few weeks. At 300 rem, radiation sickness is universal, and some fatalities start to appear, rising to 50 percent at 450 rem and 80 percent at 600 rem. Almost no one survives doses of 1,000 rem or more.

These, however, are the effects of prompt doses, which is to say doses that occur on a time scale much shorter than the weeks-to-months time scale for cellular reproduction and bodily self-repair. The situation is much like drinking alcohol or any other chemical toxin. A man could drink a martini a night for years and suffer no obvious ill effects, his liver having adequate time to cleanse his body after each drink. Drinking a hundred martinis in a single night, though, would kill him. Similarly, radiation causes damage to living organisms by inducing chemical reactions within cells that create toxic substances that can kill or otherwise derange individual cells. Below a certain dose rate, the self-repair capabilities of individual cells can act fast enough to reject the radiation-induced toxin and save the cell. At significantly higher rates, human body tissues acting as a whole are able to generate replacement cells for those that have become casualties, before the loss of those cells causes problems for the body as a whole. It is only when dose rates occur at a pace that overwhelms these self-repair mechanisms that severe health impacts occur.

Now, in addition to causing prompt radiation sickness and death when dealt out in overwhelming amounts, smaller chronic doses of radiation can increase the statistical probability of cancer in humans and other animals. This is because the radiation-induced toxin introduced into a cell via a radiation hit can be a carcinogen. The exact relationship between such chronic doses and delayed cancer effects is not universally agreed upon, but it has been studied in far greater detail than the effects of any of the chemical carcinogens present in the human environment. For example, prior to 1960, large-scale radiation treatment to spinal bone marrow was used in Britain to deal with ankylosing spondylitis. Numerous follow-up studies were done on those so treated to look for radiation-induced leukemia. In the largest of these studies 14,554 adult patients were tracked for twenty-five years following their treatments, which ranged from 375 to 2,750 rem each. Of the group studied, sixty individuals died from leukemia, which compares unfavorably with an expected death rate from leukemia of six for a random group in the contemporary British population. Nevertheless, despite the huge doses, the fatality rate of the irradiated group was less than 0.5 percent. On the basis of this and hundreds of other such studies, the authoritative National Academy of Sciences-National Research Council study known as the "Biological Effects of Ionizing Radiation (BEIR) report"[17] estimated the statistical probability of fatal cancer within thirty years induced by chronic doses of radiation totaling 100 rem in individuals over the age of ten (see Table 5.1).

So, according to BEIR estimates, the likelihood of fatal cancer is 1.8 percent within thirty years for every 100 rem received. So, if a female astronaut gets a dose of 50 rem over the course of a two-and-one-half-year Mars mission, and after her return lives thirty years until she dies, the chance of her getting a fatal cancer due to that exposure would be $50/100 \times 1.81\% = 0.905\%$. (The chance of her getting a fatal cancer within one year would be 1/30th this amount, or 0.03 percent. The risk of radiation-induced cancer occurring during the mission itself is almost negligible.) If the astronaut were a male, the chance would be

TABLE 5.1
Estimates of Cancer Risk Due to Chronic Radiation Doses Totaling 100 rem

Type of cancer	Probability of Fatal Cancer within 30 Years
Leukemia	0.30%
Breast	0.45%
Lung	0.40%
GI, including stomach	0.30%
Bone	0.06%
All other	0.30%
Total	1.81%

slightly less, 0.68 percent, because of the elimination of breast cancer risk. Assuming these astronauts don't smoke, the chance of them dying of cancer if they did not go to Mars would be about 20 percent. Therefore, by taking the dose associated with the trip, they increased their cancer risk from 20 percent to somewhat less than 21 percent.

Now, in the above example I cited a chronic (not prompt) dose of 50 rem delivered over the course of a two-and-one-half-year Mars mission. The question then is, how do the mission profiles available to a present-day piloted Mars mission affect the radiation dose the crew may be expected to receive?

There are two kinds of radiation that can affect astronauts on a Mars mission: solar flares and cosmic rays.

Solar flares are composed of floods of protons that burst forth from the Sun at irregular and unpredictable intervals on the order of once per year. The amount of radiation dose a solar flare would deliver to a completely unshielded astronaut can be hundreds of rem in the course of several hours, which as we have seen would be enough to cause radiation sickness or even death. However, the particles composing solar flares individually each have energies of about one million volts, and can be stopped relatively easily by a modest amount of shielding. For example, if we look at the three largest solar flares recorded in history, those of February 1956, November 1960, and August 1972, we find that the dose they would have delivered to an astronaut protected only by the hull of an interplanetary spacecraft like our hab (which with its hull, furniture, miscellaneous engineering systems, fittings, and other objects has about 5 grams per square centimeter of mass spread around its periphery to shield its occupants) would have averaged about 38 rem, while if the astronaut had gone into an onboard pantry storm shelter (where the Mars Direct hab has about 35 grams per square centimeter of shielding see Figure 5.1) he could have been shielded by stacked provisions reducing the dose to about 8 rem.[18,19,20] If he had been sitting in the hab *on Mars* during an event representing the average of these flares, he would have taken about 10 rem if outside the shelter, or 3 rem within the shelter. (The Mars surface doses are much lower because the planet's atmosphere and surface shields out most of the flare.)

Cosmic-ray doses are different. Because they are composed of particles with energies of billions of volts, it takes meters of shielding to stop them, which basically makes cosmic-ray shielding impossible during interplanetary flight. On Mars, however, cosmic-ray shielding can readily be provided by the planet itself which masks out all cosmic rays coming from below, and by the use of sandbags to block out at least part of the cosmic-ray dose impinging upon the hab from above.

FIGURE 5.1
A schematic of the Mars Direct hab. In the event of a solar flare, the airlock could double as a storm shelter for the crew.

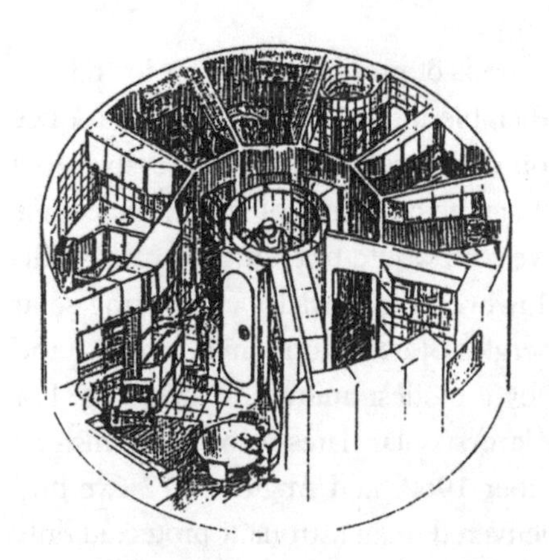

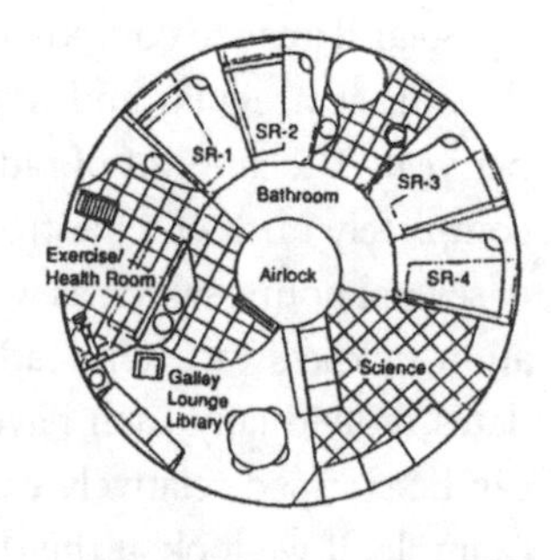

Also, unlike solar flares, cosmic rays don't occur in huge occasional floods. Rather they occur as a fairly constant but thin rain of radiation. An astronaut in the hab during flight through interplanetary space will take a cosmic-ray dose varying between 20 rem and 50 rem per year, depending upon where the Sun is in its eleven-year sunspot activity cycle. The biggest cosmic-ray doses will occur during the period of minimum solar activity, because during the period known as "solar max" the Sun's magnetic field expands and actually manages to shield the whole solar system to some extent against cosmic rays coming in from interstellar space. So as an average, however, 35 rem per year of cosmic-ray doses can be expected during interplanetary flight. If the crew was unsheltered on the surface of Mars, the cosmic-ray dose would be about 9 rem per year, while under shelter (a sandbag-roofed hab) it would be about 6 rem per year. Since the crew will be spending most, but not all of its time on Mars in the hab, 7 rem per year of cosmic rays is probably a reasonable average for that phase of the mission.

If we put all this data together with the flight profiles of the conjunction and opposition missions, and assume that a solar flare equal to the average of the three worst in history occurs at a rate of once per year during the mission, we obtain the predicted radiation doses shown in Table 5.2.

As discussed in the previous chapter, the Mars Direct mission would use the conjunction trajectory, whose estimated round-trip mission radiation dose varies between 41 and 62 rem, depending upon whether the Sun is at solar min or solar max phase of its eleven-year cycle. So,

TABLE 5.2
Radiation Dose Experienced on Mars Missions

	Conjunction	Opposition
Cosmic rays in transit	31.8 rem	47.7 rem
Solar flares in transit	5.5 rem	9.6 rem
Cosmic rays on Mars	10.6 rem	0.8 rem
Solar flares on Mars	4.1 rem	0.3 rem
Total average dose	52.0 rem	58.4 rem

the 50-rem estimate for a round-trip Mars mission is realistic, reflecting an average value between solar min and solar max conditions. We can also see that the worst expected solar flare dose on the Mars Direct mission is about 5 rem, far below the 75 rem threshold for any prompt radiation sickness effects.

Looking at Table 5.2, note also how silly the arguments for the opposition-class mission are from the standpoint of radiation dose reduction. Despite its much greater mass and cost, and much lower mission value (due to its limited stay on Mars), the total radiation dose received on the opposition-class mission is *more* than the conjunction mission, and its expected prompt dose from solar flares is 75 percent higher. But basically, the chronic doses experienced on either of these trajectories are predictable, and are negligible when compared to all the other risks that must be accepted in manned space flight. The only real risk from radiation is the possibility of a freak solar flare delivering a prompt dose far in excess of anything that has been measured over the past fifty years. The chance of this occurring is much higher on the opposition-class mission due to its close-in pass to the Sun. Thus, there is no radiation-dosage rationale for choosing the opposition mission plan over a Mars Direct-style conjunction or even minimum energy trajectories. Quite the contrary, from the perspective of radiation hazards, the opposition trajectory is the worst choice possible.

By the way, contrary to the scaremongering of certain people who would like to obtain large research budgets in this area, there is nothing extraordinary about cosmic-ray radiation doses compared to other types of radiation doses. Cosmic rays deliver about half the radiation dose experienced throughout life by people on the surface of the Earth, with those living or working at high altitude receiving doses that are quite significant. For example, a trans-Atlantic airline pilot making one trip per day five days a week would receive about a rem per year in cosmic-ray doses. Over a twenty-five-year flying career, he or she would get more than half the total cosmic-ray dose experienced by a crew member of a two-and-one-half-year Mars mission.

In fact, because cosmic ray dose rates in low Earth orbit are fully 50 percent as much as those in interplanetary space, some half-dozen astronauts and cosmonauts (Waltz, Foale, Krikalyov, Solovyov, Polyakov, and Avdeyev) participating in Mir or ISS missions have already received cosmic radiation doses equal to, greater than, or even double those that would be received by members of a human Mars mission, and *none* have exhibited any radiological health effects.

So, once again, using only chemical propulsion, not warp drive, we can fly a crew to Mars and return them home with radiation doses limited to 50 rem or so. While such doses are not to be recommended to the general public, they represent a small fraction of the total risk of not only space travel, but such common recreations as mountain climbing or sailboarding. Radiation hazards are not a show stopper for a piloted Mars mission.

ZERO GRAVITY

Another dragon barring the path to Mars is the menace of zero gravity. Long-duration exposure to zero gravity carries the risk of serious deterioration of human muscles and bone tissue, we are told, and, therefore, before astronauts go to Mars we must undertake a long-term program of experimentation with human subjects exposed to extended periods of zero gravity on board the Space Station. This program will require several decades, many billions of dollars in "microgravity life science research," and a few dozen human beings willing to sacrifice their health to "scientific research."

I find this argument bizarre. Now, it is certainly true that spending long periods in zero gravity will cause cardiovascular deterioration, decalcification and demineralization of the bones, and a general deterioration of muscular fitness due to lack of exercise. Zero gravity also depresses some aspects of the body's immune system. These effects are well documented from the experiences not only of the U.S. Skylab astronauts, who spent up to three months at a time on-orbit, and crews on the International Space Station, whose standard rotation lasts six months, but of Soviet cosmonauts, some of whom have spent stints in zero gravity on their *Mir* space station of almost eighteen months—nearly three times the duration of the trans-Mars or trans-Earth cruises required to perform the Mars Direct mission. In all cases, near total recovery of the musculature and immune system occurs after reentry and reconditioning to a one-gravity environment on Earth. The demineralization of the bones ceases upon return to Earth, but actual restoration of the bones to preflight condition appears to be a very extended process. The Soviets have experimented with various countermeasures to zero gravity, including intensive exercise, drugs, and elastic "penguin suits" that force the body to exert significant physical effort in the course of routine movement. As might be expected, programs of intensive (three hours a day) exercise have proven effective in reducing general muscular deconditioning, and to some extent cardiovascular deterioration, but countermeasures taken to date have shown little benefit in slowing bone demineralization. It should be understood that while these effects are all quite tangible and definitely not desirable, they are not too extreme; in no case have such zero-gravity "adaptations" prevented astronauts or cosmonauts from satisfactorily performing their duties while they are in the zero-gravity environment, and after even the longest flights, crew members recovered enough to become basically functional again within 48 hours after landing. Indeed, within a week of landing, the members of the 84-day Skylab 3 crew were able to play strong games of tennis. The recovery time to functionality upon Mars arrival after a six-month zero-gravity exposure should be swifter, because the crew will only have to deal with reacclimation to Mars' 0.38g environment after landing, instead of the 1g shock experienced after reentry on Earth. The point, however, is that an awful lot of research has already been done in this area, and we know what the effects are. Given that is the case, we can rightly ask whether it is necessary, or even ethical, to subject further astronaut crews to such experimentation solely for the purpose of more exhaustive research on zero-gravity health deterioration effects. I don't think it is. In fact, given what we know today, I'd have to classify the proposed program of continued experimentation on humans with long duration zero-gravity health effects as unethical and worthless, and I know a lot of astronauts who agree with me on that point. It just doesn't make sense to expose dozens of astronauts to a larger zero-gravity dose than a Mars mission might provide in order to "ensure the safety" of a much smaller crew who actually fly there. Doing so is like training bomber pilots by having them fly their planes through real flak. If you are willing to accept the health consequences of long-duration exposure to zero gravity, you might as well take your licks in the process of actually getting to Mars.

But in fact we don't need to fly to Mars in a zero-gravity mode at all. A Mars-bound spacecraft can be provided with artificial gravity. This can be done by spinning the spacecraft, using essentially the same centrifugal force physics that allows a small child to swing a bucket round and round without losing a drop of water. The equation that governs this effect can be written:

$$F = (0.0011)\,W^2R$$

where F is the centrifugal force measured in Earth gravities, W is the spin rate in revolutions per minute (rpm), and R is the length of the spin arm in meters. I offer this equation because by looking at it you can see that for a given level of gravity produced, the larger W is, the smaller R can be. For example, if Mars' normal gravity is desired ($F = 0.38$), then if W is 1 rpm, R is 345 meters. But if W is 2 rpm, then R is 86 meters; if W is 4 rpm, R is 22 meters; and if W is 6 rpm, R is 10 meters. Thus, there are two ways to produce artificial gravity; you can either go with fast spin rates and short spin arms, or slow spin rates and long spin arms. By "spin arm" I mean the distance between the location of the crew and the center of gravity of the spacecraft about which they are being rotated. If the spacecraft is to be a single rigid structure, it can easily be made to spin by having small rocket thrusters at each of its ends fire sideways in opposite directions. However if significant amounts of artificial gravity are desired on a rigid spacecraft the only viable option is to go for the fast-spin/short-arm technique. In the 1960s, NASA conducted experiments with humans on rotating structures and found that, after some initial disorientation, humans could adapt to living, functioning, and moving about on structures with rotation rates as high as 6 rpm.[21]

Rapid-spin/short-arm artificial gravity systems are the easiest for the engineer to design and implement, but they also have some disadvantages. For example, if R is 10 meters, then a 2-meter-tall person standing in such a gravity field will have his head at $R = 8$ meters, and thus experience only 80 percent as much gravity at his head as at his feet. This much difference is tangible and can be disconcerting, at least at first. On the other hand, if the spin arm was 100 meters long, the 2-meter-tall person would feel 98 percent as much gravity at his head as at his toes, and probably would not notice any difference. In addition, if the crew member were to walk rapidly about the ship, he would experience Coriolis forces due to the interaction between his attempt to move in a straight line and the fact that the ship (the floor he is walking on) is not only moving but changing direction rapidly. Once again, at 6 rpm these effects are quite noticeable but at 2 rpm they are negligible. Thus, if you want the artificial gravity to feel like dry land on Earth (desirable, but not necessarily a requirement—sailors adapt quite well to very erratic gravity/Coriolis-force environments experienced on pitching ships at sea), the best way to do it is to employ a slow spin rate together with a long spin arm. Such very long spin arms can be provided if the spacecraft is split up into several parts that can be connected to each other across long distances (hundreds to thousands of meters) by cables, or "tethers."

While excellent in principle, in the past such tethered artificial gravity systems were generally frowned upon because in traditional "Battlestar Galactica"–type spacecraft designs, the only thing massive enough to serve as a useful counterweight on the other end of the tether from one functional part of the spacecraft would be another functional part of the spacecraft. In other words, if you wanted to provide artificial gravity to the crew habitat on one end of the tether, you probably would have to split your ship in half, and put a good portion of your propellant tanks on the other end. Such a configuration might work fine on paper, but would be an invitation to disaster in practice. If the tether should snag when you reeled it in, a large fraction of your mission critical hardware, such as your Earth-return propellant, would be permanently inaccessible, and, as a consequence, your mission would be lost. In the Mars Direct plan, however, this is not a problem. Because the crew is flying to Mars in a relatively lightweight habitat and not in an interplanetary Battlestar, their spacecraft is light enough to be counterbalanced on the tether's other end by the burnt-out upper stage booster that threw them toward Mars (Figure 5.2). This item is not mission critical—it's just junk and never needs to be reeled in. A similar

FIGURE 5.2
Tethered artificial gravity system requires two objects swinging around a mutual center of gravity. For Mars Direct, the hab (on the right) is counterbalanced by the spent upper stage (on the left).

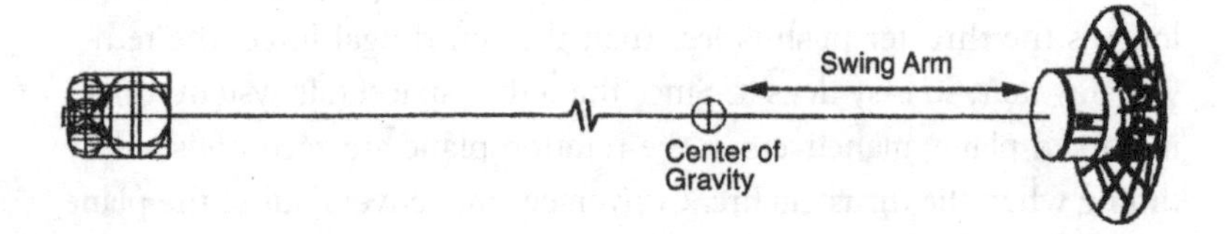

tether scheme can be employed between the burnt-out upper stage of the ERV propulsion system and the ERV cabin during the flight home. Thus, except for some brief phases just before trans-Mars and trans-Earth injection, just before aeroentry at Earth and Mars, and just after aerocapture at Mars, the crew of a piloted Mars mission never need be subjected to a zero-gravity regime.

The tether used should be of a hefty interconnected multistrand variety, designed to remain intact even if several of the strands should be individually cut in numerous different places by micrometeorites or other space debris. Such fail-safe tethers have been designed and demonstrated by aerospace engineers Robert Forward and Bob Hoyt. The tether should also not be used as a wire for the transmission of large quantities of electric power. In the failed tethered satellite mission flown by the Space Shuttle in February of 1996, a power surge in the multikilowatt tether/power system caused the tether to melt itself and break.

I've been asked how a rotating spacecraft will perform necessary maneuvers, such as the midcourse correction ΔVs of 20 meters per second or so that are typically necessary on interplanetary missions. Actually, it's not that hard. Maneuvers have been performed on spinning spacecraft before. *Pioneer Venus Orbiter* and the *Pioneer Venus Probe Carrier* were spinning, interplanetary spacecraft with precise targeting requirements at Venus. They used repeated, timed thruster firings to create a net ΔV in any direction needed.

The Mars Direct mission tethered assembly would do much the same. For example, if you want to create a ΔV in any direction that is within the spin plane of the spacecraft, you fire a thruster repeatedly along the line of the tether while the tether is pointing in the direction desired. Since the tether is taut, thruster firings that push the habitat toward the upper stage have the effect of reducing tether tension. As long as the thruster push is less than the centrifugal force, the tether will stay taut, so easy does it. Since the tether-spacecraft system rotates in a fixed plane, maneuvers in the rotation plane are accomplished by timing when the thrusters fire. Conversely, maneuvers out of the plane are accomplished with continuous, very low-thrust burning perpendicular to the plane of rotation.

A piloted Mars spacecraft has so much power (several kilowatts at least) that effective voice and essential flight telemetry data communication with Earth can be achieved with an omnidirectional antenna. So, while the ship would use a high-gain antenna that could actively track the Earth as the ship rotates for high data rate video transmission, this item is not really mission critical. If the spin plane of the assembly is oriented such that it always faces the Sun, then any power-generating solar arrays used by the spacecraft need not be controlled by active gimbals either. Navigation scanning sensors are available that can operate just fine at rotation rates much higher than even 6 rpm, so these can be fixed to the habitat as well. In other words, none of these instruments requires a counterrotating platform to operate successfully on a tethered spacecraft.

In short, using artificial gravity on the Mars Direct spacecraft is thoroughly practical, and completely kills the zero-gravity dragon. At a conference a few years back I quizzed a NASA official who advocated a multidecade program investigating zero-gravity health effects on humans prior to a piloted Mars mission. "Why not just employ artificial gravity?" I asked. "We can't do that," he said, "all our data is going to be for zero gravity." Get the picture?

HUMAN FACTORS

One of the more bizarre dragons that mar the charts of Mars navigators goes by the name of "The Human Factors Problem." Some people assert that the psychological problems associated with a round-trip piloted Mars mission are unique and probably a show stopper. Either very fast ships that reduce the round-trip time to weeks, or else very large and luxurious ships that can accommodate large crews with ample social and physical space, must be used for the mission, they claim. Unless such concessions to the modern American suburban lifestyle are provided, they declare, the crew will surely go crazy. Unfortunately, since neither the ultrafast space-warper nor the Club Med interplanetary cruise ship options are in fact feasible, these concerned parties recommend that any Mars mission be postponed until substantial sums have been spent in areas of psychological research to solve "The Human Factors Problem." (Once again we hear the chorus of the now familiar song, "Oh, you can't go to Mars till you give us dough . . .")

Let's consider this argument. In the type of human Mars mission that we propose, a crew of four will spend six months on an outbound mission leg more or less confined to the interior of a two-deck hab containing a private room for each crew member as well as some common social areas (recreational space walks or "EVAs" are possible, especially if the mission is conducted in zero gravity, but we'll put that option aside for now). The total interior floor space is about 101 square meters (1,083 square feet), somewhat small living space for a four-person apartment by American standards, but rather large compared to the accommodations available to a middle-income apartment dweller in Tokyo. After the six-month voyage, the crew will land with their hab on Mars, and live there for a year and a half, during which time they will have extra living quarters available to them in the Earth return vehicle cabin on-site, as well as in the pressurized rover. Moreover, during the long surface stay, the crew will be heavily occupied outside the hab conducting broad-ranging field exploration. Finally, during the last six months of the trip, the crew will be confined to the ERV's cabin, which has about half the living space of the hab. During the entire voyage, normal telephone conversations with people on Earth will not be possible because of the time lag in radio signal transmission. Instead, voice, video, or text and still-image messages will have to be sent, with round-trip delays for reply ranging from seconds all the way up to forty minutes.

Well, it's true that the above mission plan imposes psychological rigors on the crew not experienced by most civilians in the course of today's daily life. But let's compare it to the stresses that many ordinary people have overcome in the past.

The space shrinks talk a lot about the trauma of Mars mission crew members "being away from home for three years." Well, my father, and my uncles, and several million other GIs were "away from home for three years" during World War II under vastly tougher conditions than those that will face the crew of Mars 1 (a dugout at the Anzio beachhead was a much more stressful environment than a habitat on the surface of Mars). In addition to the constant threat of death by enemy action, front line soldiers also had to endure hard labor, low pay, cold, heat, insects, diseases, lice infestations, terrible rations, and sleeping on cold, wet ground in the snow or rain, sometimes for months on end. In addition, the vast majority of soldiers were enlisted men who had to endure the constant, brutal insults of the military discipline system, day and night being treated like dirt by 90-day wonders and other officers whose collective conceit was that rank made them superior beings. In contrast to these conditions, the crew of the first Mars mission may face risks, but not armies and fleets of people and war machines doing everything in their power to kill them.

The Mars crew will not have to endure extended hard physical labor. Insects, lice, and diseases are not part of their program. Their food will be good, and they will sleep in dry clothes in nice warm beds. During the interplanetary cruise phases of their mission, they may share some of the GI's boredom, but this burden will be greatly lightened by a hefty on-board supply of books, games, writing supplies, and other materials to support various hobbies or amusements, as well as by the knowledge shared by all members of the crew that when they get back to Earth their fortunes are made. Compared to the constantly degraded GI, the psychological boost enjoyed by Mars-bound astronauts of knowing that they are golden people, celebrated as heroes by millions on Earth cannot be overstated. During the war, the standard method of communication with those back home for the GI was by V-mail, with a round-trip communication time of several weeks. By comparison, the fact that the astronauts might have to wait up to forty minutes to hear back from their folks hardly seems cause for tears.

The point I'm making is simply this: If you abstract yourself from America's contemporary comforts and look at human history, everywhere you will see people chosen essentially at random—whether as front line soldiers, refugees in hiding, prisoners, submariners, explorers, trappers, or pre-twentieth-century merchant seamen—all of whom had to, and by and large did, competently endure long-term conditions of isolation, privation, and psychological stress greatly exceeding what the select crew of a piloted Mars mission will face. Humans are tough. We have to be. We're the survivors of the saber-tooth tigers and the glaciers, of tyrannical empires and barbarian invasions, of horrible famines and devastating plagues. You name it, you've got ancestors who have faced it, and overcome it. The same can certainly be said of the handpicked and highly trained crews of the first human missions to Mars.

The human psyche will not be the weak link in the chain on a piloted mission to Mars. On the contrary, it is likely to be the strongest.

DUST STORMS

The fourth dragon, the Martian dust storm, is actually the oldest, and so has lost a few of its teeth, especially since its primary potential beneficiaries, the Mars atmospheric scientists, lack the keen commercial instincts of other critics. However, it still scares some. Moreover, since this particular dragon is more of an exaggeration than an illusion, it's worth our while to address it here.

The existence of powerful dust storms on Mars was suspected by telescopic observers from the nineteenth century on, and the robotic spacecraft exploration programs conducted by the United States and Soviet Union since the 1960s have provided ample data to confirm the hypothesis. Mars' orbit is eccentric. During its southern hemisphere's summer, it comes about 9 percent closer to the Sun than the year-round average, while during the southern winter, it is 9 percent farther from the Sun than average. This combination of expected summer season heating with the extra warmth generated by being closer to the Sun than usual causes the planet's southern hemisphere to have extreme seasonal temperature variations (and the out-of-phase northern hemisphere to have mild seasons). During the extra-cold southern winter, large amounts of carbon dioxide precipitate out of the atmosphere on the southern polar cap (which has a dry ice layer covering its water ice) and adsorb into the Antarctic regolith. This extra layer of frozen and adsorbed carbon dioxide is then blasted back into the atmosphere when the strong heating of the early southern summer hits the south polar regions. This sudden addition of gas to the planet's atmosphere is so large that it actually raises the planet's atmospheric pressure by about 12 percent in the course of a few months (the full-year winter-summer pressure swing being nearly twice that), in the process causing huge winds that pick up and transport a considerable amount of dust. These dust storms, therefore, originate in the early southern summer near the south pole, and then spread north, occasionally going so far as to engulf the entire planet. Wind speeds in these storms have been clocked between 50 and 100 km/hour. The storms, which recur occasionally throughout the southern summer, gradually die out with the approach of the southern autumn. As in the case of weather on Earth, there is some randomness in all this: In some years dust storm activity is almost nonexistent, while in others it dominates the entire planet for virtually the whole southern summer. However, in general, clear weather can be expected in the north during that hemisphere's spring, summer, and fall.

That's the story, and it sounds pretty formidable. Indeed, in November 1971 when the U.S. *Mariner 9* orbiter and the Soviet *Mars* 2 and 3 lander probes reached Mars, a global dust storm was in progress. For four months, the surface of the planet was entirely blocked out by dust and *Mariner 9* couldn't see a thing. This didn't hurt *Mariner 9*'s mission very much—it just waited in Mars orbit until things cleared up and then proceeded to image the entire planet without difficulty. However, in the case of the Soviet landers, the story was very different. They had been preprogrammed to target for landing sites near 45° south latitude, and that's where they went, parachuting right into the heart of the maelstrom. Both were destroyed.

However, while parachuting into a Martian dust storm is a bad idea, the story is very different if you are already on the ground when the dust storm hits. The Martian atmosphere is only about 1 percent as thick as the Earth's, and, therefore, the dynamic pressure created by a 100 km/hour Martian wind is only equal to that of a 10 km/hour (6 knot) breeze on Earth. The *Viking* landers *1* and 2 and the *Spirit* and *Opportunity* rovers all operated for years on the surface (their design life was 90 days), and all were subjected to many dust storms during their stay. Despite this, no damage to the *Vikings* or the MER rovers or any of their instruments was detected. Furthermore, while the dust storm can block visibility of the surface from orbit, local visibility on the surface is not seriously impaired. The dust does reduce light levels, much as an overcast day does on Earth, but to an observer on the surface the surrounding area is not fogged out. If a surface installation were powered by solar panels, some problems could be expected from dust storms reducing light levels. However, since photovoltaic panels can convert light to electricity even after it has been scattered by dust (a clear optical view of the Sun is not necessary), power loss would not be total. Instead, during a typical bad dust storm, one might expect solar power electric output to fall by about 50 percent. Thus, provided that the power system is designed to ensure sufficient power for minimal life-support functions for the duration of the dust storms, things should be okay. Of course, if either a nuclear reactor or a radioisotope generator provides base power supply, or if a large power reserve is available in the form of locally produced chemical propellants (which can be burned in a chemical combustion engine to turn a generator), this problem becomes moot.

Some people have voiced concern that dust deposited by storms could obscure solar panels or other optical surfaces, such as windows or instruments. This problem was not observed on *Viking* or the MER rovers. Indeed, *Spirit* and *Opportunity* repeatedly had their solar panels cleaned by high winds. Apparently, the total quantity of dust actually suspended by the storms is rather small. However, in the case of a human Mars mission, dust deposition certainly wouldn't be much of a problem. If a solar panel becomes covered with dust the solution is simple; send someone outside with a broom!

So, to sum up, the only real hazard represented by dust storms is to objects that are dominated by aerodynamic forces (because they have a lot of sail area compared to their weight), such as balloons or parachute suspended landers. If a lander does not use a parachute for landing (high-altitude drogues are okay), and the Mars Direct landers do not need to, it should be able to punch its way through a dust storm as easily as an airplane can fly through a cloud. Of course, most pilots would prefer to land under conditions of complete visibility, and this is why the Mars Direct plan has the spacecraft brake into orbit prior to landing. If the weather is bad at the landing site when the hab arrives, the crew can just wait on orbit like *Mariner* 9 until the skies clear. Interestingly, however, for the decade 2016 to 2025, it is possible to choose Earth-to-Mars trajectories during every launch year that have the ships arrive at Mars during the clear weather season.

Dust storms won't keep us from Mars.

BACK CONTAMINATION

The last of the five dragons infesting the maps of would-be Mars explorers is not only illusory, but hallucinatory. This is the alleged threat of back contamination.

The story goes like this: No Earth organism has ever been exposed to Martian organisms, and therefore we would have no resistance to diseases caused by Martian pathogens. Until we can be assured that Mars is free of harmful diseases, we cannot risk infecting the crew with such a peril that could easily kill them, or if it didn't, return to Earth with the crew to destroy not only the human race but the entire terrestrial biosphere.

The kindest thing that can be said about the above argument is that *it is just plain nuts*. In the first place, if there are or ever were organisms on or near the Martian surface, then the Earth has already been, and continues to be, exposed to them. The reason for this is that over the past billions of years, millions of tons of Martian surface material has been blasted off the surface of the Red Planet by meteor strikes, and a considerable amount of this material has traveled through space to land on Earth. We know this for a fact because scientists have collected nearly a hundred kilograms of a certain kind of meteorite called "SNC meteorites,"[22] and compared the isotopic ratios of their elements to those measured on the Martian surface by the *Viking* landers. The combinations of these ratios (things like the ratio of nitrogen-15 to nitrogen-14), as well as the fact that the gas trapped in the rock matches the Martian atmosphere, represent an irrefutable fingerprint proving that these materials originated on Mars. Despite the fact that in general each SNC meteorite must wander through space for millions of years before arrival at Earth, it is the opinion of experts in the area that neither this extended period traveling through hard vacuum, nor the trauma associated with either the initial ejection from Mars or reentry at Earth would have been sufficient to sterilize these objects, if they had originally contained bacterial spores.[23] Indeed, chemical investigations of the famous SNC meteorite ALH84001 (see Addendum) have shown that portions of it never rose above 40°C at any time during its entire interplanetary journey, and therefore, if there had been bacteria within it when it departed Mars, they easily would have survived the trip. Furthermore, on the basis of the amount we have found, it has been estimated that these Martian rocks continue to rain down upon the Earth at a rate of about 500 kilograms per year. So, if you're scared of Martian germs, your best bet is to leave Earth fast, because when it comes to Martian biological warfare projectiles, this planet is smack in the middle of torpedo alley. But don't panic—they're not so dangerous. In fact, to date the only known casualty of the Martian barrage is a dog who was killed by one of the falling rocks in Nakhla, Egypt in 1911. Statistically, the hazard presented to pedestrians by furniture being thrown out onto the street from upper-story windows is a far greater threat.

The fact of the matter, however, is that life almost certainly does not exist on the Martian surface. There is no (and cannot be) liquid water there—the average surface temperature and atmospheric pressure will not allow it. Moreover, the planet is covered with oxidizing dust and bathed in ultraviolet radiation to boot. Both of these last two features—peroxides and ultraviolet light—are commonly used on Earth as methods of sterilization. No, if there is life on Mars now, it almost surely must be ensconced in exceptional environments, such as a heated hydrothermal reservoir underground.

But couldn't such life, if somehow unearthed by astronauts, be harmful? Absolutely not. Why? Because disease organisms are specially keyed to their hosts. Like any other organism, they are specially adapted to life in a particular environment. In the case of human disease organisms, this environment is the interior of the human body or that of a closely related species, such as another mammal. For almost four billion years the pathogens that afflict humans today waged a continuous biological arms race with the defenses developed by our ancestors. An organism that has not evolved to breach our defenses and survive in the microcosmic free-fire zone that constitutes our interiors will have no chance of successfully attacking us. This is why humans do not catch Dutch elm disease, and trees do not catch colds. Now, any indigenous Martian host organism would be far more distantly related from humans than elm trees are. In fact, there is no evidence for the existence of, and every reason to believe the impossibility of, macroscopic Martian fauna and flora. In other words, without indigenous hosts, the existence of Martian pathogens is impossible, and if there were hosts, the huge differences between them and terrestrial species would make the idea of common diseases an absurdity. Equally absurd is the idea of independent Martian microbes coming to Earth and competing with terrestrial microorganisms in the open environment. Microorganisms are adapted to specific environments. The notion of Martian organisms outcompeting terrestrial species on their home ground (or terrestrial species overwhelming Martian microbes on Mars) is as silly as the idea that sharks transported to the plains of Africa would replace lions as the local ecosystem's leading predator.

If I appear to be spending excessive time on beating this idea, it's partly as a result of a NASA planning meeting for the planned (robotic) Mars sample return mission during which someone seriously proposed that, to allay alleged public concerns, any sample acquired on Mars be sterilized by intense heat before returning it to Earth. While an extremely unlikely find, the greatest possible treasure a Mars sample return mission could provide would be a sample of Martian life. Yet, certain of those attending the meeting would destroy it preemptively (and a great deal of valuable mineralogical information in the sample as well). The proposal was so grotesque that I countered by asking the assembled scientists, "If you should find a viable dinosaur egg, would you cook it?" The question is not entirely out of line; after all, dinosaurs are our comparatively close relatives and they did have diseases. In fact, every time you turn over a shovelful of dirt you are returning a sample of the Earth's disease-infested past to menace the current biosphere. Nevertheless, neither paleontologists nor gardeners generally wear decontamination gear.

Just as the discovery of a viable dinosaur egg would represent a biological treasure trove but no menace, so a sample of live Martian organisms would be a find beyond price, but certainly constitute no threat. In fact, by examining Martian life, we would have a chance to differentiate between those features of life that are idiosyncratic to terrestrial life and those that are generic to life itself. We could thus learn something fundamental about the very nature of life. Such basic knowledge could provide the basis for astonishing advances in genetic engineering, agriculture, and medicine. No one will ever die of a Martian disease, but it might be that thousands of people are dying today of terrestrial ailments whose cure would be apparent if only we had a sample of Martian life in our hands.

THE LUNAR SIREN: WHY WE DON'T NEED LUNAR BASES TO GO TO MARS

We now come to a totally different kind of mythical creature blocking the path to Mars, one appearing not in the guise of a threatening monster or fearsome dragon, but in the alluring dress of a lovely goddess. This is Diana, the lunar siren, whose seductive song has probably done as much to wreck would-be Mars ventures to date as all five dragons combined.

According to Diana's followers, it is a point of religious belief that we cannot venture human expeditions to Mars until after the goddess has been appeased by the construction of a substantial array of temples—

that is, bases—on the lunar surface. This is a commendably original basis for a pagan religion, and it really shows how far we've come since the days of the Roman Empire, but the fact of the matter is that it has no basis in reason.

Yes, it is quite true that due to its low gravity and negligible atmosphere, it would be much easier to send a rocket from the surface of the Moon to Mars than to launch it from the surface of the Earth. Furthermore, it is also true that Moon rocks are almost 50 percent oxygen by weight, so that once technologies are developed that can break down the iron and silicon oxides that make up most of the Moon's materials, a copious supply of liquid oxygen could be made available for spacecraft refueling on the lunar surface. Unfortunately, fuel to burn in this oxygen, such as hydrogen or methane, is essentially unavailable on the Moon. Nevertheless, since the oxygen content of various rocket propellant mixtures varies from 72 percent to 86 percent by weight, the Moon can in principle be made into a base that could support a substantial fraction of required space transportation logistics.

But this analysis neglects some basic facts about solar system transportation. You see, before the spacecraft can refuel at the Moon, it has to get to the Moon. Now the ΔV required to go from low Earth orbit (LEO) to the lunar surface is 6 km/s (3.2 km/s for trans-lunar injection, 0.9 km/s to capture into low Lunar orbit, and 1.9 km/s to land on the airless Moon.). On the other hand, the ΔV required to go from LEO to the Martian surface is only about 4.5 km/s (4 km/s for trans-Mars injection, 0.1 km/s for post-aerocapture orbit adjustment, and 0.4 km/s to land after using the aeroshield—but no parachute—for aerodynamic deceleration). Put briefly, from a propulsion point of view, *it is much easier to go from LEO directly to Mars than it is to go from LEO to the surface of the Moon.* So, even if infinite quantities of free rocket fuel and oxygen were sitting right now in tanks on the lunar surface (and they aren't), it would make absolutely no sense to send a rocket there to refuel itself for a voyage to Mars. Basically, refueling at the Moon on your way to Mars is about as smart as having an airplane flying from Houston to San Francisco stop over for refueling in Saskatoon. Putting the lunar refueling node in lunar orbit doesn't change things very much. You still have to perform almost as much ΔV to move the spacecraft from LEO to lunar orbit as you do to send it to Mars. Add in the supplies required to support the making of oxygen on the Moon along with the hardware and fuel to haul large quantities of it to lunar orbit (you have to ship hydrogen or methane to the lunar surface to use to lift oxygen to orbit) and it quickly becomes apparent that the whole scheme is nothing but a logistics nightmare that would enormously increase the cost, complexity, and risk required to mount a piloted Mars mission.

So, the Moon is not useful as a Mars transportation base. Well then, say Diana's followers, you still need to use the Moon as a test bed and training site to prepare for a Mars mission.

But lunar conditions are so dissimilar from those on Mars that the Arctic (or Utah for that matter) would do just as well for crew training, and at far lower expense. (In fact, the Mars Society, www.marssociety.org, a nonprofit organization that I lead, has established Mars practice bases in both the Canadian Arctic and the southern Utah desert for exactly that purpose, and has operated them on a total budget of less than $2 million, raised primarily from private sources, for the past ten years.) Mars has an atmosphere and a 24-hour day, with daytime temperatures varying between –50°C and +10°C. The Moon has no atmosphere, a 672-hour day, and typical daytime temperatures of about +100°C. While the Earth's gravity is 2.6 times that of Mars, Mars' gravity is 2.4 times that of the Moon. Furthermore, the types of resource utilization that one would undertake on Mars (exploitation of the atmosphere in gas-based chemical reactors and extraction of permafrost from soil) are completely different from the high-temperature rock-melting techniques that will be employed on the Moon. In addition, the types of geologic investigations needed on Mars, given its complex hydrologic and volcanic history, will much more closely resemble those that can be done on Earth than those that can be done on Luna. We won't learn how to live on Mars by practicing on the Moon.

The Moon does have some uses, most notably as a platform for astronomy using a coordinated array of optical telescopes to obtain super-high-resolution views of the universe at large (an "optical interferometer"). It makes sense, therefore, to gain maximum benefit by ensuring that the same set of hardware used to accomplish Mars missions is designed in such a way that it can also be used to support transportation of humans and equipment to the Moon. As discussed in chapter 3, this is the case with the Mars Direct mission design. Therefore, in much the same way as Apollo lunar hardware could be used as an afterthought to create the Skylab space station, so an ancillary benefit of the Mars Direct mission is that it will give us the capability of setting up lunar observatories whenever we want them.

However, what needs to be clearly understood is that a lunar base is neither necessary nor desirable as an asset to support piloted missions to Mars. With respect to the path to Mars, it is a fatal siren, a diversion into a dead end. The late NASA administrator Thomas O. Paine knew all about this trap. In one of the last speeches of his life he put it this way: "As Napoleon Bonaparte once said explaining his winning strategy for war with Austria: 'If you want to take Vienna, take Vienna!' Well, if you want to go to Mars, go to Mars!"

Well said, Tom. On to Mars.

6: EXPLORING MARS

We are not sending a crew to Mars to set a new altitude record for the Aviation Almanac. We are going to Mars to explore a planet; to determine if it ever harbored life in the past and to survey its potential as a future home for a new branch of human civilization. Sending a few robotic probes, no matter how sophisticated, will never get the job done. Nor will even a few piloted excursions to the Red Planet's surface, especially if the crews are limited to lingering near their short-term base. No, to learn about Mars, we'll have to get about Mars, and in a major way.

With a surface area of 144 million square kilometers, the Red Planet has as much terrain to explore as all the continents and islands of Earth put together. Moreover, the Martian terrain is incredibly varied. It includes canyons, chasms, mountains, dried river and lake beds, flood runoff plains, craters, volcanoes, ice fields, dry-ice fields, and "chaotic terrain," to name just a few surface features. The U.S. Geological Survey currently records no less than thirty-one types of Martian terrain on its "Simplified Geologic Map," and all these before real high-resolution imaging of Mars has truly been done. Some of the Martian terrain features, such as the 3,000-kilometer-long Valles Marineris, are of continental extent, and the thorough exploration of even a single such feature will require continental scale mobility.

The dry riverbeds discovered on Mars by *Mariner 9* are proof that Mars once had a warm, wet climate, suitable for the origin of life. This was possible in Mars' early years, because in its youth the planet's carbon dioxide atmosphere was much thicker, endowing it with a very strong greenhouse effect. Venus has a thick carbon dioxide atmosphere today, which has turned that planet into a baking hell. At Mars' greater distance from the Sun, though, a thick carbon dioxide greenhouse is just what is needed to create the temperate conditions required for the development of life. Most Mars scientists currently believe that such conditions persisted on Mars for a period of time considerably longer than it took life to evolve on Earth. Current theories on the origin of life regard the emergence of life as a natural development of progressive self-organization by matter that should inevitably occur wherever the appropriate physical and chemical conditions exist. If that is indeed the case, life should have appeared on Mars, because during the period of life's origin on Earth, Earth and Mars were similar environments. Over geologic time, Mars lost its greenhouse and became the frigid, arid world it is today, and this climatic deterioration almost certainly has driven life from its surface and possibly into extinction. Nevertheless, microscopic organisms can leave macroscopic fossils. We have found some on Earth, called bacterial stromatolites, that date back 3.7 billion years, making them contemporary with Mars' tropical era. Even if Martian life died out entirely, its fossil remains could still be there. Today, all we know about the chances for the evolution of life is that it occurred on one planet, our own. We have no way of knowing whether that development was a one-in-a-trillion freak chance or whether it was a dead sure bet. Freak chances can occur in a single sample experiment, but never twice in a row. If we were to find either living organisms or simply fossils on Mars, we would know that the universe belongs to life.

Thus the search for life, either extant or fossilized, will be the highest priority for early Mars explorers, as around its result turns the question of whether life is a universal or unique phenomenon. But the results of the *Viking* missions showed that if extant life does exist on Mars, it is rare, and its finding will take more than a bit of searching.

Likewise, the experience of professional paleontologists on Earth has shown that the hunt for fossils will require much footwork, because the creation of a detectable fossil is a very low-probability event. Imagine the sequence of events necessary to create a fossil. First, when the organism dies it must be immediately isolated from the environment. If not, it will soon decay, or perhaps be scavenged by its former pals who want for themselves whatever it was made of. Then it must sit hidden in isolation from the environment for millions or billions of years, only to be revealed just before you happen to come along looking for it. (If it is exposed any significant time before your arrival, the environment will destroy it before you get to see it.) Recall that triceratops, and, more recently, bison once roamed the plains of North America in herds of tens of millions, yet American hikers today don't run much risk of accidentally tripping over their fossilized skeletons. No, if you want to find yourself a fossilized dinosaur, or Martian stromatelite, you better be prepared to do quite a bit of traveling. And if you want proof that they're *not* there, you'll have to travel even more, because the ability to demonstrate a convincing negative result will depend upon a search of virtually the planet's entire surface. In the end, the mobility requirements of Martian exploration are profoundly simple: To explore a planet you need mobility on a planetary scale. It's a simple but often overlooked point.

So, how will the crew of our first piloted Mars mission get around? The battery-powered Lunar rover used during the Apollo program had a one-way range of about 20 kilometers, giving it a sortie range of 10 kilometers from the landing site. A manned Mars expedition equipped with equivalent transportation would be able to explore only about 300 square kilometers, regardless of the length of its surface stay, and nearly *half a million* such missions would be required to examine the entire surface of Mars *just once*. Even if it were considered sufficient simply to examine a variety of points of interest, the limited mobility afforded by such a vehicle would be a severe impediment and vastly increase the cost of mounting a serious manned Mars exploration program. For example, Table 6.1 shows a list of points of interest in the Coprates triangle area, surrounding a landing site at 0° latitude and 65° west longitude. Because it is near the equator (and thus comparatively warm and sunny year 'round) and has such a large variety of interesting targets for investigation nearby, this site is one of the leading candidates as the landing zone for the first human expedition to Mars.

We can see that if surface mobility were limited to a 100-kilometer range (ten times the Apollo lunar rover's), at least twelve landings would be needed to visit all the sites listed. If the mission had a surface mobility of 500 kilometers, then only four missions would be required to visit all fourteen sites, and these four missions could access eight times the surface area as that available to the twelve missions conducted by crews with a 100-kilometer range.

Now, manned Mars missions are likely to cost billions of dollars each. It is true that this cost can be reduced by introducing technologies such as nuclear thermal propulsion or cheaper launch vehicles. However, while such efforts are to be heartily encouraged, it must be pointed out that introduction of any of these technologies will cost bil-

TABLE 6.1
Surface Features of Interest in the Exploration of Mars

Feature	Distance from Base (km)	Direction
Ophir Chasma	<300	southwest
Juventae Chasma	<300	southeast
Slope and bedrock material	<300	south
Cratered plateau material	<300	east
Chaotic material	<300	east
Degraded crater material	<300	south
Hebes Chasma	600	west
Center of Lunae Planum	650	north
Northern plains	1,200	northwest
Kasei Vallis	1,300	north
Viking 1 landing site	1,400	northeast
Paleolake site	1,500	northeast
Volcanic flows	2,000	west
Pavonis Mons	2,500	west

lions, and only reduce Mars mission costs by around a factor of two. The expansion of surface mobility, on the other hand, is likely to be cheaper, and can potentially increase mission exploration effectiveness a hundredfold or even more.

It is clear that there is nothing more important in determining the cost-effectiveness of a program of human Mars exploration than the degree of mobility provided on the surface of the planet.

MARS CARS

Options abound for building Mars cars. Wheels, treads, half-tracks, and even motorized legs are all viable choices for getting around. What matters most is the power that actually moves the vehicle.

The only cars used in space to date have been the Apollo Lunar rovers, which were unpressurized electric vehicles powered by batteries. If we employed the latest advanced lithium-ion batteries (such as are used in camcorders) and gave them enough charge to power the rover for ten hours, such a system could be made to produce about 10 watts of power for every kilogram of its weight. If instead of using batteries we employed hydrogen/oxygen fuel cells such as those on the Space Shuttle to provide the electric power, the system power/mass ratio could be raised to about 50 W/kg. That's certainly an improvement, but it pales against a much more familiar household technology.

Internal combustion engines can have power/mass ratios of 1,000 W/kg. That's twenty times higher than that of a hydrogen/oxygen fuel cell, one hundred times that of the battery-driven system. A combustion engine delivers far more power with far less mass than anything else (that's the main reason why they are preferred for the vast majority of vehicle applications on Earth), and that has great implications for our Mars cars. For a given life-support system mass, the vehicle's range will be directly proportional to its speed, which is in turn proportional to the power. But if you try to match one of the competing option's power level with that of the combustion engine's, the competing option's weight will rapidly become excessive. Imagine a rover equipped with 50 kW (about 65 hp) of power. The mass of the required internal combustion engine would only be about 50 kilograms, while a set of fuel cells delivering that much power would weigh in at 1,000 kilograms. The combustion-powered car could thus take along 950 kilograms of additional science equipment and consumables compared to a fuel-cell-powered vehicle of equal power, and again have much greater endurance, capability, and range. Furthermore, the fact that the combustion-powered vehicle is virtually power unlimited allows sortie crews to undertake energy-intensive science at a distance from the base that would otherwise be impossible. For example, a combustion vehicle sortie crew could drive to a remote site in a pressurized rover and generate 50 kW to run a drilling rig to try to reach the Martian water table. Rover data transmission rates are also proportional to power, and can therefore also be much higher, which in turn increases both crew safety and sortie science return. Furthermore, combustion engines enable the small, lightweight power plants needed to drive fast, nimble single-person all terrain vehicles (ATVs). Just as they do on Earth, such versatile ATV-style systems would offer many advantages to explorers operating in the Martian outback. Combustion engines can also be used to provide high power for either main base or remote site construction activity (bulldozers, etc.). The bottom line is that the greater power density of combustion-powered engines provides for greater mobility with much smaller, lighter, and far more capable vehicles, and that translates into a more potent and cost-effective Mars exploration program all around. If you want to do anything serious on Mars, you will need to employ combustion-powered vehicles.

But there's a hitch. The use of combustion-powered vehicles is very fuel intensive. For example, I estimate that a 1-tonne pressurized ground rover would require about 0.5 kilograms of methane/oxygen bipropellant to travel one kilometer. Thus, an 800-kilometer round-trip excursion would consume about 400 kilograms of propellant. Traveling at an average rate of 100 kilometers a day, this would only represent an eight-day sortie. In the course of a 600-day surface stay, many such excursions would be desired to make effective use of the available time. If the rover is employed in this way for just 300 of the 600 days, 15 *tonnes* of propellant will be used. Having to import that much mass from Earth just to support rover operations would be a logistics disaster. If you want to be able to take advantage of combustion-powered vehicles on Mars, you have to be able to make their propellant on Mars.

The type of combustion engine used in a Mars car could be any of the common cycles in use on Earth today, including internal combustion, diesel, or gas turbine. However, if you try to burn a pure rocket propellant combination, such as methane/oxygen in a combustion engine, the result will be an engine that burns too hot to allow for the type of reliability and long life we need in a car motor. Diluting the burning mixture with atmospheric carbon dioxide drawn in by a fan gets around this problem. The carbon dioxide acts as an inert buffer, bringing down the flame temperature in the same way that nitrogen in air does on Earth.

The range of a ground rover powered by chemical combustion will depend critically upon the energy/mass ratio of the propellant we use. While in principle any bipropellant combination could be used, transportation logistics dictate that at least most of the propellant used be manufactured on Mars out of indigenous materials. A list of potential combinations is given in Table 6.2.

The Martian atmosphere is 95 percent carbon dioxide, and thus the hydrogen/carbon dioxide (H_2/CO_2) and hydrazine/carbon dioxide (N_2H_4/CO_2) combinations given in Table 6.2 can function as air-breathing engines, much in the manner that internal combustion and jet engines do on Earth. In these cases, therefore, the energy/mass ratio given is that of the energy release per unit mass of the non-carbon dioxide fuel, since the carbon dioxide does not have to be carried by the vehicle. It can be seen that from the point of view of energy/mass ratio, the hydrogen/carbon dioxide engine is superior to all other options considered. However, storing hydrogen presents formidable problems that probably makes the use of such a system on a ground rover impractical. Taking that into consideration, the high energy density of methane/oxygen would appear to make it the preferred option. This is fortunate, because it turns out that methane/oxygen is the easiest propellant to make on Mars. It also is the best option to use as the propellant for rocket vehicles taking off from Mars. As we have seen, the Mars

TABLE 6.2
Potential Bipropellants for Use in Mars Mobility Vehicles

Bipropellant	Energy Density	
	W-hr/kg	*W-hr/liter*
Hydrogen/carbon dioxide	25,833	416
Hydrazine/carbon dioxide	1,329	1,111
Hydrogen/oxygen	3,750	1,312
Carbon monoxide/oxygen	1,816	2,144
Methanol/oxygen	2,129	2,093
Methane/oxygen	2,800	2,380

Direct plan uses methane/oxygen as the propellant for the ERV. Therefore, the same in-situ propellant plant (ISPP) that makes our rocket propellant also can be used to make our rover propellant.

It may be observed however, that the energy density figures for the methanol/oxygen combination are not too bad. This is interesting, because methanol/oxygen is a good combination to use in fuel cells (buses in Vancouver are currently powered by such systems), and after methane, methanol is the second easiest fuel to manufacture on Mars. While methanol is greatly inferior to methane as a rocket fuel, its noncryogenic nature and ease of handling (it handles like water—windshield-wiper fluid is 1/3 methanol) commends it for general use as a portable source of power for astronauts operating on the surface. Thus, if we are willing to accept the added system complication of employing two different fuel manufacturing systems, making methane for the return rocket and methanol for surface systems, the option of utilizing methanol/oxygen fuel cells for rover propulsion could also be considered.

So, the rover will either burn methane/oxygen diluted with carbon dioxide, or employ a methanol/oxygen fuel cell. The waste product of either system will be carbon dioxide and water. The carbon dioxide is of no value—more can always be obtained from the Martian air—so it will be vented as exhaust. The water is another matter, however. Properly designed Mars cars will therefore carry condensers that will allow them to recapture the water portion of their engine combustion products. (This is not hard to do. U.S. Navy dirigibles operating in the 1920s did exactly the same thing. They needed their exhaust water for ballasting purposes.) At the conclusion of a given rover sortie, the condensed waste water will be brought back to the base to be combined with carbon dioxide and synthesized back into methane/oxygen propellant by the base's chemical plant. If 90 percent of the water is recaptured, this system will allow the rovers to reuse the same propellant supply 10 times over.

How about rover life support? Well, using the same ISPP units responsible for propellant manufacture, unlimited quantities of oxygen can be readily produced on the Martian surface out of the carbon dioxide that comprises 95 percent of the Martian atmosphere. However, nitrogen and argon combined only compose about 4.3 percent of the Martian atmosphere, and thus buffer gas for breathing will be much harder to come by. (You can use carbon dioxide as a buffer gas for engines, but not for breathing. In concentrations above 1 percent it is toxic to humans.) It is therefore imperative that the habitats and pressurized vehicles operate at the lowest buffer gas partial pressures possible. For the surface habitat I recommend the 5 psi (3.5 lbs oxygen, 1.5 lbs nitrogen) atmosphere used by NASA astronauts in the long duration Skylab missions flown in the 1970s.

Apollo crews, however, operated on two-week missions in an atmosphere consisting of 5 psi oxygen and no buffer gas. Since the maximum rover excursion will be of this order, this is what I recommend for the pressurized rovers. There are major advantages to doing things this way. Such a low-pressure rover would require no airlock, and so could be built much lighter than would otherwise be possible. When they want to leave the rover ("go EVA"), the astronauts inside would simply don their spacesuits, purge the pure oxygen atmosphere in the rover cabin, and then open the hatch and walk outside. Since no nitrogen is in the air mixture, this depressurization could be done very quickly; without nitrogen in your blood, you can't get the bends. Assuming a rover interior volume of 10 cubic meters, 3.3 kilos of oxygen would be lost each time the rover was depressurized in this way. If part of the rover's interior atmosphere were pumped into a compressed oxygen cylinder prior to valving, oxygen losses would be reduced further, but in any case the losses could easily be made up by in-situ production of oxygen at the base. The low-pressure rover would allow the use of a low-pressure (3.8 psi oxygen, no buffer gas, as in Apollo) spacesuit for EVAs, with no prebreathing period to prepare for the EVA required. Such a suit would be the lightest and most flexible possible, and thus enhance the quality of surface field science performed. (The current Shuttle spacesuits are virtually miniature spaceships, and are much too heavy to be used on Mars.) Since the oxygen is replaceable, a simple once-through system in which exhaled air is vented directly to the environment (in the manner of scuba gear) would be feasible, allowing a great simplification in spacesuit design. Such a simplification would not only further the goal of reduced spacesuit mass, but would dramatically enhance spacesuit serviceability, reusability, and reliability, making possible a Mars surface mission incorporating not tens, but thousands, of EVAs.

Assuming a breathing rate of 5 gallons a minute, each astronaut using such a low-pressure oxygen "scubasuit" would expend 1.3 kilograms of oxygen in the course of a four-hour EVA. A methanol/oxygen fuel cell could tap off some of this expended oxygen, and use it in combination with a small supply of methanol to provide the spacesuited astronaut with a copious personal power supply. Thus, if two astronauts were to perform two EVAs each per rover excursion day, venting the rover twice in the process, 12 kilograms of oxygen would be used up. If the rover were to be operated in this manner every day of the 600-day surface stay, a total of 7 tonnes of oxygen would be used up. Wasting this much oxygen would be a burden if it had to be transported from Earth. If it were produced on Mars, though, it would require only twenty-four days of operation of an ISPP plant driven by a 60 kWe reactor.

MAKING PROPELLANT ON MARS

It should be evident by this point in the discussion that both the ability to get to Mars affordably and to do anything meaningful once we get there depend critically upon one key technology: the in-situ production of propellant out of the Martian atmosphere. But is this really possible? Absolutely. In fact, all of the chemical processes used in the Mars Direct plan have been in large-scale use on Earth for over a century.

The first step in propellant production is to acquire the required raw materials. Since the hydrogen component of the bipropellant mixture represents only about 5 percent of the total propellant weight, it can be imported from Earth. Heavily insulating the hydrogen tanks with multilayer insulation can reduce in-space boiloff of liquid hydrogen to less than 1 percent per month during the six-to-eight-month interplanetary transit without any requirement for active refrigeration. Since the hydrogen raw material is not going to be directly fed into an engine, it can be gelled with a small amount of methane to prevent leaks. Gelling the hydrogen cargo will also reduce boiloff further (as much as 40 percent) by suppressing convection within the tank.

The only raw materials thus required from Mars are carbon and oxygen, which are the two most plentiful elements in the 95 percent carbon dioxide Martian atmosphere and can be acquired "free as air" anywhere on the planet. The atmospheric pressure measured at the two *Viking* sites varied over a Martian year between 7 and 10 millibars (1 bar is Earth sea-level atmospheric pressure, or 14.7 psi; 10 millibar, or mbar, is 1 percent Earth sea-level atmospheric pressure), with a year-round average of about 8 mbar observed at the higher altitude *Viking 1* landing site on Chryse Planitia. Pumps that can acquire gas at this pressure and compress it to a workable pressure of 1 bar or more were first demonstrated by the English physicist Francis Hawksbee in 1709. Even better pumps are available today. However, you don't even need a pump to compress carbon dioxide. Instead, you can use a sorbant bed that will act like a sponge, soaking up carbon dioxide. All you need to do is take a jar and fill it with either activated carbon or zeolite, and then expose it to the Martian atmosphere at night. Given the chill (–90°C) nighttime temperatures, the material bed will soak up to 20 percent of its weight in carbon dioxide. Then when day comes, you warm the bed up to 10°C or so, and the carbon dioxide will outgas. You can generate very high pressure carbon dioxide gas this way, with essentially no moving parts and very little power expenditure. In fact, you can even use the waste heat generated by other components on your propellant maker to drive the outgassing process. At my lab at Martin Marietta we built such a system and it worked quite well.

Now, to ensure quality control in the propellant production process, no substances of unknown composition, to wit, Martian dust, should enter the chemical reactors. This can be accomplished by first placing a dust filter on the bed inlet or pump intake to remove the vast majority of the dust, and then compressing the Martian air to about 7 bar pressure. When carbon dioxide gas is brought to this pressure and then allowed to equilibrate to ambient Martian temperatures, it will condense into a liquid state. (We don't see liquid carbon dioxide on Earth because the pressure is too low.) Any dust that managed to evade the pump filters will go into solution, or precipitate to the bottom of the CO_2 tank, while the nitrogen and argon constituents of the air will remain gaseous and thus can be removed, to be either discarded, or, better yet, kept for use as life-support system buffer gas. If carbon dioxide is then vaporized off the holding tank, it will be distilled 100 percent pure, as all dust will be left behind in solution. Distillation purification processes working on this principle have been widely used on Earth since the mid-1700s, when Benjamin Franklin introduced a desalination device for use by the British Navy.

Once pure carbon dioxide is obtained, the entire process becomes thoroughly controllable and predictable, as no unknown variables can be introduced by Mars. With the design of adequate quality control in the carbon dioxide acquisition process, the rest of the chemical production process can be duplicated on Earth under precisely the same conditions that will be present on Mars, and reliability guaranteed by an intensive program of ground testing. Very few of the other key elements of a piloted Mars mission (engines, aerobrakes, parachutes, life-support systems, orbital rendezvous or assembly techniques, etc.) can be subjected to an equivalent degree of advance testing. This means that, far from being one of the weak links in the chain of a Mars mission, the in-situ propellant process can be made one of the strongest.

Once carbon dioxide is acquired, it can be rapidly reacted with the hydrogen brought from Earth in the methanation reaction, which is also called the Sabatier reaction after the chemist of that name who studied it extensively during the latter part of the nineteenth century.

The Sabatier reaction produces methane and water from carbon dioxide and hydrogen and is written as:

$$CO_2 + 4H_2 \rightarrow CH_4 + 2H_2O \quad (1)$$

This reaction is *exothermic*, that is, it releases heat, and will occur spontaneously in the presence of a nickel or ruthenium catalyst (nickel is cheaper, ruthenium is better). The equilibrium constant which determines the completeness of the reaction is extremely strong in driving the reaction to the right, and production yields of greater than 99 percent utilization with just one pass through a reactor are routinely achieved. In addition to having been in wide-scale industrial use for about a hundred years, the Sabatier reaction has been researched by NASA, the U.S. Air Force, and their contractors for possible use in Space Station and Manned Orbiting Laboratory life-support systems. The Hamilton Standard company, for example, developed a Sabatier unit during the 1980s for use on the Space Station, and subjected it to about 4,200 hours of qualification testing.

The fact that the Sabatier reaction is exothermic means that no energy is required to drive it. Furthermore, the reactors used are simple steel pipes, rugged and compact, that contain a catalyst bed. In fact, on the basis of results obtained from the lab programs at Martin Marietta and Pioneer Astronautics, I believe that the Sabatier reactor setup required to perform all the methane production needed for the Mars Direct mission could consist of no more than a set of three reactors, each 1 meter long and 12 centimeters in diameter.

As the reaction (1) is run, the methane so produced is liquefied either by thermal contact with the super-cold input hydrogen stream or (later on after the liquid hydrogen is exhausted) by the use of a mechanical refrigerator. (Methane is liquid at about the same "soft cryogenic" temperatures as liquid oxygen.) The water produced is condensed and then transferred to a holding tank, after which it is pumped into an electrolysis cell and subjected to the familiar electrolysis reaction, which splits water into its components, hydrogen and oxygen:

$$2H_2O \rightarrow 2H_2 + O_2 \quad (2)$$

The oxygen so produced is refrigerated and stored, while the hydrogen can be recycled back to the Sabatier reaction (1).

Electrolysis is familiar to many people from high-school chemistry, where it is a favorite demonstration experiment. However, this universal experience with the electrolysis reaction has created a somewhat misleading mental image of an electrolysis cell as something composed of Pyrex beakers and glassware strung out across a desk top. In reality, modem electrolysis units are extremely compact and robust objects, composed of sandwiched layers of electrolyte-impregnated plastic separated by metal meshes, with the assembly compressed at each end by substantial metal end caps bolted down to metal rods running the length of the stack. Such solid polymer electrolyte (SPE) electrolyzers have been brought to an extremely advanced state of development for use in nuclear submarines, with over twenty million cell-hours of experience to date. Testing has included subjecting cells to depth charging and loads of up to 200 gs. Both the Hamilton Standard and the Life Sciences companies have developed lightweight electrolysis units for use on the Space Station. These units have the capacity to perform the propellant production operation for the Mars Sample Return ISSP mission. The SPE units that Hamilton Standard has supplied for use by Britain's Royal Navy have the correct output level to support the propellant production requirements of the manned Mars Direct mission. These units have operated for periods of up to 28,000 hours without maintenance, about four times the utilization required for Mars Direct. The submarine SPE electrolysis units are very heavy, as they are designed to be so for ballasting purposes. SPE electrolysis units designed for space missions would be much lighter.

If all the hydrogen is expended cycling the propellant production process through reactions (1) and (2), then each kilogram of hydrogen brought to Mars will have been transformed into 12 kilograms of methane/oxygen bipropellant, with an oxygen-to-methane mixture ratio of 2:1. Burning the bipropellant at such a ratio would provide a specific impulse of about 340 seconds. This would be okay, but the optimum oxygen-to-methane combustion mixture ratio is about 3.5:1, as this provides for a specific impulse of 380 seconds and a hydrogen-to-bipropellant mass leveraging of 18:1. This is the level of performance we need to reach for the optimal design of the manned Mars Direct mission.

To obtain this optimal level of performance, an additional source of oxygen must be procured beyond that made available by the combination of reactions (1) and (2). One possible answer is the direct reduction of carbon dioxide.

$$2CO_2 \rightarrow 2CO + O_2 \quad (3)$$

This reaction can be accomplished by heating carbon dioxide to about 1,100°C, which will cause the gas to partially dissociate, after which the free oxygen so produced can be electrochemically pumped across a zirconia ceramic membrane by applying a voltage, thereby separating the oxygen product from the rest of the gas. The use of this reaction to produce oxygen on Mars was first proposed by Dr. Robert Ash at JPL in the 1970s, and since then has been the subject of ongoing research by both Ash (now at Old Dominion University), and Kumar Ramohalli and K. R. Sridhar (at the University of Arizona). The advantage of this process is that it is completely decoupled from any other chemical process, and an infinite amount of oxygen can be produced without any additional feedstock. The disadvantages are that the zirconia tubes are brittle, and have small rates of output, so that very large numbers would be required for the manned Mars Direct application. It also requires about five times as much power per unit oxygen produced as does water electrolysis. Improved yields have recently been reported at the University of Arizona, so the process may be regarded as promising, but still experimental.

An alternative that would keep the set of processes employed firmly within the world of Gaslight Era industrial chemistry would be to run the well-known (to chemical engineers) water-gas shift reaction in reverse. That is, recycle some of the hydrogen produced in the electrolysis unit into a third chamber where it is reacted with carbon dioxide in the presence of an iron-chrome or copper catalyst to produce carbon monoxide and water as follows:

$$CO_2 + H_2 \rightarrow CO + H_2O \quad (4)$$

This reaction is mildly endothermic, but will occur at 400°C, which is well within the temperature range of the Sabatier reaction. If reaction (4) is cycled with reactions (1) and (2), the desired mixture ratio of methane and oxygen can be produced with all the energy required to drive reaction (4) provided by thermal heat output from the Sabatier reactor. Reaction (4) can be carried out in a simple steel pipe, making the construction of such a reactor quite robust. The disadvantage of reaction (4) is that in the temperature range of interest it has an equilibrium constant of only about 0.1, which means that in order to make it go it is necessary to run a condenser and a membrane separator to remove water and carbon monoxide from the reactor on an ongoing basis, and then recycle the unreacted hydrogen and carbon dioxide back to the reactor using a pump. (Water and CO are the products on the right-hand side of equation (4); so long as they are continuously removed, chemical principles dictate that the reaction will keep moving to the right, producing water and CO so as to try to maintain the appropriate equilibrium concentration in the reactor.) Such a system was first demonstrated by myself and Brian Frankie at Pioneer Astronautics in 1997, with subsequent improvements achieving nearly complete conversion of carbon dioxide and hydrogen to CO and water. By running such a reverse water gas shift (RWGS) unit in parallel with a Sabatier/electrolysis cycle, the desired methane/oxygen propellant ratio needed to perform the Mars Direct mission could readily be achieved.

A more elegant solution would be simply to combine reactions (1) and (4) in a single reactor as follows:

$$3CO_2 + 6H_2 \rightarrow CH_4 + 2CO + 4H_2O \quad (5)$$

This reaction is mildly exothermic, and, if cycled together with reaction (2), would produce oxygen and methane in a mixture ratio of 4:1, which would give the optimum propellant mass leveraging of 18:1 with a large extra quantity of oxygen also produced that could function as a massive backup to the life-support system. In addition, salvageable carbon monoxide would also be produced that could conceivably be used in various combustion devices or fuel cells. If all the carbon monoxide and oxygen produced is included, the total propellant mass leveraging obtained could be as high as 34:1!

In a project conducted for NASA between 2005 and 2007, Pioneer Astronautics demonstrated this cycle in action, with an end-to-end demonstration system that took simulated Martian air at 8 mb from a chamber, pumped it up to 3 bar pressure, then applied the combined reaction (5) to produce methane, CO, and water, then electrolyzed the water to produce oxygen for use and hydrogen for recycle, distilled the methane from the CO, and liquefied the final methane product. Led initially by Tony Muscatello and completed under the guidance of Douwe Bruinsma after Tony left Pioneer to take a post at NASA Kennedy Space Center, the machine was shown to be able to produce methane and oxygen in any ratio desired, operating under automated control for periods of up to five days nonstop.

Still another method of obtaining the required extra oxygen is simply to take some of the methane produced in reaction (1) and pyrolyze it into carbon and hydrogen.

$$CH_4 \rightarrow C + 2H_2 \qquad (6)$$

The hydrogen so produced would then be cycled back to attack more Martian carbon dioxide via reaction (1). After a while, a graphite deposit would build up in the chamber in which reaction (6) was being carried on. (This reaction is actually the most common method used in industry today to produce pyrolytic graphite.) At such a time, the methane input to the reactor would be shut off, and instead the chamber would be flushed with hot carbon dioxide gas. The hot carbon dioxide would react with the graphite to form CO, which would then be vented, cleaning out the chamber.

$$CO_2 + C \rightarrow 2CO \qquad (7)$$

Such a plan, incorporating two chambers, with one carrying out pyrolysis while the other is being cleaned, has been suggested to me as the simplest solution to the extra oxygen problem by Jim McElroy and his group of researchers at Hamilton Standard.

Now, it is sometimes the case that it is very easy to write down a chemical synthesis system as a series of equations on paper, but a very different matter to build a unit that can perform it as a matter of practice. That, however, is not the case here. I know, because I've led a number of projects that have built Mars ISPP units from scratch. The first, and in certain ways the most dramatic of these began in the fall of 1993, when David Kaplan and David Weaver, representing NASA's Johnson Space Center, approached me and asked if Martin Marietta could demonstrate a working model of the kind of Mars ISPP system I had been promoting at conferences and in papers. There was a catch, however. NASA could provide only $47,000 in funds to support the project, a very small budget to develop and demonstrate a new aerospace technology, and the work would have to be completed by January 1994. This was quite a challenge—at Martin Marietta $47,000 will ordinarily buy you a report containing a couple dozen viewgraphs. I was convinced, however, that the technology involved was fundamentally simple, and that the project, whose accomplishment seemed farfetched on the budget and schedule proposed, was fundamentally feasible. After much discussion with Martin Marietta management, the challenge was accepted. In October 1993, Martin Marietta was put on contract to undertake the work, with David Kaplan serving as the JSC program manager, Steve Price as the project manager at Martin Marietta, and me serving as principal investigator and lead engineer.

The design of the system was done during October 1993, and most of November was spent waiting for parts to arrive in the mail. By the end of November, all the required components were in hand, and construction, sized full scale to the requirements of a Mars Sample Return mission, was begun in earnest.

The Sabatier reactor was built from scratch, filling a metal pipe 36 centimeters long and 5 centimeters in diameter with a ruthenium catalyst obtained from a chemical supply company. (We found out later that this was ten times the volume we needed for our system, but we were on a tight schedule that would not allow us to build anything twice. So, overdesigning seemed like the way to go.) The electrolyzer, standing just 25 centimeters tall and weighing only 3 kilograms, water included, was ripped from a Packard Instrument laboratory hydrogen supply unit. Nichrome heaters, used to warm the Sabatier reactor up to its operating temperature (after which the heat from chemical reactions would keep it hot without electricity) were obtained and wrapped around the Sabatier reactor. A condensing system was built to separate the methane product from the water product, and the whole system plumbed into a cycle, with pressure and temperature sensors and gas flow meters inserted at strategic points and wired to a computer data display to allow system monitoring and control. By the end of the second week of December, the system was complete and ready for operation. (See plates.)

On December 15, the system was turned on for the first time, with just the Sabatier reactor running. By the end of the second hour of operation, the water level in the condenser vessel had risen noticeably, indicating that the system was working. Subsequent laboratory analysis of the effluent gas from the Sabatier reactor showed that it was operating at 68 percent efficiency in converting input hydrogen and carbon dioxide into methane and water.

On subsequent days, adjustments were made to the system to improve performance. On the 22nd of December, 85 percent conversion efficiency was achieved, with hydrogen for the Sabatier reactor being supplied by the electrolyser. On January 5, running for the first time with full system integrated operation, we reached 92 percent efficiency. Finally, on January 6, 1994, the full system ran in its fully integrated mode all day, achieving a 94 percent conversion efficiency in the process.

With the conclusion of the January 6 run, all test objectives had been achieved, and there was still enough money left in the kitty to write the report.[24]

Following that success, additional small chunks of money from first JSC, then the Jet Propulsion Laboratory, allowed further improvements and elaborations of the Martin Marietta (subsequently Lockheed Martin) system. Sorption beds were added, allowing the unit to acquire its carbon dioxide from a Mars atmospheric reservoir held at Martian pressures. The Sabatier reactor had its efficiency increased to 96 percent and was miniaturized by a factor of 10, and a 2-kilogram Stirling cycle refrigerator was added, allowing us to liquefy all the product oxygen and store it in a cryogenic dewar. Automated control systems were also added, allowing the system to run 10 days at a time without operator intervention. The total mass of all working components in this system, which was sized to produce 400 kilograms of propellant to support a

Mars Sample Return mission, was about 20 kilograms, and the total power needed was less than 300 watts.[25] In 1996, I left Lockheed Martin to found my own company, Pioneer Astronautics, where we have developed numerous additional machines, demonstrating the reverse water gas shift, methanol, benzene, ethylene, and propylene production, and combined Sabatier/electrolysis/RWGS systems. Meanwhile, the old team at Lockheed Martin, led now by Larry Clark, has continued to advance the Sabatier/electrolysis system there to ever greater efficiency and more flightlike configurations. Studies indicate that when scaled up to the size needed to accomplish the Mars Direct mission, the mass leverage of all of these units would be even more pronounced, as the percent of system mass represented by parasitic elements such as flow meters and pressure sensors would fall into the noise.

We can make rocket fuel and oxygen on Mars.

STAYING IN TOUCH WITH THE BASE

Using combustion-powered ground vehicles, the first Mars explorers will be able to wander far from their base, but if they do, how will they maintain communication? Mars is, after all, a little more than half the diameter of the Earth, so the horizon is correspondingly closer. If the terrain on Mars were as flat as Kansas, the horizon would only be about 40 kilometers (25 miles) away—and Mars is most definitely not like Kansas. So, if the excursion team wants to go anywhere on Mars, they're going over the horizon. That rules out line-of-sight radio transmissions. How will they manage to stay in touch with the base?

One answer is to have a communication relay satellite stationed in Mars orbit, 17,065 kilometers above the equator. At that altitude, the satellite will be flying at a velocity of 1.45 km/s, and will take 24.6 hours to orbit Mars. Since this is the length of a Martian day, the satellite will keep pace with the planet as it turns, and to an observer on the ground will not appear to move at all. Such an "aereosynchronous" satellite is the exact analog on Mars of the geosynchronous satellites currently used extensively to support communication on Earth. If our Mars expedition lands at the equator, the satellite will hang directly overhead all day and all night, supporting communication from the base to anyone or anything within a region with a radius about the base of nearly 5,000 kilometers, covering nearly half the surface of the planet.

But communications satellites cost money, and, more importantly, are subject to failure. If the satellite should go on the blink while the exploration team is 400 kilometers out from the base, what then?

The backup plan is to use ham radio. You see, Mars has an ionosphere, a layer of charged particles in the high upper reaches of its atmosphere, that can be used to reflect radio signals, enabling global surface-to-surface communication in the short-wave radio bands, just as on Earth. We know a lot about the properties of the Martian ionosphere from measurements taken by *Mariner* 9, the *Viking* orbiters and landers, and the European *Mars Express* probe. It extends upwards from an altitude of about 120 kilometers, and is composed of an ion population consisting of 90 percent O_2+ and 10 percent CO_2+, matched by an equal number of free electrons produced via photo-ionization. During the day the electron density reaches a peak concentration of about 200,000/cm^3 at an altitude of about 135 kilometers. During the night, this density falls off to a peak value of about 5,000/cm^3 at an altitude of about 120 kilometers. These numbers are lower than the electron density in Earth's ionosphere by about a factor of 25. However, since the maximum usable frequency for shortwave radio goes as the square root of the electron density, the maximum usable frequency available on Mars is only lower than that obtainable on Earth by about a factor of 5. So, while on Earth, hams can talk to each other with frequencies as high as 20 MHz, the best you can do on Mars is about 4 MHz during the day and 700 kHz at night. The latter figure is kind of low if you want to transmit images or engage in other kinds of high data rate transmission, but it is more than adequate for engineering telemetry or voice communication. In fact on Earth this frequency band—AM radio—is favored for the commercial transmission of talk radio and other forms of (alleged) communication.

Furthermore, while Mars' shortwave communication bands are positioned at a somewhat lower frequency than those of the Earth, this disadvantage (using higher frequencies enables a higher data rate) is more than counterbalanced by the fact the Martian ionosphere is much less afflicted with radio noise. On Earth, shortwave radio transmission power requirements are driven up by radio noise caused by distant thunderstorms and large numbers of other hams, military users, and commercial radio stations. All of these problems would be absent on Mars.

Some ham setups in current use may conjure up images of heavy, unwieldy equipment unsuitable for mobile communications. However, advanced shortwave technology highly suitable for use by Mars explorers has been developed on Earth for military purposes. One such example is the Advanced Miniature High Frequency System (AMHFS) developed by Defense Systems Inc. The AMHFS is a two-way transmitter/receiver system, with each unit having a mass of 0.8 kilograms and a volume of 0.7 liters (smaller than a quart-sized container)—small enough to be carried not only in rovers, but by individual astronauts during EVA. Based upon its terrestrial performance, this system could transmit globally over the sunlit side of Mars at a rate of 2.4 kb/s using 10 W of radiated power, or 30 W electric. This speed of transmission is adequate for engineering telemetry, e-mail, real-time low-quality voice, or high-quality voice packet transmission. To achieve high-quality real-time voice (as in terrestrial telephones) twenty times this data rate would be required, and thus 600 W of power, which could be easily generated on the rover. However, these power requirements may be sharply reduced if Mars' ionosphere is really as quiet as theory predicts. In any case, the AMHFS uses an adaptive sounding technique that automatically searches the radio spectrum to find the maximum usable frequency in real time, and then causes the two units in communication to perform a handshake acknowledging the link and verifying that data has been transmitted correctly. Thus, even if ionospheric conditions are unpredictable or variable during the transmission, the AMHFS can adapt to find and maintain the best communication channel. The AMHFS uses its electronics to compensate its antennae size for the wavelength chosen for communication. Thus, the same 6-meter whip antenna can be used for transmission at 0.5 MHz as at 5 MHz. The antennae used are very lightweight and are typically simply helical springs, or "stacers," that can be released to pop into place for deployment.

The use of shortwave radio for communication has an additional benefit for Mars explorers. The same system can be used for exploration via deep-ground-penetrating radar. A 3 MHz radio signal has a wavelength of 100 meters. In the dry Martian environment, such signals, if directed downward, could be expected to penetrate about 10 wavelengths, or 1,000 meters, into the ground. It now appears that Mars has an underground liquid water table that may be found within 500 meters and 1,000 meters of the surface. Even if this is not true globally, it is almost certainly true in some localities, as proven by the observation made by the *Mars Global Surveyor* spacecraft of the appearance of a water erosion feature on the side of a crater between 2001 and 2005, which could only have been created by a transient outflow of water from an underground source during *MGS*'s time on orbit. Indeed, such underground liquid water reservoirs may well be common, as geothermal heat will necessarily cause pockets of subsurface ice to melt and form hot subsurface reservoirs. (Mars is geologically alive. It is estimated that some of the volcanic features in Tharsis may be less than 200 million years old. From the point of view of Mars' 4.5-billion-year planetary age, they might as well have erupted yesterday.) A rover team driving along with a shortwave radio could send radar pings into the ground. If liquid water exists underground within a kilometer or so of the surface, its much higher electrical conductivity than the surrounding dry soil or ice will cause the radio signal to be strongly reflected back to the rover's receiver, and the time delay between transmission and reception will tell the crew how deep the reservoir is. If they should discover a heated pool relatively close to the surface it will be time to get out the drilling rig. Water is, after all, the staff of life.

NAVIGATION ON MARS

In addition to maintaining communication with the home base, Mars explorers will also need to navigate. While good maps of Mars are available from orbital imaging, the essential problem for a Mars rover crew will be determining their own location. This is critical not only for documenting the location of various scientific finds, but, more importantly, to prevent getting lost. On the deserts of Mars, as in the North African desert during World War II, getting lost means dying. A radio beacon at the base could help a crew find its way home, but its range would reach at most only to the nearby horizon (just 40 kilometers away, remember). Upon approaching the limits of the base beacon's range, a departing rover crew could station a second beacon on a hilltop, and then another, and another, to mark a return path. Such techniques are, however, quite limiting, and as in the story about the bread-crumb trail being eaten by birds, are subject to catastrophic failure if one of the beacons composing the trail should cease functioning. What other navigation techniques will be available to the rover crew?

Well, to an aerospace engineer, the first thing that comes to mind is to use navigation satellites. If a satellite is stationed in low polar orbit about Mars, its latitude at any given moment will be known. If you put a radio beacon on the satellite (beginning with the *Mars Global Surveyor* launched in 1996, all Mars orbiters actually have had such beacons), the rover crew can listen for it, and when they compare the time of closest approach with the satellite's itinerary recorded in the rover's computer, they can determine their latitude. In addition, the satellite's rate of approach to the rover crew's position will be fastest if the rover is sitting directly below the satellite's ground track but much slower if they are stationed far to the side. Measuring the Doppler shift caused by the approach and recession rate of the satellite beacon can allow the rover crew to determine how far east or west they are from the satellite's north-south ground track. Once again, comparing this information with computer records specifying the satellite's longitude as a function of time would allow the crew to determine their longitude.

These high-tech techniques are quite accurate. A similar approach is used on Earth in the Argos satellite system to track the movement of falcons and elk (except that in this case the beacon is on the elk and the receiver and required calculations are done on the satellite) with precision of about one kilometer. However, there are a number of problems. The satellite is in a roughly 2-hour orbit with Mars turning beneath it. Therefore, an observer on the ground will encounter the satellite only once during the day and once at night, so that a single satellite will only allow position fixes to be taken once every 12 hours. This can be remedied by having more satellites orbiting in a number of north-south planes spaced around the planet, but then you start running into real money. And what if the satellite beacon, or the rover receiver, or the rover computer, should fail? What then? Are there more reliable low-tech navigation techniques that can serve as a backup?

On Earth, the magnetic compass has long served as the mariner's key navigation tool. Unfortunately, compasses won't work on Mars because the planet has virtually no magnetic field. However, the time-honored techniques of celestial navigation can be used on the Red Planet, with much greater facility, in fact, than has ever been possible on Earth.

If you've ever practiced celestial navigation, you'll know that determining latitude is easy, while determining longitude is hard. To determine latitude all you need is a sextant to measure the angle between the celestial pole and the horizon. That angle is your latitude, period. In Earth's Northern Hemisphere this measurement is easy to do, because the celestial pole is marked within 1° accuracy by the pole star, Polaris. The direction of Polaris also tells you which way is north—with better accuracy than any compass. Does Mars have a conspicuous pole star? Not really, but its celestial pole, located at 21.18 hours right ascension, 52.89° north declination, is still pretty easy to find, as it lies almost exactly halfway between the two bright stars Deneb and Alpha Cephei. So, with a sextant and a clear night (a more common occurrence on desert Mars than on the rainy, misty old Earth) you can readily determine your latitude on Mars.

What about longitude? On Earth, with the aid of an accurate clock set on some standard time, such as Greenwich Mean Time, you can determine your longitude by measuring the time of sunrise and comparing it to the value given in an almanac for the sunrise time for that day at the Greenwich meridian (the prime meridian, 0° longitude) at your latitude. For example, if the almanac says that the Sun will rise at 6 A.M. at your latitude on the prime meridian on March 21, and you observe the Sun rising at 7 A.M. on your Greenwich Mean Time clock, you would know you are at 15° west longitude, since the Earth turns at a rate of 360 degrees in 24 hours, or 15 degrees per hour.

This works reasonably well on Earth, but it works much better on Mars, because on Mars in addition to the Sun, the two rapidly moving asteroidlike moons Phobos and Deimos can also be used as longitude beacons. As seen from the surface of Mars, Phobos, the inner moon, would have a visual magnitude of –10, which makes it about 300 times brighter than Venus as seen from the Earth at its brightest, while Deimos would have a magnitude of –7, about twenty times brighter than Venus. Except during dust storms, both of these satellites should be clearly visible from the Martian surface both by day and by night. Both moons are in almost precisely equatorial orbits, so by measuring their angular distance from the zenith when they are at their highest in the sky, you can use these moons to determine your latitude, even during the middle of the day. Phobos orbits Mars every 7 hours 39 minutes, while Deimos has a period of 30 hours 18 minutes. Between the Sun, Phobos, and Deimos, the Martian navigator will have lots of sunrise and moonrise events to chose from to compare against his almanac and clock, and with each rising, longitude can be determined. In fact, using a bit of math which would be A-B-C for a trained navigator; it is possible for an observer on Mars equipped with a sextant, a clock, and an almanac to determine his latitude and longitude simultaneously whenever any two of the three objects (Sun, Phobos, and Deimos) are visible in the sky.

By the way, on Earth we define one nautical mile as one minute (1/60th of a degree) of latitude. This is about 1.82 kilometers. However, if we define a Martian nautical mile as equal to one minute of latitude on Mars, it turns out that a Martian nautical mile equals one kilometer almost exactly (well, 983 meters). So, on Mars, navigators and metric system buffs will finally be able to get along!

KEEPING TIME ON MARS

There has been a fair amount of discussion in the literature of possible timekeeping systems for use on Mars. Having just discussed surface navigation, now would be a good time to address this matter.

As we've seen, the Martian day lasts 24 hours and 39.6 minutes of terrestrial time. Timekeeping systems proposed to date have usually preserved terrestrial clock units, with an additional partial hour thrown in just after midnight.[26] Alternatively, totally novel, usually decimal-based clocks using a completely new set of time units have sometimes been proposed.[27]

It should be clear from the previous section's discussion that a clock employing unequal hours would be a nightmare for those attempting navigation or astronomy on the Martian surface. On the other hand, a decimal or other novelty clock would probably be disorienting and in any case would require a complete overhaul of Mars' existing system of surface geographical coordinates (which employs the same base-60 degrees, minutes, and seconds system that is used to map the Earth).

The practical answer is simple—just divide the Martian day into 24 Martian hours, each composed of 60 Martian minutes, each of which in turn is composed of 60 Martian seconds. The conversion factor between Martian days, hours, minutes, and seconds and their terrestrial equivalents would thus be 1.0275 across the board. A time of day on Mars, say 6:00 A.M., would have exactly the same physical significance with regard to the orientation of the planet toward the Sun as it does on Earth. All the equations of celestial navigation used on Earth would then remain precisely valid. That is, regardless of whether you are on Mars or Earth, one hour of time would equal 15 degrees of longitude, one minute of time would equal 15 minutes of longitude, and one second of time would equal 15 seconds of longitude.

Such a clock solves all the practical problems associated with daily timekeeping on Mars. In fact, it is in implicit use already among mission planners at Jet Propulsion Lab today who, for example, might describe a future Mars orbiter's path as a 6:00 A.M.–6:00 P.M. orbit, meaning a satellite that travels tracking the dawn–dusk terminator on Mars. That is, "6:00 A.M." as they use it is really a Mars local time in the sense described above, and the twelve hours separating it from "6:00 P.M." are Martian hours. It is unfortunate that such a clock annoys physicists who regard the terrestrial second as the sacrosanct unit of physical time. They really shouldn't worry—Martian crystallographers and others who require a high degree of precision in quoting their measurements of frequencies will still be able to quote the measurements in terms of terrestrial seconds. The standard International System of Units can remain intact. However, for purposes of operating on Mars, the terrestrial second is no more useful a unit of timekeeping than the terrestrial day, and must yield to its Martian counterpart.

TELEROBOTICS: EXTENDING THE REACH OF THE CREW

For safety reasons, while two members of the crew (a scientist and a mechanic) are out on a rover excursion, the other two will generally remain behind at the hab base. Thus, if the rover crew should get into trouble, the reserve crew can ride to the rescue in a backup vehicle (such as one of the open rovers). In general there will always be at least two people stationed at the base, and, in between rover excursions (which might typically last one to ten days each), the entire crew of four will be there. Granted, there are many useful activities that the crew can undertake at the base—analyzing samples, conducting various scientific and engineering experiments, and engaging in construction and necessary maintenance of equipment. Nevertheless, since the primary function of the initial Mars missions will be exploration, it would be extremely beneficial if personnel stationed at the base could use part of their time to explore. This they will be able to do, provided that the expedition is supplied with a contingent of telerobots.

Martian telerobots would be small wheeled or treaded roving vehicles equipped with TV cameras, microscopes and other scientific instruments, manipulator arms, and a radio. Controlled from the Mars base via either shortwave radio or through an aereosynchronous relay satellite, these telerobots could be driven rapidly by remote control, as the radio link time delay would be insignificant (the Earth–Mars radio time delay, up to 40 minutes round trip, severely impairs telerobotic operations conducted from Earth). The telerobots could be deployed by the rover crews as they travel, allowing the base crews to explore in greater detail various sites that rover crews identify as interesting but had no time to investigate at length themselves. The telerobots could also be used to probe into places too small or risky for humans, such as caverns or narrow crevices.

However, the base crews could also deploy some telerobots themselves, lofting them up on balloons and then bringing them down to land thousands of kilometers away. (A balloon on Mars could be expected to fly 2,000 kilometers in a single day.) The flight path of the balloons can't be controlled of course, but provided that the wind patterns of Mars have been mapped out in advance by missions like MAP the path the balloon-borne telerobot takes could well be predicted. As the telerobot flies, its cameras can be used to send real-time aerial images to the base crew, who, looking through the telerobot's eyes, will be able to choose the best time and place to land the system. Upon landing, the telerobot could either release the balloon, thereby committing itself to its selected landing region for life, or, if the winds are light, it could attempt to secure the balloon's anchor line to a rock formation. In the latter case, the telerobot could then leave the balloon and explore the area for a few hours, but then reattach itself to the balloon, release anchor, and take off to visit still another more distant site.

A potentially even more powerful option would be telerobots that can fly themselves wherever they need to go. One method of doing this would be to employ a concept known as a "gashopper." In the gashopper, a set of solar panels run a small pump that acquires and liquefies by compression CO_2 from the Martian atmosphere, which is then stored in a tank at about 10 bar pressure. Once this is done, the solar power would be redirected to drive a set of resistive heaters to heat up a bed of high-heat capacity material contained within a steel vessel to a temperature of perhaps 800°C. The vehicle could then take flight by allowing the pressurized CO_2 to flow through the high temperature heat storage vessel, where it would be vaporized and then released out a nozzle to produce rocket thrust. At Pioneer Astronautics, we have built and flown such gashoppers configured as both winged rocketplanes and ballistic rockets capable of vertical takeoff and landing, using magnesium oxide pellets as a thermal storage medium. Using higher performance heat storage media, such as beryllium pellets or liquid lithium, ballistic gashoppers operating on Mars would be able to achieve hops of 20 km length, while winged aircraft could travel 150 km in a single flight. The best design would probably be something like the British Harrier fighter, capable of vertical takeoff and landing with long distance flight enabled by wings. After each landing, the gashopper could release a small teleoperated rover to engage in local exploration for a

FIGURE 6.1
Gashopper prototype during flight-testing at Pioneer Astronautics in July 2005. (Pioneer Astronautics)

few weeks, while the flight vehicle replenishes its CO_2 propellant supply from the atmosphere. Then, when the tanks have been refilled, and the engine brought back up to temperature, the rover could be recalled and flown to a new site for further exploration.

Neither cliffs nor canyons, nor even small mountains will stand in the way of the flying telerobots. Deployed and controlled without time delay from the first manned Mars base camp, they will make vast regions of the planet accessible to scientific exploration.

Having a telerobot deployed at a distant site is the next best thing to being there. But in this case next best is a distant second best; to truly explore Mars we will need to send actual human explorers all over the planet. How can this be done? To some extent, this goal can be met by sending each new Mars Direct mission to a new landing site, opening up a new region for exploration. But while necessary in the short run to get a significant expanse of Mars investigated, in the long run such a strategy is inefficient, as it prevents follow-on missions from using assets left behind by preceding missions. At some point, then, after an initial set of exploratory missions, landings of successive missions should be concentrated at a single site, to build up a major base. Such a base will have, among other things, the resources required to maintain much larger teams of astronauts on Mars, and to support the operation of piloted rocket propelled flight vehicles that will give these explorers truly global reach in their investigations of the Red Planet. It is to the development and use of such a base that we will turn to as the subject of our next chapter.

FOCUS SECTION—A CALENDAR FOR THE PLANET MARS

Martian colonists will need a calendar that is tied to physical and seasonal conditions on the Red Planet—using Earth dates just won't do. If I tell you it is February 1, you know that it is freezing in Minneapolis and high summer in Sydney, but what does it tell you about conditions on Mars? In fact, the need for a Martian calendar and timekeeping system is already upon us because of current and planned unmanned exploration missions. You know the season on Earth now and can predict it with ease for any date named in the future, but without a Martian calendar you'll be hard pressed to do the same there. So, we might as well remedy this right now.

Here is the problem: Mars has a year consisting of 669 Martian days, or "sols." As we have seen, the correct method to measure time within these days is to use units 1.0275 times longer than their terrestrial counterparts. But equipartitioned months don't work for Mars, because the planet's orbit is elliptical, which causes its seasons to be of unequal length.

In order to predict the seasons, a calendar must divide the planet's orbit not into equal division of days, but into equal angles of travel around the Sun. If we want months to be useful units and choose to retain the terrestrial definition of a month as a twelfth of a year, then a month really is 30 degrees of travel about the Sun. But what to name them? Using the current terrestrial month names could be confusing, and a totally new system would be completely arbitrary. There is, however, a set of names available that has long been universally known to humanity and that has real physical significance not only for Mars, but for any planet in our solar system: the signs of the zodiac. All the constellations of the zodiac lie in the plane of motion of all the planets. Ancient astrologers, having a geocentric point of view, named the months for whatever zodiacal constellation the Sun appeared to be located in as viewed from Earth. An interplanetary culture, though, must adopt a heliocentric, or Sun centered, point of view. Therefore, I have chosen to name the Martian months for whatever constellation Mars would be found in as seen from the Sun. For Martian colonists then, the sign of the month would be seen high in the sky during the midnight hours of a given month. It is currently the custom among planetary scientists to start a planet's year with the vernal equinox (the beginning of spring, March 21, in the Earth's northern hemisphere), and so, consistent with that custom, the Martian year begins with the month of Gemini and ends with Taurus. The complete Martian year is given in Table 6.3.

In order to convert Earth dates to Mars dates, I have invented a

TABLE 6.3
The Martian Year

Month	Number of Sols	Begins on Sol#	Noteworthy Features
Gemini	61	1	Gemini 1, Vernal equinox
Cancer	65	62	
Leo	66	127	Leo 24, Mars at Aphelion
Virgo	65	193	Virgo 1, Summer solstice
Libra	60	258	
Scorpius	54	318	
Sagittarius	50	372	Sagittarius 1, Autumnal equinox
Capricorn	47	422	Dust storm season begins
Aquarius	46	469	Aquarius 16, Mars at Perihelion
Pisces	48	515	Pisces 1, Winter solstice
Aries	51	563	Dust storm season ends
Taurus	56	614	Taurus 56, Martian New Year's Eve

FIGURE 6.2
The Mars Areogator.

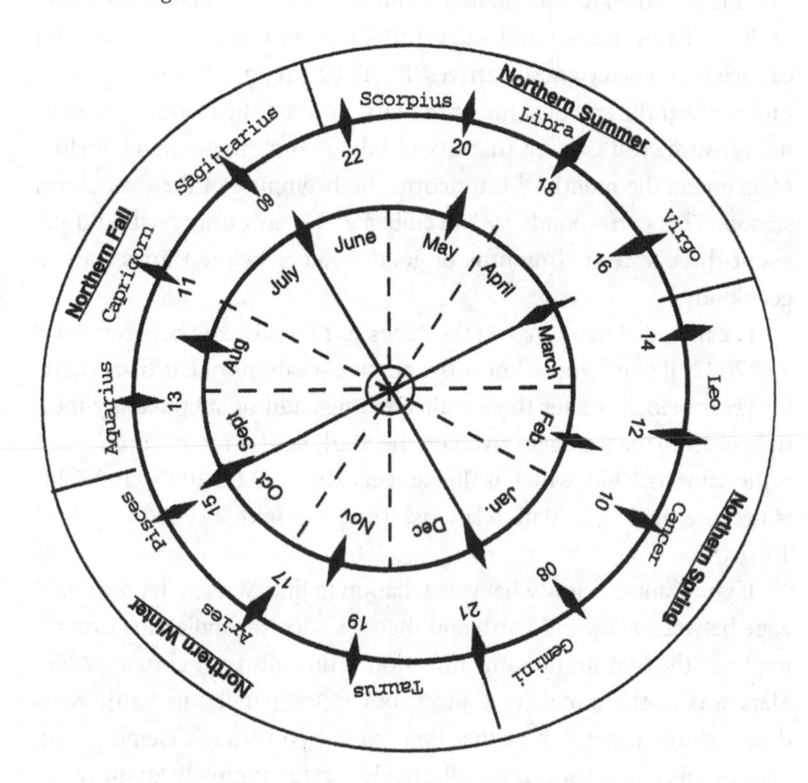

device which I call an Areogator, a copy of which is provided in Figure 6.2. You can use it to find the month (and therefore season) on Mars during any chosen month on Earth, or vice-versa; the relative positions and angles of Earth and Mars about the Sun; as well as to determine where in the sky Mars will be as seen from Earth, or vice-versa, any given time in the past or future.

Let's say you want to know the position of Mars during a given year, 2012, for example. Place a penny, representing Mars, on the diamond on Mars' orbit labeled "12," and a nickel, representing Earth, on the diamond on Earth's orbit at the beginning of January. These will be the comparative positions of Earth and Mars around January 1, 2012. You can see that it will be early Leo, late spring in the northern hemisphere on Mars at that time. Now, to move forward in time, just move your

Mars marker (the penny) forward one diamond and your Earth marker (the nickel) forward one diamond. Move each three diamonds more, until the Earth reaches mid August, the time of *Curiosity*'s arrival. You can see that when *Curiosity* arrives, it will be late in the month of Scorpius, or near the end of summer in Mars' northern hemisphere. Counting forward, you can see that it will take two more diamonds before Mars enters the month of Capricorn, the beginning of the dust storm season. This corresponds to November 2012, so *Curiosity* should get about three terrestrial months of good weather before things start to get cloudy.

I've included markings on the Areogator for all years between 2008 and 2022. If you want to know the relative positions of Earth and Mars for years before or after those indicated, just add or subtract any multiple of 15 to the numbers given on the markings (in other words, 1975 is the same as 1990, which is the same as 2005, 2020, 2035, etc.). This is because the Earth–Mars relationship repeats with a synodic cycle of 15 years.

If you want to know what constellation to find Mars in, lay a straight edge between Mars and Earth and then visualize a parallel line extending from the Sun in the same direction. Thus, during February 2008, Mars was in the month of Cancer, but a line parallel to Earth–Mars drawn through the Sun at that time would go through Gemini, and, because the constellations are effectively located infinitely far away relative to solar system dimensions, this is the constellation that Mars was seen in by observers on Earth at that time. At the same time, any astronomers located on Mars would have seen Earth in Sagittarius.

You'll notice that the diamond markers on Mars' orbit are not equally spaced. This is because as Mars travels around in its elliptical orbit, it speeds up and slows down. For those interested in making their own Areogators, the correct locations of the diamond markers are at 0°, and plus or minus 28.8°, 56.5°, 82.4°, 106.2°, 129.0°, 149.6°, and 170.2° from perihelion (the closest position of Mars to the Sun). Perihelion occurs in the middle of the month of Aquarius, in the same direction as Earth is from the Sun on September 1.

Now for a complete system of dating, it is necessary to know not only the month of the year, but also what year it is in some absolute sense. You can see that the beginning of the month of Gemini also corresponds to Mars' position around January 1 of the years that match the characteristics of 2021 (1946, 1961, 1976, 1991, 2006, 2021, 2036, etc.).

The latest such year that precedes all space probes sent to Mars is 1961. I have therefore chosen to start the Martian calendar with that year. Based on this system, I've calculated some of the great dates in Martian history. These are shown in Table 6.4.

TABLE 6.4
Great Dates in Martian History

Occasion	Terrestrial Date	Martian Date
Calendar begins	January 1, 1961	Gemini 1,1
Mariner 4 flyby	July 15, 1965	Libra 25, III
Mariner 6 flyby	July 31, 1969	Sagittarius 16, V
Mariner 7 flyby	August 5, 1969	Sagittarius 20, V
Mariner 9 arrival in orbit	November 14, 1971	Pisces 20, VI
Mars 2 & 3 landings	December 2, 1971	Pisces 38, VI
Viking 1 arrival in orbit	June 19, 1976	Leo 41, IX
Viking 1 landing	July 20, 1976	Virgo 6, IX
Viking 2 landing	September 3, 1976	Virgo 49, IX
Mars Observer disappears	August 21, 1993	Libra 16, XVII
Pathfinder lands	July 4, 1997	Libra 59, XX
MGS arrives on orbit	September 12, 1997	Sagitarius 12, XX
Mars Polar Lander crashes	Dec 3, 1999	Aquarius 40, XXI
Mars Odyssey arrives on orbit	October 24, 2001	Aquarius 46, XXII
Mars Express arrives on orbit	December 24, 2003	Taurus, 5, XXIII
Spirit lands on Mars	January 3, 2004	Taurus 14, XXIII
Opportunity lands on Mars	January 24, 2004	Taurus 34, XXIII
MRO arrives on orbit	March 10, 2006	Cancer 8, XXV
Phoenix lands on Mars' North Pole	May 25, 2008	Leo 61, XXVI
Curiosity lands	August 23, 2012 ?	Scorpius 47, XXVIII

For those interested in calculating exact dates, the equation to use is:

Mars Year = 1 + 8/15(Earth Year – 1961)

To use this equation, you first have to put the Earth date in decimal form. For example, July 1, 1973, is 1973.5. The equation will then give you the Mars year in decimal form. In the case of July 1, 1973, the answer will be Mars year 7.667. This means that it is year VII on Mars, and, taking the fraction 0.667 and multiplying it by 669 (the number of days in a Martian year) gives you sol number 446. Using Table 6.3, you can see that this corresponds to Capricorn 25.

It is my firm belief that we now possess the technology that could allow a human landing on Mars within ten years of any time a decision is made to launch the program. As I write this it's 2011, and if we launch in October 2022, the first human crew will arrive April 9, 2023. On Mars the date will be Leo 15, XXXIV, the height of the northern Martian spring. The weather will be at its best, with clear skies and low winds, and a landing will be called for. It'll be about time.

7: BUILDING THE BASE ON MARS

The purpose of the first several human missions to Mars will be to explore, to survey, and to answer above all the question of whether or not the Red Planet ever harbored life. But as Mars becomes increasingly well-explored, this question will be answered, one way or the other, and another question will become paramount—not whether there was life on Mars, but whether there *will* be life on Mars. As we've seen, Mars is unique in the solar system, and in this chapter and the next, we'll see that it is not only more varied than any of our other planetary neighbors, but that it is the only planet other than Earth possessing the full array of materials and energy sources necessary to support not only life, but a new branch of human civilization.

Mars is not just a destination for explorers or an object of scientific inquiry, it is a *world* compared to which all other known extraterrestrial bodies are utterly bleak poverty zones. On Mars the resources exist that could allow travelers to grow food, make plastics and metals, and generate large quantities of power. There is no element in large-scale use by human society today which cannot be found in adequate quantities on Mars, and its environmental conditions, in terms of radiation, sunlight availability, and day-night temperature swings, are all well within limits acceptable to different stages of human settlement on its surface.

The resources of Mars could someday make the Red Planet a home not just for a few explorers, but for a dynamic society of millions of colonists building a new way of life in a new world.

Useful materials are not really resources, however, until you develop the technologies to exploit them. If humans are ever to settle Mars, or even to establish a permanent scientific facility of any size, a set of new resource utilization technologies will have to be developed and demonstrated on the Red Planet. To do this we will need a substantial base on the planet where an intense program of agricultural, civil, chemical, and industrial engineering research can be carried out. The base will also give us the capability to support the operation of rocket propelled flight vehicles with global reach, thereby greatly amplifying our ability to discover mineral resources and scientific wealth throughout the planet.

Thus, after a certain number of exploration missions have been carried out, an optimum site for development can be selected and the Mars program will shift from exploration into a second phase, that of base building. While the initial Mars Direct exploration missions will make use of the Martian air to provide fuel and oxygen, in the base-building phase this elementary level of local resource utilization will be transcended as the crew of a permanent Mars base masters an increasing array of techniques that will transform Martian raw materials into useful resources. In establishing a sizable Mars base, we will learn how to extract native water and grow greenhouse crops on Mars; how to produce ceramics, glasses, metals, and plastics; how to construct habitats and inflatable structures; and how to manufacture all sorts of useful materials, tools, and structures. While the initial exploration phase can be accomplished with small crews of just four members each operating out of Spartan base camps spread over vast areas of the Martian surface, building a base will require a division of labor entailing a larger number of people, perhaps on the order of fifty individuals, equipped with a wide variety of equipment and substantial sources of power. In short, the purpose of the base-building period is to develop a mastery of those techniques required on Mars to produce food, clothing, shelter, and everything else needed to make colonizing the Red Planet possible.

FOUNDING THE BASE

Under the Mars Direct plan, crews open new territories on Mars every other year to exploration and possible settlement. Eventually, one of these outposts will be considered the best location for the first permanent Mars base. Once that location is identified, all new crews will land their spacecraft at the designated site. In the Mars Direct plan, the habitat used to house the crew on the outbound leg of the mission is landed and left behind on the planet. Thus, as the sequence of missions proceeds, each mission will add a habitat to the base infrastructure. The habitats landed at the base site (which will be preselected for trafficability) can have wheels attached to their landing gear legs, and then, with the aid of a cable and windlass, be rolled together to be either mated up directly or connected with the aid of inflatable tunnels. Alternatively, second-generation habs can be built whose landing gear legs can articulate not only up and down (as all landing gear must), but also side to side, thus allowing the six-legged habs to walk much as the Martians did in H. G. Wells's *War of the Worlds*! Either way, using one of these techniques, an initial Mars base of some size can be rapidly built up as an interconnected network of Mars Direct style "tuna can" habitats.

While housing in tuna cans will be sufficient for the iron men and women of the first Mars exploration crews, the prospect is marginal for supporting a large scientific population at a permanent Mars base and utterly hopeless as the basis for a program of Mars colonization. An early task, then, both necessary for the base's own self-development and for all that will follow will be the development of large habitable structures. This will employ the same live-off-the-land approach we employed to get to the planet, as these new structures can be assembled out of native materials.

VAULTS OF BRICK

In a series of papers published in the late 1980s, engineer Bruce MacKenzie analyzed this problem in some detail and came to the conclusion that the optimum native material for building the first large structures on Mars is brick.[28] This low-tech concept may seem somewhat surprising at first, but there's actually quite a lot of merit to the proposal. Making brick is quite simple. For that reason some of Earth's first cities were built of brick, and for the very same reason brick may be the literal building block of mankind's first settlements on Mars. To manufacture brick you simply take finely ground soil, wet it, put it in a mold under mild compression, dry it, and then bake it. High temperatures are not really required—in many parts of the world sun-baked bricks are still used—an oven temperature of 300°C can produce pretty good bricks, especially if some scrap material such as torn parachute cloth is mixed into the mud to add cohesion. (You may recall the biblical description of the Egyptians mixing straw with mud to make brick. This was good engineering, an early example of composites manufacture.) Even the 900°C kiln temperature needed to make first-rate modern bricks can readily be produced on Mars, using either a solar reflector furnace or the waste heat from the base's nuclear reactor. True, water is needed for the process, but if the oven is constructed correctly, nearly all the water used can be recovered from the steam produced as the brick is dried at 200°C prior to baking. On Mars, excellent raw material for brick manufacture is available nearly everywhere in the form of a finely ground, iron-rich claylike dust that covers most of the surface to a depth of at least several tens of centimeters. Mixed with water, the same ruddy dust can be used to produce mortar to make the bricks stick together. In fact, in experiments done at Martin Marietta in the late 1980s with Martian soil simulant, chemist Robert Boyd showed that by simply wetting and drying Martian soil, "duricrete" material could be created that is over half as strong as terrestrial concrete.[29] Viking results show that Martian soil contains very high amounts of calcium (about 5 percent) and sulfur (2.9 percent), while analysis of SNC meteorites, which are known to have come from Mars, has shown that these are present on the Red Planet in the form of gypsum ($CaSO_4 \cdot 2H_2O$). On Earth, gypsum is the raw material used to make plaster, and it can be baked to produce lime. This can be added to mortar to produce conventional Portland cement, with a resulting significant improvement in tensile strength.

Structural materials have different kinds of strength, tensile and compressive, reflecting their ability to resist stretching and crushing, respectively. A rope or cable can have a great deal of tensile strength, but no compressive strength. A steel girder has plenty of both kinds of strength. Brick walls and columns, on the other hand, have plenty of compressive strength but are quite weak in tension. They are very difficult to crush but are almost useless for holding things together. Nevertheless, brick and mortar structures built three thousand years ago in ancient Egypt still stand today. Constructions made of brick can prove equally durable on Mars provided that Martian architects adopt the central rule governing nearly all ancient architecture: Keep brick structures in compression.

To build a pressurized structure out of bricks on Mars, you excavate a trench and then within it build a Roman-style vault, or better yet, a series of Roman-style vaults or even a Roman-style atrium as shown in Figure 7.1. The vaults are covered with soil, thereby putting a large downward load upon them, and only then are pressurized with breathable air (produced either by the oxygen-making chemical units described in chapter 6 or by the greenhouses described later in this chapter). How much soil covering is needed depends upon how much air pressure is used. If we stick with our proposed Martian standard of 5 psi (3.5 psi oxygen and 1.5 psi nitrogen, as in *Skylab*), the vaults will experience a pressure force trying to explode them upward of about 3.5 tonnes per square meter. Assuming that Martian soil has an average density four times that of water, this would mean that a layer of dirt about 2.5 meters deep on top of the vault would be enough to keep the whole structure compressed. (Remember gravity on Mars is only 0.38 that of Earth. If terrestrial gravity held sway we could get away with just one meter.) A dirt layer this thick would also provide massive radiation shielding, reducing the cosmic-ray dose experienced by those living in such a subsurface structure to roughly terrestrial levels. In addition, the soil would provide excellent thermal insulation, causing the large temperature swings experienced on the surface during the Martian day-night cycle to go essentially unnoticed by those below and greatly reducing total power requirements to heat the habitat. The brick and soil construction would probably leak air, albeit very slowly. This can be remedied, however, by using a thin layer of plastic sealant either sprayed on the walls or attached to them in the form of wallpa-

FIGURE 7.1
Roman style vaults either singly or in series (a) can be used to construct large subsurface pressurized habitats, including even spacious atriums (b) on Mars. (Designs by MacKenzie, 1987)

(a)

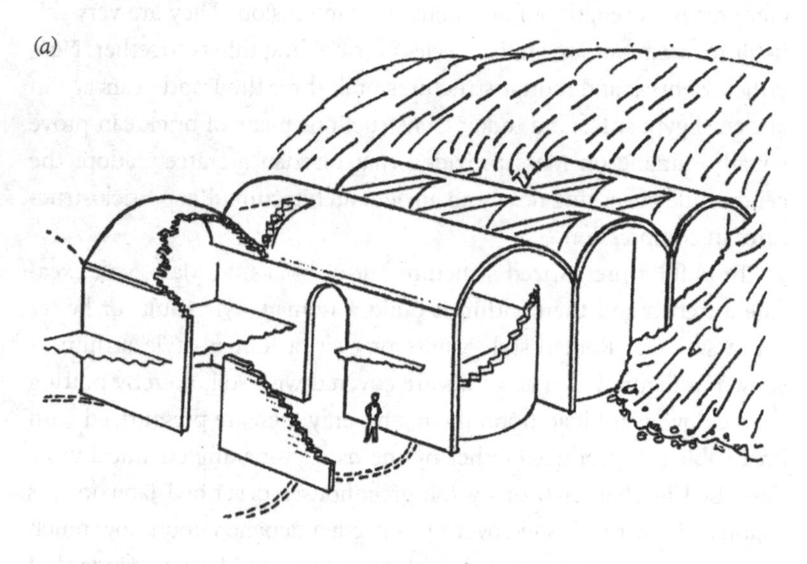

(b)

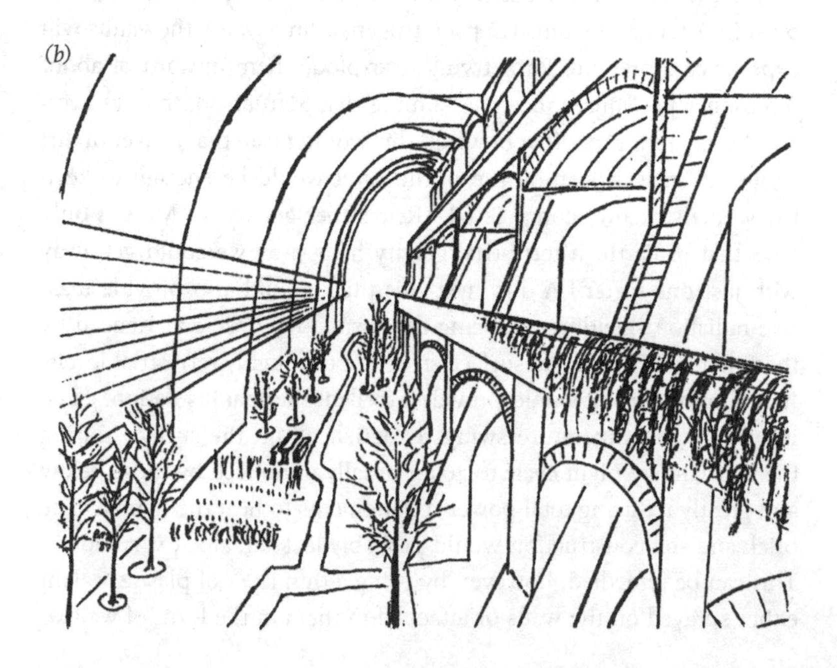

per. Slow leaks should tend to seal themselves over time, however, as the relatively moist air leaking from the structure causes leak-blocking permafrost or ice to form in the diffusion paths of the surrounding soil. As can be seen in Figure 7.1, using these relatively simple, fundamentally ancient techniques, pressurized structures the size of shopping malls can be constructed on Mars.

AT HOME IN A DOME

Habitation in a subsurface shopping mall is a big improvement over living in Mars Direct-style tuna cans (my teenage daughter Rachel would probably jump at the chance to live in a mall), but ultimately on Mars we can do better. We don't have to burrow underground to protect ourselves from radiation (as on the Moon) because the Martian atmosphere is dense enough to shield people living on the surface against solar flares. The planet's surface will be open to us, and, even during the base-building phase, large inflatable structures made of transparent plastic protected by thin, hard-plastic ultraviolet- and abrasion-resistant geodesic domes could be readily deployed, rapidly creating large domains for both human habitation and eventual crop growth. It should be noted in passing that even without the problems of solar flares and a month-long diurnal cycle, such simple transparent surface structures would be impractical on the Moon as they would create unbearably high temperatures inside. On Mars, in contrast, the strong greenhouse effect created by such domes would be precisely what is necessary to produce a temperate climate inside.

During the base-building phase, domes of this type up to 50 meters in diameter and containing the 5 psi atmosphere necessary to support humans could be deployed. If made of high-strength plastics such as Kevlar (with a fabric yield stress of 200,000 psi—twice as strong as steel) such a dome one millimeter thick would be three times as strong as it needs to be to resist bursting and weigh only about 8 tonnes (including its subsurface hemisphere) with another 4 tonnes required for its unpressurized Plexiglas shield. (A habitation dome made of ripstop Kevlar fabric is unlikely to fail catastrophically. Even if someone shot a large-caliber bullet through a 50-meter diameter dome, it would take over two weeks for the air to leak out, leaving plenty of time for repair.) In the early years of settlement, such domes could be imported prefabricated from Earth. Later on they could be manufactured on Mars, along with larger domes. (With the mass of the pressurized dome increasing as the cube of its radius, and the mass of the unpressurized shield dome increasing as the square of the radius: 100-meter domes would mass 64 tonnes and need a 16-tonne Plexiglas shield, etc.)

The key problem with using domes is their foundations. The natural shape for a pressure-containing flexible container to assume is a sphere, as this spreads the load out everywhere equally. While a spherical shape is simple and robust, it does pose a daunting problem when used as the basis for a dome shelter, for you must undertake an enormous amount of excavation work to construct the dome. Imagine partially burying a beach ball in the sand such that its lower hemisphere is buried and its upper hemisphere exposed. To do so, you have to dig out a pit equal in size to the lower hemisphere. While that may seem trivial at the beach, it certainly wouldn't be a trivial amount of excavation on Mars when you're planning to erect a 50-meter dome. In the latter case, you would again excavate a pit and put your sphere in place, but you would then shovel the excavated material back inside to fill up the lower half of the sphere's interior. The result would be a grand space 50 meters across and 25 meters from its dirt floor to the top of the dome (Figure 7.2a)—beautiful, but a lot of work because it would have entailed digging out and then replacing about 260,000 tonnes of material. Finding a natural crater of about the right size would give you a big head start, but it's very unlikely that nature will provide you with one, let alone two or more, at your desired base site.

There is a way around this problem, though, one that relies on using an upper and lower hemisphere with a different radius of curvature. Place a dime on top of a quarter and you'll see what I mean. The quarter has a larger radius and, correspondingly, a larger radius of curvature than the dime. The arc described by the bottom half of the quarter is much flatter than that described by the bottom half of the dime. So, to resolve our excavation problem, instead of using a true hemisphere for the subsurface section of the dome, we could use a partial spheri-

FIGURE 7.2
Methods of construction of domes on the Martian surface: (a) burying half of a spherical dome: (b) burying a dome whose lower half has twice the radius of curvature as the upper half: (c) anchoring a "tent" type dome; (d) a spherical housing complex located entirely above ground, employing Kevlar suspended decks. (Artwork by Michael Carroll)

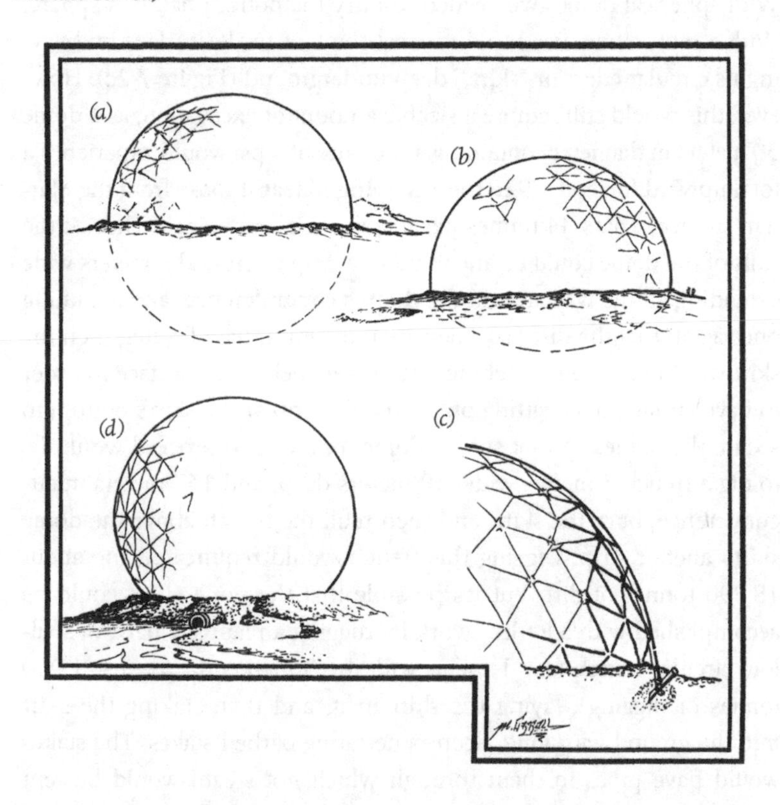

cal section that has a larger radius of curvature than the upper hemisphere (Figure 7.2b), resulting in far less excavation. For example, if the above-surface dome is a true hemisphere 50 meters in diameter (a 25-meter radius of curvature), the subsurface section could be a spherical section with a 50-meter radius of curvature. In this case, instead of having to dig a hemispherical hole 25 meters deep to house the dome, a shallow basin only 3.35 meters deep would suffice, and the total amount of soil moved would be reduced from 260,000 tonnes to about 6,500 tonnes. This latter figure makes the proposition a whole lot more practical. If digging and moving equipment capable of loading one typical dump truck (20 cubic meters) with soil every hour were employed, this excavation job could be done in forty eight-hour shifts.

Another alternative is to employ a hemispherical tent for the dome. With spherical domes we needed to bury the bottom half of a sphere. With a tent, all we need to do is seal the tent to the surface by burying its circular edge, or "skirt," deep underground (Figure 7.2c). However, this would still require a sizable amount of excavation, as a dome 50 meters in diameter containing a pressure of 5 psi would experience a total upward force of 6,926 tonnes trying to tear it loose from the Martian surface. That's 44 tonnes per meter of circumference. Thus, if the skirt of the dome could be anchored to a strip of ground 3 meters wide extending all the way around the dome's circumference, and assuming once again that the dirt has a density four times that of water, then the skirt would have to be anchored 10 meters below the surface in order to have enough mass sitting on an anchor strip at the skirt's bottom to secure the dome. To root such a dome, one way to proceed would be to dig a trench 3 meters wide, 10 meters deep, and 157 meters in circumference, bury the skirt, and then refill the trench above the dome skirt's anchor strip. Digging this trench would require moving about 18,800 tonnes of dirt. But it's possible that the same effect could be accomplished with a lot less work by digging a relatively narrow, shallow circular trench (say 1 meter wide by 3 meters deep—just 1,900 tonnes of digging), laying the skirt in it, and then staking the skirt into the ground with long, deep-penetrating barbed stakes. The stakes would have pipes in them through which hot steam would be sent down deep underground, where it would eventually freeze into a solid and extremely strong ring of permafrost rooting the stakes, and thus the dome, firmly in place.

A fourth alternative would be again to use a sphere, but not bury it. Instead, we could suspend decks to a series of Kevlar cables circling the sphere at various latitude parallels, as shown in Figure 7.2d. For example, if a 50-meter-diameter sphere were employed, the first deck could be 4 meters above the sphere's bottom, the next at 7 meters, then 10, then 13, and so on, with decks spaced every 3 meters until the fifteenth deck, 46 meters above the surface. The total habitable area within such a structure would be enormous, about 21,000 square meters. Because of the nature of the construction, it should not be heavily loaded, so lightweight dividers made of a material such as sound-deadening plastic foam would be used to subdivide the decks into apartments, labs, cafeterias, gymnasiums, auditoriums, or whatever. Access to the structure could be via a tunnel leading to an airlock in the sphere's "south pole." Piling dirt around the base of the sphere would help distribute the loads created by the weight of the sphere as it weighs down on Mars. Installation of a central brick column could reinforce the load-carrying capacity of each deck and facilitate the inclusion of an elevator into the structure. Because this freestanding sphere rises so high above the Martian surface, the unpressurized geodesic Plexiglas dome required to shield it would have to be much larger than for the other concepts, but it would still weigh only about 16 tonnes.

We can see that setting up large habitable domes on the surface of Mars requires the mastery of new nontrivial civil-engineering techniques in a novel environment. Thus, early Martian architecture may well resemble Roman architecture, with simple, subsurface brick vaults predominating. However, once the required manufacturing and civil-engineering techniques are mastered, networks of 50- to 100-meter domes could rapidly be produced and deployed, opening up large areas of the surface to both shirtsleeve human habitation and agriculture. Within surface-rooted domes (Figure 7.2a,b,c), people could live in relatively conventional houses (except that they wouldn't need roofs) made of—what else—brick. If agriculture-only areas are desired, the domes could be made much lighter, as plants do not require more than about 0.7 psi atmospheric pressure. Indeed, because of their lower pressure and reliability requirements, it is likely that Martian domes would find their first application in support of greenhouse agriculture, evolving later to enable large open-air surface settlements.

MANUFACTURING PLASTICS

As a family friend pointed out to Dustin Hoffman's character in *The Graduate*, the key materials in modern life are made of plastics. Get into plastics and your future is assured, my boy. Well, because Mars, like Earth, possesses abundant supplies of native carbon and hydrogen, opportunities to get into the plastics industry abound there as well.

The key to plastics manufacture on Mars is the production of synthetic *ethylene*, which itself can be done as an extension of the reverse water-gas shift (RWGS) reaction discussed in chapter 6 as a means for making oxygen. You may recall the RWGS reaction:

$$H_2 + CO_2 \rightarrow H_2O + CO \quad (1)$$

We can use this reaction to produce all the oxygen we need on Mars by hitting Martian atmospheric carbon dioxide with hydrogen, discarding the carbon monoxide and electrolyzing the resulting water, storing the oxygen so released, and then recycling the hydrogen to make more water, and thus more oxygen, and so forth. But let's say we do things a little differently. Instead of feeding hydrogen and carbon dioxide in a ratio of 1:1 suggested by equation (1), let's feed them together with a ratio of 3:1. Then we have:

$$6H_2 + 2CO_2 \rightarrow 2H_2O + 2CO + 4H_2 \quad (2)$$

(Yes, I know I could divide all the proportions in equation (2) in half and it would still be the same, but bear with me.) So now we take the water produced by equation (2) and condense it out. Maybe we electrolyze it; maybe we don't. That all depends on whether we would rather have water or hydrogen and oxygen. The key thing, however, is what we do with the rest of the products after the water has been removed. If we choose, we can send the remaining mixture of carbon monoxide and hydrogen into another reactor, where in the presence of an iron-based catalyst they can be reacted in accordance with:

$$2CO + 4H_2 \rightarrow C_2H_4 + 2H_2O \quad (3)$$

Bingo. C_2H_4 is *ethylene*, a great fuel and the key to the petrochemical and plastics industries. Reaction (3) is strongly exothermic, and so like the methane-making Sabatier reaction discussed in chapter 6 can be used as a heat source to provide the energy needed to drive the endothermic RWGS. It also has a high equilibrium constant, making the achievement of high ethylene yields possible. Side reactions typically occur, however, producing propylene (C_3H_6), which is fine, as it is also an excellent fuel and valuable plastic-making stock. Waxy higher hydrocarbons may be produced as well, however, which is not so fine as they can cause problems unless they are distilled out of the product. However, while more complicated, this system has important advantages over a simple Sabatier reactor. In the first place, ethylene has only two hydrogen atoms per carbon, while methane has four. Thus, using ethylene for fuel instead of methane cuts the hydrogen importation or water-mining requirements to make fuel in half. In the second place, ethylene has a boiling point (at one atmosphere pressure) of −104°C, much higher than methane's boiling point of −183°C. In fact, under a few atmospheres pressure ethylene is storable without refrigeration at Mars average ambient temperatures, whereas methane's critical temperature is below typical Mars nighttime temperatures. Therefore, ethylene can be liquefied on Mars without the use of a cryogenic refrigerator, whereas methane cannot be. This cuts the required refrigeration power for an ethylene/oxygen propellant production system about in half relative to that of a methane/oxygen production system. It also greatly reduces the need to insulate the ethylene fuel tanks, and makes handling the resulting fuel a lot simpler all around. In the third place, the density of liquid ethylene is 50 percent greater than liquid methane's, allowing the use of smaller and therefore lighter fuel tanks on Mars ascent vehicles or ground rovers employing ethylene instead of methane fuel. Fourth, ethylene has other uses besides rocket or rover or welding fuel. It is used as an anesthetic, as a ripening agent for fruits, and as a means of reducing the dormant time of seeds. These features will all be very useful to the developing of a Mars base.

But as wonderful as all this is, it's small potatoes compared to ethylene's and propylene's starring roles as the basic feedstocks for a range of processes to manufacture polyethylene, polypropylene, and numerous other plastics. These plastics can be formed into films or fabrics to create large inflatable structures (including habitation domes) as well as to manufacture clothing, bags, insulation, and tires, among other things. They can also be formed into high-density, stiff forms to produce bottles and other watertight vessels both enormous and minute, tableware, tools, implements, medical gear, and innumerable other small but necessary objects, boxes, and rigid structures of every size and description, including those that are both transparent and opaque. Lubricants, sealant, adhesives, tapes can all be manufactured—the list is nearly endless. The development of an ethylene-based plastics manufacturing capability on Mars will thus offer enormous benefits in opening up all sorts of possibilities and capabilities necessary for the human settlement of the Red Planet.

Plastics are, of course, among the most central materials of modern life. They can be made on Mars because of the ubiquitous presence there of carbon and hydrogen. This should give pause to those who believe the prospects for settlement on the Moon are superior to those on Mars. The Moon has no significant quantities of accessible carbon and hydrogen; outside of ultra cold, permanently shadowed polar craters, they exist there only in parts per million quantities, somewhat like gold in seawater. The manufacture of cheap plastics will never be possible on the Moon. In fact, on the Moon, for a long time to come, plastics would literally be worth their weight in gold.

MAKING CERAMICS AND GLASS

Clay-type minerals are also ubiquitous in the Martian surface soils, making the manufacture of ceramics for pottery and similar purposes a straightforward enterprise. The most common material measured by the *Viking* landers on Mars, though, was silicon dioxide, SiO_2. Making up about 40 percent of both the *Viking 1* and 2 soil samples by weight, silicon dioxide is the basic constituent of glass, which thus can readily be produced on Mars using sand-melting techniques similar to those that have been used on Earth for thousands of years. Unfortunately for the Martian glass industry, however, the second most common compound (about 17 percent in the *Viking* samples) is iron oxide, Fe_2O_3. This poses a problem. If you want to make optical-quality glass, the sand used as feedstock must be nearly iron free, and such sand may well be hard to come by on Mars. So, if you want to make optical glass on Mars, you first need to remove the iron oxide. This can be done by hitting the iron oxide with hot carbon monoxide "waste" from your

RWGS reactor, thereby reducing it to metallic iron and carbon dioxide, and then removing the iron metal product with a magnet. Tiresome, I admit, but you get to keep the iron for other uses, such as making steel, as will be discussed shortly. In fact, since the base will almost certainly need much more steel than optical glass, after the base foundry has been functioning for a while there should be no shortage of iron-denuded material for the glass makers to work with. It should be noted, however, that optical-quality glass is not required to make many important glass products, including fiberglass, an excellent material for constructing various types of structures.

TAPPING WATER

> In the Martian mind, there would be one question perpetually paramount to all the local labor, women's suffrage, and Eastern questions put together—the water question. How to procure water enough to support life would be the greatest communal problem of the day.
>
> —Percival Lowell, *Mars*, 1895

Percival Lowell may have been wrong about a lot of things, but he was certainly prescient in his comment regarding water on Mars. From making rocket fuel, rover fuel, and oxygen, to producing plastics, bricks, mortar, and pottery; to growing crops, to sealing leaks and hardening soil with artificial permafrost, all the opportunities to open Mars to human exploration and settlement we have discussed so far depend upon water. While the logistics of transporting water to Mars are hopelessly unattractive, for our first several missions we can afford to make water, bringing only its 11 percent hydrogen component from Earth to combine with the oxygen in Mars' carbon dioxide atmosphere. But once the base-building phase begins, we must move beyond this. The increased propellant requirements deriving from the expanded level of human activity, the multitude of additional civil and chemical engineering uses, and above all the agricultural requirements that arise during the base-building phase, all conspire to increase water demand on Mars far beyond the point where hydrogen transport from Earth remains a viable option. If human civilization is ever to grow on Mars, we must find a way to obtain indigenous water.

So, if we are wise, we will position our base near where water is likely to be found. This should be readily possible. If you look at Mars today, you'll see a large region of depressed topography in the Martian Arctic that contains very few craters. It is believed that in Mars' early history, this vast basin was filled with water, which shielded the surface from meteor impact during the planet's first billion or so years. The last remnant of this ancient ocean is the northern polar cap, which is made of water ice (about two million cubic kilometers of it, by current estimates[30]). The European *Mars Express* orbiter has also discovered craters in the Martian north that are filled with water ice.[31] But these are just the known sources of pure water. Mapping the planet from orbit using its gamma ray and neutron spectrometers, NASA's *Mars Odyssey* spacecraft has found continent-sized regions in both hemispheres of Mars where the surface soil is 40 to 60 percent water by weight. However, we see from orbital images that the north boasts many more dry river beds and outflow channels than the south. It is likely that when these channels flowed their last, deposits of ice or permafrost were left at their mouths. These deposits may still exist, hidden from our view by a layer of dust. Measurements of atmospheric humidity taken from orbit also leave no doubt that the northern hemisphere is wetter than the southern, with the wettest time of the year being the northern spring. The existence of much larger amounts of water in the northern hemisphere's past is also significant to future Martian colonists for another reason; hydrological activity is key to the formation of a large variety of mineral ores. If Horace Greeley had lived on Mars, his advice to young Martians seeking their fortune would have been simple—Go north.

There are a number of possible ways to get water on Mars. The first, most attractive, but most problematical method is simply to find it. As discussed in chapter 6, there may be subsurface, geothermally heated pools of liquid water on Mars. Such pools could be detected within a kilometer of the surface by rover crews equipped with ground penetrating radar. The rover crews won't have to search randomly. Low-resolution radar investigations conducted from orbit or from aircraft or balloon-borne probes can identify in advance the best places to look. Other clues could come from detection of methane vents, indicating subsurface hydrothermal activity (and possibly life!), or images such as those provided by *Mars Global Surveyor*, revealing transient outflows of water from the sides of cliffs or craters in the recent past. If we find such a pool and drill down to it, the hot pressurized water should come shooting out of the ground like a Texas oilfield gusher. Once it hits the low-pressure, cold Martian air, the water won't stay warm for long. Depending upon its speed of ejection, it will probably freeze into ice crystals and fall back to the ground before it has gone a hundred meters. In no time at all a snow volcano could form, possibly of considerable size. Extracting the water in such a spectacular way would be rather wasteful though, because such a hydrothermal well could also represent a significant source of power. But, as far as water access is concerned, next to siting the base over a hot artesian well, this is about as good as it gets.

Of course, things might not work out so well. Subsurface liquid water within drilling range may not be found. What then? Well, the next best thing would be to find brines. Saturated salt solutions can be liquid at temperatures as low as –55°C, which means that even without geothermal heat, such liquid brines, protected from evaporation by a modest layer of soil or ice, could exist on Mars today, very close to the surface. In addition to being a good source of water, brines would be of great interest as candidate sites for finding extant Martian life. No brines have been identified on Mars as yet, but the *Spirit* and *Opportunity* rovers both found plentiful salts on the edges of ancient lakes, and some scientists believe that light-colored features surrounding certain basins imaged on Mars from orbit may represent very large salt deposits left behind on the shore lines of vanished Martian seas.

After brines, the next most interesting source of water on Mars would be ice. There are large deposits of water ice on Mars' north polar cap, but that's not where we are going to build our base. We see no large permanent deposits of ice south of 70° north latitude, but theory indicates that poleward of 40° N, underground ice should be stable within a meter of the surface. But there can be local anomalies. Where I live in Colorado, it can be winter on the north side of the house while it is summer on the south side, and it is not uncommon even on a blistering, mid-August day to come across snow nestled in a shady depression of a hill's northern side. On the basis of such experience, one therefore might suspect that in some cold crevice, lava tube, cavern, or shady slope on the north face of some hill on Mars there is ice to be found, in regions where planetary-scale climate models say it can't be. This has proven to be the case. Observations taken by the *Mars Reconnaissance Orbiter* reported in 2009 reveal exposed pure water ice several feet deep in five relatively new craters located between 43 and 56 degrees north latitude. (Three of the locations are in the Cebrenia quadrangle. These locations are 55.57° N, 150.62° E; 43.28° N, 176.9° E; and 45° N, 164.5° E. Two others are in the Diacria quadrangle: 46.7° N, 176.8° E and 46.33° N, 176.9° E.) This discovery proves that pure water is available on Mars at mid latitudes.

Still, such pure ice deposits in a nonpolar region are not to be found everywhere. Martian explorers would much more frequently encounter permafrost, or frozen mud. This can contain plenty of water, but those who would harvest it may wish to bring dynamite. Permafrost at Martian temperatures can be pretty tough stuff. In fact for some applications it's the ideal material for construction on Mars. A permafrost brick would be much stronger than a hot fired red clay brick, and you don't need an oven to make one or use mortar to get one brick to adhere to the next. Instant rock, just add water. Instant water, just melt the rock.

So much for the heroic forms of Martian water-prospecting and mining. Let's take a look now at some more mundane, industrial-style methods.

Martian soil has some water in it. We know that for a fact because at both *Viking* landing sites, random samples of soil scooped from the top 10 centimeters of the surface emitted about 1 percent of their weight in water when heated to 500°C. That's not too bad, but in fact the test was unfairly skewed, because surface soil is the driest there is; the samples were heated for only 30 seconds; and furthermore, the samples were held in an unsealed vessel at 15°C for days before the test. Since 15°C is much warmer than average for Mars, the odds are very high that a significant amount of water was lost from the samples via outgassing prior to the test. On the basis of the *Viking* results, it would be a good bet that *average* Martian soil is at least 4 percent water. This has since been confirmed by the *Mars Odyssey* orbiter. But some soils are likely to be much wetter than this average. For example, there are salts on Mars that typically contain up to 10 percent chemically bound water that can be released by heating to appropriate temperatures. Clays, which are common on Mars, also have excellent water adsorption capacities. For example, smectite clays have been found in SNC meteorites. Simectite clay is known commonly as "swelling clay," because it can absorb several tens of percent of its weight in water and will swell in the process. The mineral gypsum ($CaS0_4 \cdot 2H_2O$) has also been found in many SNC meteorites. It's likely that gypsum is quite common on Mars, because sulfur and calcium concentrations measured at both *Viking* landing sites were much higher (forty and three times, respectively) than their averages in soils on Earth. Gypsum can be over 20 percent water by weight.

Whether 4 percent or 20 percent, to get this water out of the soil, all that is needed is heat. This can be done in one of two ways—either you bring the soil to the heater or the heater to the soil. The first of these methods is illustrated in Figure 7.3. A truck loaded with some relatively wet soil dumps its load onto a conveyor belt leading into an oven. The oven heats the soil to 500°C or so, causing the adsorbed water to outgas. The steam so produced is collected in a condenser, while the dehydrated dirt is dumped. The resulting "slag heap" is an inconvenience, but the energetics of this system aren't too bad. If soil with a 4 percent water content is used as a feedstock, the energy required to run the system is about 3 kWh (kilowatt hours) of heat for every kilogram of water produced.[32] At that rate a 100 kWe (kilowatt-electric) reactor could produce 900 kg/day of water if its electricity is used to power the oven, or up to 18,000 kg/day of water if the reactor's waste heat is used to bake the dirt. (Thermoelectric generators used on today's space nuclear power sources are only 5 percent efficient at turning their power into electricity; the other 95 percent comes out as "waste heat.")

Alas, but there is that annoying waste pile of dried dirt. We could make 18,000 kg/day of water, but we'd be piling up 462,000 kg/day of desiccated slag. That might be acceptable—it's only about 120 cubic meters, or six truckloads, of material. Maybe we can use the slag for something; maybe we can just dump it in a nearby crater.

FIGURE 7.3
Truck, oven, and slag pile system for extracting water from Martian soil. (Artwork by Michael Carroll)

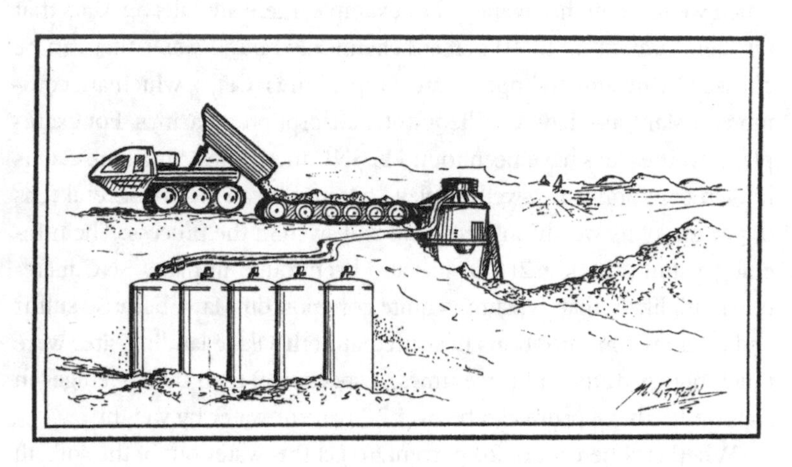

But, if you don't want to move all that dirt around, the alternative is to bring the heater out to the field. One way that's been suggested to do this is to have a mobile oven that wheels along, ingesting soil, baking it, condensing the steam, and ejecting the desiccated dirt as it travels.[33] You probably wouldn't want to use a nuclear reactor for such a system, but, instead, a radioisotope thermoelectric generator (RTG) such as has been used on *Voyager, Viking, Galileo,* and other outer solar system spacecraft. The standard RTG puts out 300 watts of electricity, enough to move the cart, along with 6 kW of waste heat, sufficient to produce 56 kilograms of water a day from 4 percent grade feedstock. Such a unit would be quite handy to small crews operating out in the field, or as an adjunct piece of equipment for early exploration missions (42 kilograms of water produced daily over the course of a single, 500-day Mars Direct mission surface stay adds up to 28,000 kilos of water), but its output is quite small relative to the needs of a large developing Mars base. Of course, we could produce all the water we need by operating a multitude of them, but all those RTGs would be expensive, and anyway we'd still be moving a lot of dirt, pebbles, and rocks around, with all the wear and tear on the involved equipment that implies. Is there a gentler approach?

One way might be for the cart to use a microwave device to heat the soil below it. This would cause the water in the soil to vaporize and rise up as steam. The cart would carry a kind of canopy with a flexible skirt brushing the ground all around it. This skirt would act as a sufficiently good seal to hold the water vapor until most of it frosted out on the canopy roof, after which it could be collected for use. The advantage of this scheme is that no digging is required, and furthermore, microwaves can be tuned so that they put most of their energy into heating the water molecules, instead of wasting power by heating water and dirt indiscriminately. Unfortunately, the rising vapor will transfer heat to the soil, so that in the end much of the heat ends up wasted anyway (although not as much as in a purely thermal heating system). The problem, however, is that the microwave power input must be electrical, not thermal. The 6,000 W of waste heat the RTG produces can't be used to drive the system, only the unit's 300 W of electrical output. Thus, even if one watt of microwave power should prove twice as efficient as thermal power in driving water out of soil, you still come out with only one-tenth the output because thermal power is twenty times more available. If the water concentration were very high, however, and the ground too strong to break (as would be the case with permafrost) this system might work better than a mobile digger, though its output would still be rather low. For example, let's assume we operate such a system over a permafrost deposit that is 30 percent water by weight. We estimate that about 1 kWe-hr would be needed to extract each kilogram of water. So, over the course of a Martian sol (24.6 terrestrial hours), the microwave cart driven by a 300-watt RTG could extract about 7.4 kilograms of water. The only way to improve on this performance would be to apply a lot more power, perhaps by connecting the cart by a long cable to the base's nuclear reactor and applying 100 kWe. In that case, 2,200 kilos of water per day could be produced, but mobility would be lost.

I think a better solution would be to put a transparent tent over a selected area of Martian terrain and warm the inside via the greenhouse effect that would occur naturally within. The greenhouse heating could be augmented by positioning large, lightweight reflectors around the tent, and moving them with the Sun to maximize the solar heating of the enclosed area. Inside the tent, the soil would be heated, not to 500°C certainly, but far above its average temperature. This would cause a fraction of the adsorbed water the soil contains to outgas, and the moisture released could be captured as frost on a cold plate kept refrigerated in one corner of the tent (just like the frost buildup in your freezer). To see how effective such a system might be, consider that the average solar incidence on Mars is 500 watts per square meter (W/m^2). If the tent is a hemisphere 25 meters in diameter, and the tent greenhouse plus reflector arrangement causes the equivalent of an extra 200 W/m^2 of heating to occur within the tent, the total effective power of the system would be 98 kW. This is enough to release 300 kilograms of water from 4 percent-grade soil in the course of an eight-hour day.

FIGURE 7.4
Mobile methods of extracting water from Martian soil: (upper left) soil eater on wheels; (center) mobile microwave system with skirt; (bottom) portable greenhouse dome with condenser. (Artwork by Michael Carroll)

This amount of water would be available within the first half centimeter of soil within the tent. Made of 0.1 mm thick polyethylene, the tent would have a mass of only 100 kilograms (and therefore weigh 38 kilograms on Mars), so it could be carried about by rover crews to a new position every day. After the tent moved on, the mined surface soil would rehydrate itself naturally, allowing the same field to be repeatedly farmed for water.

A completely different approach would entail extracting water from the Martian atmosphere. The problem here is that the air on Mars is very dry—under typical conditions you have to process one million cubic meters of Martian air to acquire one kilogram of water. In a classic paper, engineer Tom Meyer and Mars scientist Chris McKay proposed a mechanical compressor system capable of doing just that.[33] The authors found that every kilogram of water produced would require about 103 kWh of electrical energy. Comparing this result to the figures for the soil-based water-extraction systems described above (about 3 kWh of thermal energy per kilogram), it certainly seems unattractive, although it should be pointed out that the compressor system will also produce a lot of useful argon and nitrogen from the atmosphere for base life support. More recently, however, Adam Bruckner, Steven Coons, and John Williams of the University of Washington undertook a study in which instead of compressing the air, they simply employed a fan to blow it against a zeolite sorption bed.[34] Zeolite is an extreme desiccant, and can be used to reduce atmospheric water vapor concentrations to a few parts per billion, well below even Martian humidity. At Martian temperatures, zeolite will adsorb up to 20 percent of its weight in water. Once the zeolite is saturated, you can bake the water out at an energy cost of about 2 kWh of thermal energy per kilogram, after which the now desiccated zeolite can be used again. Since all you have to do is move the air, not compress it, the mechanical fan power is much less than the pump power needed by the Meyer and McKay system, perhaps requiring another 2 kWh of electrical energy per kilogram of water processed. Energy costs are thus comparable to the systems based upon soil water extraction. The main problem with any atmosphere-based water-extraction system on Mars, however, is that it must be rather large to achieve a useful level of output. For example, a system pairing an intake duct with a 10 square meter cross-sectional area and a fan capable of generating an intake air speed of 100 meters per second (close to 200 miles per hour) would produce just 90 kilograms or so of water per day. However, since the machine does not need to be mobile, the 8 kWe needed to run the fan could readily be provided by the base power supply. This, taken together with the facts that no digging or prospecting is required, the system is susceptible to complete automation, and the raw material, air, is infinitely renewable, may ultimately make such atmospheric water extraction systems quite attractive.

All in all, while there may not be enough water available on Mars to support Lowell's visions of water-bearing canals criss-crossing the planet, there is certainly enough available to support a Mars outpost. No doubt much of the water tapped from Mars' arid environs will go to adding a touch of green to the Red Planet.

GREEN THUMBS FOR THE RED PLANET

Given the costs of interplanetary transportation, it is obvious that if significant human populations are to settle on other worlds, they will eventually have to grow their own food. In this respect, Mars stands at an enormous advantage to the Earth's Moon and every other known extraterrestrial body. Of the four main elements comprising organic matter—hydrogen, carbon, nitrogen, and oxygen—all are readily available on Mars. It's been argued that asteroids are likely to contain carbonaceous material, and some evidence has been presented from recent lunar probes indicating that the Moon may harbor ice deposits in permanently shaded areas near its south pole. But these arguments miss the point, because the biggest problem with the Moon, as with all other airless planetary bodies and proposed artificial free-space colonies (such as those proposed by Gerard O'Neill[35]) is that sunlight is not available in a form useful for growing crops. This is an extremely important point and it is not well understood. Plants require an enormous amount of energy which can only come from sunlight. For example, a single square kilometer of cropland on Earth is illuminated with about 1,000 MW of sunlight at noon, a power load equal to an American city of one million people. Put another way, the amount of power required to generate the sunlight responsible for the crop output of the tiny country of El Salvador exceeds the combined capacity of every power plant on Earth. Plants can stand a drop of perhaps a factor of five in their light intake compared to terrestrial norms and still grow, but the fact remains: The energetics of plant growth make it inconceivable to raise crops on any kind of meaningful scale with artificially generated light. That said, the problem with using the natural sunlight available on the Moon or in space is that it is unshielded by any atmosphere. (The Moon has an additional even more intractable problem with its twenty-eight-day light/dark cycle, which is completely unacceptable to plants.) Thus, plants grown in a thin-walled greenhouse on the surface of the Moon or an asteroid would be killed by solar flares. In order to grow plants safely in such an environment, the walls of the greenhouse would have to be made of glass 10 centimeters thick, a construction requirement that would make the development of significant agricultural areas prohibitively expensive. Using reflectors and other light-channeling devices would not solve this problem, as the reflector areas would have to be enormous, essentially equal in area to the crop domains, creating preposterous engineering problems if any significant acreage is to be illuminated.

Mars' atmosphere, on the other hand, is sufficiently dense to protect crops grown on the surface against solar flares. On Mars, as we have seen, large inflatable greenhouses protected by geodesic domes could be readily deployed, rapidly creating huge temperate-environment domains for crop growth. Martian sunlight levels, at 43 percent those of Earth, are entirely adequate for photosynthesis, which in fact could be accelerated relative to Earth by filling the domes with higher concentrations of carbon dioxide than are available on Earth. We have seen that a 1-mm thick Kevlar reinforced dome fabric would be needed to support a 50-meter diameter habitation pressurized to 5 psi. However, plants require only 0.7 psi, or 50 mbar of atmospheric pressure with 20 mbar nitrogen, 20 mbar oxygen, 6 mbar water vapor, and less than 1 mbar of carbon dioxide comprising the atmosphere. A fabric only 0.2 mm thick would be sufficient for a 50-meter dome if it were used as a greenhouse only. Such a dome, enclosing about 2,000 square meters (a half acre) of crop land, would have a fabric mass of about one tonne, but its Plexiglas shield mass would still be 4 tonnes. (The Plexiglas geodesic dome shield mass could be cut almost in half if the upper hemisphere of the dome is shaped as a lens, instead of a conventional half sphere. Lens shaping of the upper hemisphere would also make construction of the shield dome easier, as it would not have to go as high. It would also drastically cut the time the plants would need to oxygenate the dome's atmosphere.) However, while plants can tolerate 0.7 psi, humans can't, and using such low-pressure domes would force those inside to wear spacesuits. Raising the dome pressure to 2.5 psi would eliminate the need for spacesuits. However, unless the base was suffering from a severe shortage of farmland, it probably makes the most sense to make the greenhouse domes suitable for operation at the same 5 psi pressure as the habitation domes. That way tunnels can be built allowing humans in shirt-sleeves to pass freely between the two types of domes without having to undergo compression/decompression operations. Moreover, common construction elements would make mass production easier and also allow humans to move into the greenhouse domes as population pressures require. The main difference between the two types of domes will be in the carbon dioxide partial pressures allowed. In the habitation domes, this will be limited to typical terrestrial levels of about 0.4 mbar. But in the greenhouses, much larger carbon dioxide levels of about 7 mbar (Mars ambient) will be employed, as this should greatly increase crop yields. (Plants on Earth suffer from carbon dioxide deprivation.) As we have seen, there are a variety of potential techniques for supplying the greenhouse with copious supplies of water. Thus, the basic prerequisites for farming—well-lit, irrigated land—can be brought into being on Mars.

How fertile is Martian land? It's hard to say, but on the basis of what we know now, Martian soil is likely to prove an excellent medium for crop growth, considerably better than most land on Earth, in fact. In Table 7.1 we show a comparison of plant nutrient elements in terrestrial and Martian soils. The data for Martian soils is based upon *Viking* results and analysis of SNC meteorites.[36]

Examining Table 7.1, we can see that with respect to the large majority of plant soil nutrients, Martian soil is richer than that of Earth.

TABLE 7.1
Comparison of Plant Nutrients in Soils on Earth and Mars

Element	Terrestrial soil (average)	Martian soil (estimated average)
Nitrogen	0.14%	Unknown
Phosphorus	0.06%	0.30%
Potassium	0.83%	0.08%
Calcium	1.37%	4.10%
Magnesium	0.50%	3.60%
Sulfur	0.07%	2.90%
Iron	3.80%	15.00%
Manganese	0.06%	0.40%
Zinc	50 ppm	72 ppm
Copper	30 ppm	40 ppm
Boron	10 ppm	Unknown
Molybdenum	2 ppm	0.4 ppm

The main question mark is nitrogen, which, due to the limits of its design, the *Viking* X-ray fluorescence instrument used to analyze the elemental composition of the soil could not assess. Nitrogen is known to exist in the atmosphere, however, so if the soil should prove nitrate poor, ammonia and other nitrate fertilizers can be synthesized. In fact, the same Sabatier reactors used to produce methane fuel can also be used to produce ammonia if nitrogen and hydrogen are used as the feedstock. Such reactors are a major source of fertilizer production on Earth. However, based upon our present understanding of planetary formation, Mars should have originated with about the same proportion of nitrogen as Earth, and most of it must still be there, undoubtedly fixed in the soil as nitrates. Natural nitrate beds should be detectable on Mars, which when mined will provide the base with fertilizer by the truckload. The other plant nutrient element in which typical Martian soils appear to be poor is potassium. This can probably be found in high concentrations in salt beds deposited on the now dry shores of Mars' ancient water bodies.

The physical properties of Martian soil are also likely to be favorable for plant growth, as the globally distributed soil layer appears to be loosely packed and porous, and well adapted mechanically to supporting plants. As discussed earlier, Martian soils are known to contain smectite clays. This is good news for future Martian farmers because smectites are highly effective at buffering and stabilizing soil pH in the slightly acidic range, and also ensure a large reserve of exchangeable nutrient ions in the soil due to their high exchange capacity.

As stated earlier, the Martian greenhouses will be pressurized at 5 psi (340 mbar), or around one-third of Earth sea-level pressure. Since Mars' gravity is one third Earth's, maintaining this air density will also make insect flight possible, facilitating pollination by bees. Initially, the domes will simply be pressurized with Martian air (95 percent carbon dioxide) with a few millibars of artificially generated oxygen included in order to make plant respiration possible. The Martian plants will thus grow in heavily carbon-dioxide–rich greenhouse environments, and photosynthetic efficiency should benefit accordingly. On Earth, in a carbon-dioxide–poor environment, plants convert sunlight into chemical bond energy with an efficiency of about 1 percent. (The net ecological efficiency of a forest or wild prairie is much lower, perhaps 0.1 percent, but that's because the dead plants are allowed to decompose. The plants themselves do considerably better, and in an agricultural setting we can take advantage of that by harvesting them before they are disassembled by bacteria.) A good guess for the efficiency of photosynthesis in a heavily carbon-dioxide–enriched environment might be about 3 percent. Assuming that the 50-meter–diameter dome is a true half-sphere, it would take plants this efficient covering its floor about 310 days to turn virtually all of the enclosed carbon dioxide into oxygen. If a lens-shaped upper dome (a radius of curvature of 50 meters instead of the natural 25) were employed, however, the time would be reduced to only eight days. The oxidant that *Viking* may have detected in Martian soil will be no problem, as it decomposes into reduced material and free oxygen on contact with water. The warm greenhouses will be moist environments, and as the moisture circulates it will quickly cause the greenhouse soils to give off their oxygen.

Mars Direct promises to land a crew on the planet within a decade. The plan makes this possible by using local Martian resources to manufacture fuel for the trip home. The crew hab is the tuna-can shaped object on the left; the conical ERV stands to the right. (Artwork Robert Murray, courtesy Lockheed Martin)

As an added benefit, Mars Direct hardware can be used for missions to the Moon, as shown here. (Artwork Robert Murray, courtesy Lockheed Martin)

The Ares heavy lift booster, composed of shuttle-derived propulsion technology, will directly launch crew and equipment to the Red Planet. (Artwork Robert Murray, courtesy Lockheed Martin)

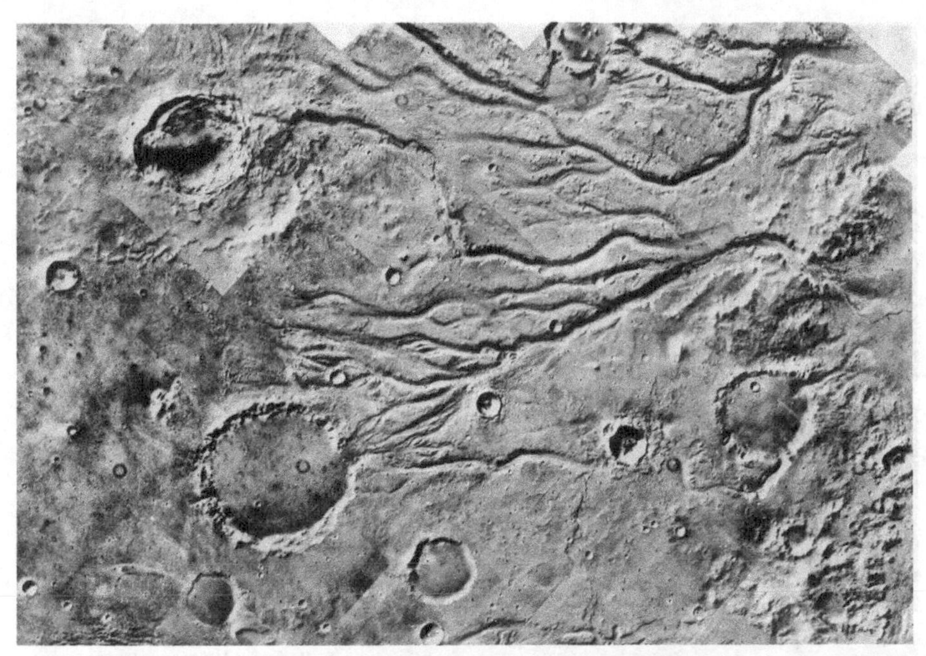

Viking *orbiter views of the surface gave clear evidence for Mars' watery past, as can be seen in this image of water-carved channels found to the west of the* Viking 1 *landing site. (Photo courtesy NASA)*

Given that Earth and Mars shared similar climates in the distant past, there is a chance that life could have evolved on the Red Planet, in this instance in the form of Martian stromatolites. (Artwork by Michael Carroll)

Mars Global Surveyor *aerobrakes into orbit. (Artwork Michael Carroll, courtesy NASA/JPL)*

Two robotic missions to the Red Planet—Mars Global Surveyor and Mars Pathfinder—*launched in late 1996. The Mars Aerial Platform mission could return information invaluable to piloted missions and a Mars Sample Return mission employing locally produced propellants could demonstrate this critical technology.*

The Mars Aerial Platform (MAP) mission will employ superpressure balloons to fly cameras over the surface of the Red Planet for hundreds of days. (Artwork Robert Murray, courtesy Lockheed Martin)

As seen from the *Pathfinder* lander, the little *Sojourner* rover examines "Yogi" with its Alpha Proton X-Ray Spectrometer (APXS). (Photo Malin Space Science Systems/NASA)

The Mars Sample Return mission will return several kilograms of soil to Earth for analysis. (Artwork Pat Rawlings, courtesy NASA/JSC)

Mars Semi-Direct step 1: Propellant production on Mars. (Artwork courtesy NASA/JSC)

In the fall of 1992, NASA broke with the "Battlestar Galactica" paradigm and adopted Mars Semi-Direct as their baseline.

Mars Semi-Direct step 2: Arrival of crew in surface hab. (Artwork courtesy NASA/JSC)

Mars Semi-Direct step 3: Crew departure. (Artwork courtesy NASA/JSC)

Mars Semi-Direct step 4: Rendezvous with ERV. (Artwork courtesy NASA/JSC)

Departure from Earth as envisioned for a traditional "Battlestar Galactica" Mars mission. (Artwork by Michael Carroll)

The first Mars missions will focus on the search for evidence of past or present life and for resources for the future. (Artwork Pat Rawlings, courtesy NASA)

Larry Clark (left) and the author perform initial checkout of the In-Situ Propellant Production plant developed by a team at Martin Marietta under contract to NASA Johnson Space Center. This demonstration unit clearly showed that making propellant on Mars is a viable concept. (Photo courtesy NASA)

NASA administrator Dan Goldin (right), shown here with the author and the Martin Marietta ISPP machine, became a supporter of Mars Direct. (Photo R. Zubrin)

The author talks Mars mission strategy with Speaker of the House Newt Gingrich. (Photo R. Zubrin)

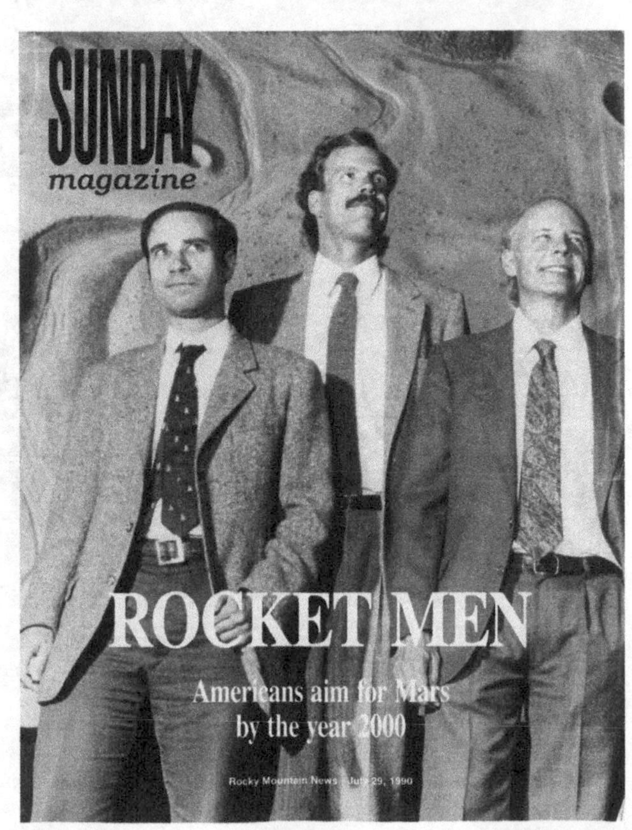

Mars Direct making press. This photo of Robert Zubrin (left), David Baker (center), and Ben Clark (right) ran as the cover of Rocky Mountain News *magazine shortly after Baker and the author began presenting the plan at conferences around the country. (Photo courtesy* Rocky Mountain News*)*

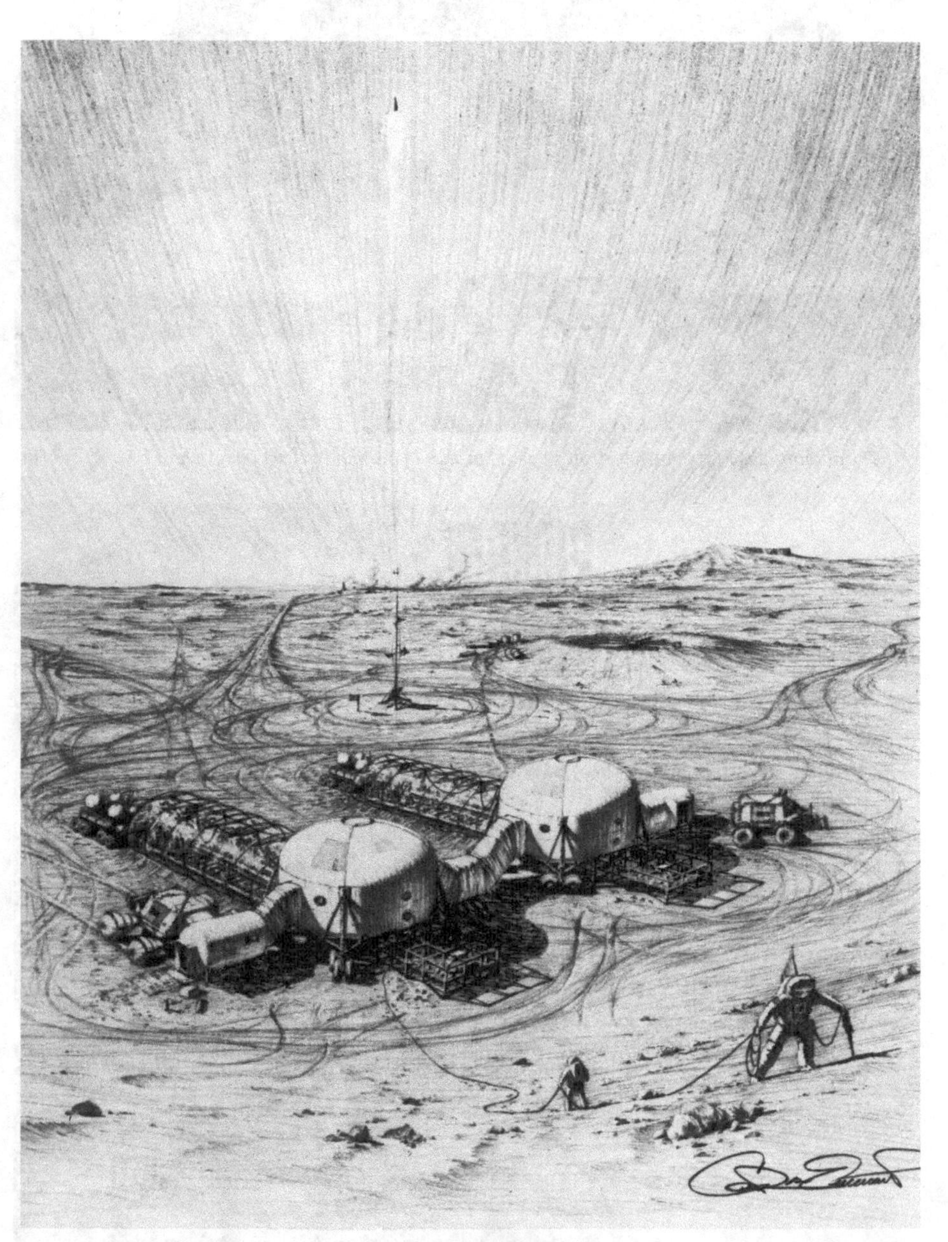

Mars Direct habitats can be mated up with the aid of inflatable tunnels to create an initial Mars base in fairly short order. (Artwork by Carter Emmart)

An Earth Return Vehicle descends to land at a growing Mars base. (Artwork by Michael Carroll)

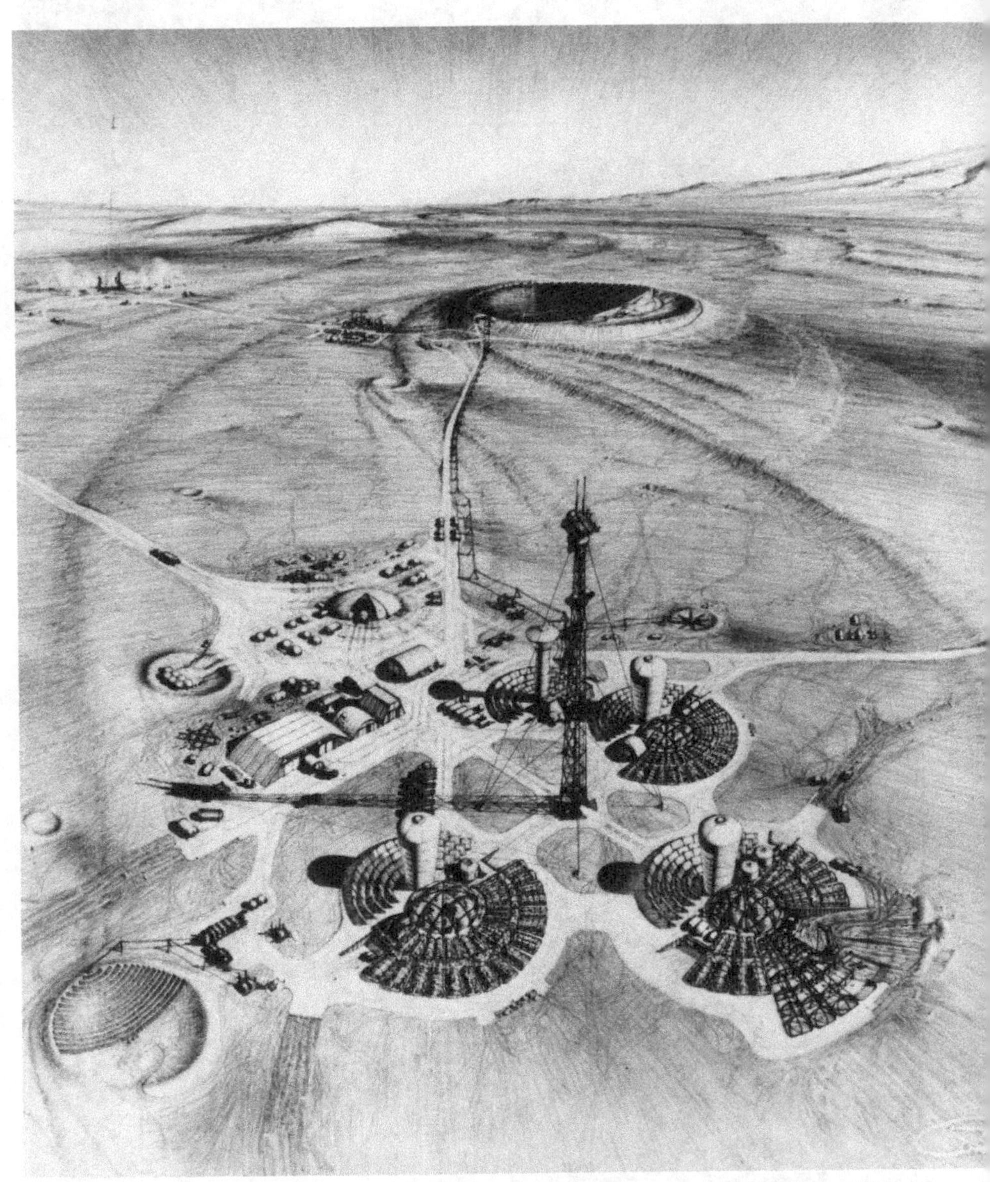

A mature base, established over a geothermal energy source, will provide a testbed for establishing technologies crucial to settling Mars. (Artwork by Carter Emmart)

Ballistic NIMF. (Artwork Robert Murray, courtesy Lockheed Martin)

NIMF Rocketplane. (Artwork Robert Murray, courtesy Lockheed Martin)

NIMFs—Nuclear Rockets using Indigenous Martian Fuels—as rocketplanes or as ballistic spacecraft would afford Mars explorers and later colonists unlimited mobility on a planetary scale.

Exploration team on a partially terraformed Mars. (Artwork by Michael Carroll)

The New Created World. (Artwork by Daein Ballard)

Liquid water once coursed over the face of Mars and, given the technological capabilities of the twenty-first century, it may once again. Several decades of terraforming could transform Mars into a relatively warm and slightly moist planet suitable some day for explorers without spacesuits, although breathing gear would still be required. Returning oceans to Mars is actually a possibility for the distant future.

We've all heard arguments advanced by vegetarians that everyone should give up eating meat, because an acre of corn can produce far more food for humans than an acre of cattle forage. These arguments are questionable on Earth, because starvation on our planet is not caused by a global food shortage but by lack of cash in the hands of the people who are starving. On Mars, however, an environment in which humans cannot simply take tillable land but must *make* it with domes and so forth, the vegetarian thesis has a fair amount of merit. Martian agriculture will have a very strong incentive to be efficient. Employing cattle, sheep, goats, rabbits, chickens, and other warm-blooded herbivores in large numbers as part of the food chain is, in fact, very inefficient. Most of the energy of the plants an animal eats goes into maintaining its body temperature, and very little ever reaches you. (Some years ago a science writer wrote several books in which he popularized the idea of goats as the key to future animal husbandry in space. They are of convenient size, omnivorous, fast breeding, milkable, and so on. Be that as it may, I'm city born, but have lived the more recent portion of my life in a rural area. I've seen what goats can do. Don't let one near your Kevlar dome. He'll eat it.) On the other hand, for almost any agricultural plant of interest, at least half of the plant is never eaten by humans. For example, in the case of corn, rice, or wheat, we don't eat the roots, stems, or leaves. Instead, these parts get plowed back into the soil with the self-consoling thought that we are keeping the ground fertile. But if that were our true objective, we'd plow the whole plant back—really, we're just wasting energy. So, if we want to be efficient, we need to find a way to use the not-directly edible parts of the plants. Time to bring in the goats? Maybe a few, to amuse the kids and keep the base security patrol busy as they leap over three-meter high fences in the light Martian gravity, but there are better ways. One is to use mushrooms. At Purdue University, for example, a NASA-funded space-agriculture research center has isolated species of mushrooms that will grow on the waste portions of plants and turn 70 percent of their material into edible protein that is as high in quality as soy (which is considerably better than goat). The fast-growing mushrooms need no light, just a dark, warm room, the waste corn stalks, and a little bit of oxygen. In other words, you can have a mushroom ranch in a closet. This, by the way, is an example of a technology developed for the extreme demands of space that can have plenty of application meeting basic human needs on Earth. But, if eating mushrooms and beans bores you, there's still hope. Cold-blooded herbivores, such as tilapia fish, are reasonably efficient in transforming waste plant material into high-quality protein. Fish farms on Mars? Why not? You don't need a very large tank to grow tilapia, and they won't escape to eat your dome.

Orchards will also be desirable to produce fruit. Therefore, wood will eventually be available as well. This wood can be used as such, to make furniture and so forth. Alternatively, it, together with other cellulose waste from agriculture, will be fed into the plastics industry, where it will significantly increase the variety of plastics that can be produced.

MARTIAN METALLURGY

The ability to manufacture metals is fundamental to any technological civilization. Mars provides abundant resources to support their production. In fact, in this respect, Mars is considerably richer than the Earth.

Steel

By far the most accessible industrial metal present on Mars is iron. The primary commercial ore of iron used on Earth is hematite (Fe_2O_3). This material is so ubiquitous on Mars that it gives the Red Planet its color, and thus indirectly, its name. Reducing hematite to pure iron is straightforward, and, as mentioned both in the Old Testament and in Homer, has been practiced on Earth for some three thousand years. There are at least two candidate processes suitable for use on Mars. The first, as discussed earlier in this chapter, uses waste carbon monoxide—reaction (1), above—produced by the base's RWGS reactor as follows:

$$Fe_2O_3 + 3CO \rightarrow 2Fe + 3CO_2 \quad (4)$$

The other uses hydrogen produced by the electrolysis of water.

$$Fe_2O_3 + 3H_2 \rightarrow 2Fe + 3H_2O \quad (5)$$

Reaction (4) is slightly exothermic and reaction (5) is mildly endothermic, so after heating the reactors to startup conditions, neither will require much power to run. In the case of reaction (5), the hydrogen needed can be obtained by electrolyzing the water waste product, so the only net input to the system is hematite. Carbon, manganese, phosphorus, and silicon, the four main alloying elements for steel, are very common on Mars. Additional alloying elements such as chromium, nickel, and vanadium, are also present in respectable quantities. Thus, once the iron is produced, it can readily be alloyed with appropriate quantities of these other elements to produce practically any type of carbon or stainless steel desired.

The widespread availability of carbon monoxide at the Mars base, due to its production as waste from RWGS reactors, opens up some interesting possibilities for novel kinds of low-temperature metal casting techniques on Mars. For example, carbon monoxide can be combined with iron at 110°C to produce iron carbonyl ($Fe(CO)_5$), which is a liquid at room temperature. The iron carbonyl can be poured into

FIGURE 7.5
Building the Mars base. (Artwork by Robert Murray, Mars Society)

a mold, and then heated to about 200°C, at which time it will decompose. Pure iron, very strong, will be left in the mold, while the carbon monoxide will be released, allowing it to be used again. You can also deposit the iron in layers by decomposing carbonyl vapor, allowing hollow objects of any complex shape desired to be made. Similar carbonyls can be formed between carbon monoxide and nickel, chromium, osmium, iridium, ruthenium, rhenium, cobalt, and tungsten. Each of these carbonyls decomposes under slightly different conditions, allowing a mixture of metal carbonyls to be separated into its pure components by successive decomposition, one metal at a time.[37]

Aluminum

On Earth, after steel, the second most important metal for general use is aluminum. Aluminum is fairly common on Mars, comprising about 4 percent of the planet's surface material by weight. Unfortunately, as on Earth, aluminum on Mars is generally present only in the form of its very tough oxide, alumina (Al_2O_3). In order to produce aluminum from alumina on Earth, the alumina is dissolved in molten cryolite at 1,000°C and then electrolyzed with carbon electrodes, which are used up in the process, while the cryolite is unharmed. On Mars, the carbon electrodes needed could be produced by pyrolyzing methane produced in the base's Sabatier reactor, as described in chapter 6. This process can be written:

$$Al_2O_3 + 3C \rightarrow 2Al + 3CO \qquad (6)$$

Aside from its complexity, the main problem with employing reaction (6) to produce aluminum is that it is very endothermic. It takes about 20 kWh of electricity to produce a single kilogram of aluminum. That's why on Earth aluminum production plants are located in areas where power is very cheap, such as the Pacific Northwest. On Mars during the base-building phase, power is not going to be cheap. At 20 kWh/kg, a 100 kWe nuclear reactor could only produce about 123 kilos of aluminum per day. Thus steel, not aluminum, will be the primary material used to build high-strength structures, although due to lower gravity, steel on Mars will weigh about the same as aluminum does on Earth! Aluminum will be saved for special applications, such as electrical wiring or flight system components, where its high electrical conductivity and/or light weight put a premium on its use.

Silicon

In the modern age, silicon has emerged as perhaps the third most important metal after steel and aluminum, as it is central to the manufacture of all electronics. It will be even more important on Mars, because by manufacturing silicon we will be able to produce photovoltaic panels, thereby continually increasing the base's power supply. The feedstock for manufacturing silicon metal, silicon dioxide (SiO_2) makes up almost 45 percent of the Martian crust by weight. In order to make silicon, you need to mix silicon dioxide with carbon and heat them in an electric furnace. The resulting reaction is:

$$SiO_2 + 2C \rightarrow Si + 2CO \qquad (7)$$

Once again, we see that the required reducing element, carbon, is a by-product of the Mars base propellant production system. Reaction (7) is highly endothermic, although nowhere near as bad as the alumina reduction reaction (6), and the energy burden involved in reducing silicon is not remotely comparable because the quantities needed tend to be much less.

For some purposes, the silicon product of reaction (7) is good enough for use. For example, you can use it to make silicon carbide, a strong heat-resistant material (it's used in tiles to protect the Space Shuttle from the heat of reentry). However, it is evident that any hematite impurities present in the reactor feedstock will also be reduced, resulting in iron impurities in the silicon product. To produce hyperpure silicon, then, good enough for computer chips and solar panels, another step is needed. This is accomplished by bathing the resulting impure silicon product in hot hydrogen gas, causing the silicon to turn into silane (SiH_4). At room temperature or above, silane is a gas, so it can easily be separated from hydrides of the other metals, all of which are solids. Then, if you want completely pure silicon, all you have to do is pipe the silane to another reactor where you decompose it under high temperatures, thereby producing pure silicon and releasing the hydrogen to make more silane. The silicon can then be doped with phosphorus or other selected impurities to produce exactly the kind of semiconductor device you need.

Alternatively, instead of decomposing the silane, you can liquefy it by refrigerating it down to –112°C. This is only about 20°C below typical Martian nighttime temperatures, so it is easily accomplished, and the resulting fluid can be stored for long periods in insulated tanks without difficulty. Why store liquid silane? Because silane will burn in carbon dioxide. Up till now, virtually all the Martian propellant combinations we have discussed, such as methane and oxygen, have required the vehicle utilizing them to carry both fuel and oxidizer in its tanks. We don't do things that way on Earth. On Earth, whether it's burning gasoline in your car or wood in your fireplace, all you do is provide the fuel. The oxidizer comes from the oxygen in the air. Since the oxidizer in general makes up about 75 percent of the reacting mixture, this latter approach is clearly a far more efficient way to go. Well, there is very little free oxygen in the Martian atmosphere; it's almost all carbon dioxide. Not many things will burn in carbon dioxide, but silane will, in accordance with:

$$SiH_4 + 2CO_2 \rightarrow SiO_2 + 2C + 2H_2O \quad (8)$$

In reaction (8), 73 percent of the propellant mass is carbon dioxide; only 27 percent is silane. Some of the products are solids, so you can't use this system in an internal combustion engine. But you could use it to fire the boiler of a steam engine, and you could use it in a ramjet engine or for rocket propulsion. Burned in accord with reaction (8), a silane/carbon dioxide rocket engine could produce a specific impulse of about 280 seconds. On the surface this is not that impressive, until you realize that you only need to carry 27 percent of the propellant with you. That is, consider a small rocket hopping vehicle that takes off and lands repeatedly, delivering its telerobot cargo to a series of chosen target sites separated by impassable terrain. It won't need to carry all its propellant. Instead, it can refuel with carbon dioxide just by running a pump each time it lands. The result is that the effective specific impulse of this system would not be 280 seconds, but 280 seconds multiplied by the ratio of total propellant to silane, which is 3.75. The result—an effective specific impulse of 1,050 seconds, unheard of in chemical rocketry.

Diborane, B_2H_6, will also burn in carbon dioxide, with a specific impulse of 300 seconds at a mixture ratio of three parts carbon dioxide to one part diborane.[37] A diborane/carbon dioxide rocket hopper could thus have an effective specific impulse of 1,200 seconds, even better than the silane/carbon dioxide system discussed above. However, boron is rare on Mars, while silicon is everywhere, and the processes required to produce diborane are rather complex. So, while small amounts of diborane may be imported to Mars early in the program to permit high-performance hopper applications (use of a diborane/carbon dioxide system may be the optimal way, for example, of performing a robotic Mars sample return mission), once a base exists capable of producing silane, this locally available product will almost certainly displace diborane.

As an aside, it has frequently been proposed that silicon be manufactured on the Moon to support the production of large quantities of solar panels there. This idea has serious flaws. Yes, it's quite true that silicon dioxide is as common on the Moon as anyone could ask for, but the carbon and hydrogen necessary to turn it into silicon metal is absent. While in the processes described above these reagents are recycled, in reality such recycling is always imperfect. If you want to produce silicon metal, or *any* other metal, on the Moon, you are going to end up having to import a lot of carbon and hydrogen. On Mars, in contrast, both of these elements are available locally.

Copper

As a final example of producing a key industrial metal at a Mars base, let us consider copper. Copper, which is absent on the Moon, has been detected in SNC meteorites at about the same concentrations that it is found in soil on Earth. This is quite low, however, about 50 parts per million. If you want to obtain useful quantities of copper, you don't extract it from soil. Instead, you must find places where nature has concentrated it in the form of copper ore. Commercially, the most important sources of copper ore on Earth are copper sulfides. As we have seen, sulfur is much more common on Mars than on Earth, and it is probable that copper ore deposits are available on Mars in the form of copper sulfide deposits formed at the base of lava flows. Once found, copper ore can easily be reduced by smelting or leaching, as has been practiced on Earth since ancient times.

The example of copper drives home the fact that, in general, the only way of accessing geochemically rare elements is by mining local concentrations of high-grade mineral ore. However, you will find ores only where complex hydrologic and volcanic processes have occurred that can concentrate these elements into local ore deposits, and, within our solar system, only Earth and Mars have experienced such processes. Because these processes have occurred on Mars, we should be able to find concentrated ore of nearly every metal, rare or common, necessary to build a modern civilization.

THE QUESTION OF POWER

It should be evident that the availability of large amounts of both thermal and electrical power is the key to being able to conduct the manufacturing processes to develop a significant Mars base. It may be unpopular to say it, but by far the best way to provide this power during the early years of base development is by importing nuclear reactors produced on Earth. On Earth today, the main sources of power for our civilization are hydroelectric, fossil fuel and wood combustion, and nuclear. Geothermal heat provides a distant fourth source of energy, and way behind it are solar power and wind, which play very minor roles. On Mars, hydroelectric dams and fossil fuel combustion are not power source options. In the long run, the prospects for generating thermonuclear fusion power on Mars are excellent, because the ratio of deuterium (the heavy isotope of hydrogen needed to fuel fusion reactors) to ordinary hydrogen found on Mars is five times as high as it is on Earth. Unfortunately, fusion reactors don't currently exist. That leaves nuclear power as the only option for the initial source of large-scale power. A nuclear reactor producing 100 kWe and 2,000 kilowatts of thermal process heat twenty-four hours a day for ten years would weigh about 4,000 kilograms—just 4 tonnes—making it light enough to import from Earth. In contrast, a solar array that could produce the same round-the-clock electrical output (but only one twentieth the thermal output) for about the same lifetime would weigh about 27,000 kilograms and would cover an area of 6,600 square meters (about two-thirds of a football field). If we wanted the same *thermal* output (for brick making and water processing), the solar array needed would weigh 540,000 kilograms and cover thirteen football fields. This is obviously far too much material to import from Earth. The advantage of nuclear power for opening Mars is enormous—so much so that the failure to date of the American political class to fund an effective space nuclear power research and development program can only be condemned in the harshest terms. If we give up space nuclear power, we will give up a world.

While the initial base power supply will need to be nuclear, once the base is well established, the equations could change. It should be possible at some point to construct solar power systems out of indigenous materials on Mars. If you are living on Mars, hundreds of tonnes of local materials may be much easier to come by than four tonnes of equipment imported from Earth.

Harnessing the Sun and Wind

There are two kinds of solar power systems that can be manufactured on Mars, dynamic and photovoltaic. Solar dynamic systems are low-tech; they work by using a parabolic mirror to concentrate sunlight on a boiler, where a fluid is heated and expanded to turn a generator turbine. These systems can be fairly efficient (about 25 percent efficiency), but to date they have not found much favor in the space program, as the fact that they rely on moving parts has caused many to consider them to be unreliable. However, at a permanently staffed Mars base, people would be on hand to maintain systems and adjust failing equipment. The reliability argument against dynamic systems becomes considerably less forceful in the context of a Mars base. Moreover, because they are low-tech assemblages of mirrors, boilers, and similar gear, it is relatively easy to see how such systems could be manufactured on Mars. The mirrors could be made of inflatable plastic, for example, covered with a very thin layer of aluminum for reflectivity. The pipes, boilers, and turbine shaft and blades could all be made of steel. To actually attain the 25 percent efficiency level, the turbines must be manufactured to tolerances that are too exact to be realistic for a Mars base, but this is hardly a show stopper. If necessary, lower tolerances and 15 percent efficiencies could easily be accepted. In addition to these advantages, the dynamic cycles also offer the attractive feature of producing a goodly amount of useful process heat, perhaps equal to four to six times their electrical output.

Solar dynamic cycles, however, require clear skies. In order for the parabolic mirrors to effectively concentrate light, the light must all come from the same place, directly from the Sun. It cannot come from diffuse sources of scattered light all over the Martian sky. Based upon *Viking* data, the kind of clear skies needed for effective solar dynamic operation can only be expected during the northern spring and summer. During the other half of the year, the solar dynamic concentrators are likely to put out very little power. Such a seasonal swing in power availability might be acceptable for some purposes. You don't necessarily have to be making metals all year long. But if solar power is to become the primary source of base power, a more dependable technology is needed.

Photovoltaic panels may potentially be such a technology. As we have seen, the key material for the manufacture of such panels, pure silicon metal, can be produced on Mars, as can the aluminum or copper for their wiring and the plastics needed to insulate the wiring. In efforts to reduce costs, simplified methods of manufacturing solar panels in large single sheets have recently been developed for use on Earth, and such methods, transported to Mars, could well make large-scale local manufacture of photovoltaic systems feasible. It's somewhat surprising, but it turns out that the performance of a photovoltaic panel on Mars is only modestly degraded when the Martian atmosphere is dusty.[38,39] Except during very bad dust storms, the dust levels that typify the northern fall and winter skies scatter most of the Martian sunlight, but block little of it. Photovoltaic panels, unlike solar dynamic reflectors, have no regard for the direction of the incident light. Thus they should work fairly well on Mars all year round. Efficiencies are low, only about 12 percent, and you get no process heat beyond the system's electric output, but that's life. The panels' performance may be significantly degraded by dust precipitating on them. This, however, can be remedied by human crews with brooms, or by manufacturing the panels with a windshield-wiper–type device attached to them.

As a further supplement to base power, wind is a possibility. Windmills have operated on Earth for centuries, and their low-tech nature makes them attractive potential items for Mars base manufacture. It's true that the great dust storms are quite intermittent, and therefore useless as a real power source. Furthermore, the air is only 1 percent as thick as Earth's, and the surface winds measured at the *Viking* sites were only about 5 meters per second (10 mph), implying negligible potential wind power. However, typical winds at altitudes well above the surface measure about 30 m/s (60 mph), which would create the same amount of power per unit of windmill area as a 6 m/s (12 mph) breeze on Earth. This would be quite acceptable for wind power generation. The key, then, to windmill practicality is how high off the surface the windmill must be placed in order to get above the stagnant surface boundary layer. At the present time this is unknown, and the answer is certain to vary locally in any case. However high this should turn out, it should be remembered that on Mars we would be erecting the windmill in a 38 percent gravity field—it may be practical to build windmill towers that Earthlings would consider outlandishly tall.

Generating Geothermal Power

Since about 1930 elementary and secondary boarding schools in the rural areas of Iceland have whenever possible been sited at locations where geothermal energy is available. In these centres the school buildings and living quarters for the pupils and staff are geothermally heated. They are also as a rule equipped with a swimming pool, and are self-supplying with vegetables (tomatoes, cucumbers, cauliflowers etc.) grown in their own hot houses. There are now many such schools in various parts of the country, and quite often they are used as tourist hostels during the summer holidays. Quite often these centres have formed the nuclei of new service communities in the rural areas.

—S.S. Einarson, Geothermal District Heating, 1973

Solar power and wind are means of potentially generating tens or hundreds of kilowatts of electricity with equipment of local manufacture. They are attractive because they can be deployed and set up almost anywhere, thereby allowing for the decentralized generation of power. This will be very useful on Mars, as the provision of some power to widely scattered assets will be necessary, and the infrastructure needed to transmit power over large distances won't be available for some time. However, the relatively modest total outputs these power sources provide make it desirable to seek a more muscular option. As British scientist Martyn Fogg has pointed out,[40] such an option is available on Mars in the form of geothermal power.

Geothermal power is generated by using the high temperatures that exist deep underground to boil a fluid such as water, and then using the steam produced to turn a generator turbine. On Earth, geothermal power is the fourth largest source of power, after combustion, hydroelectric, and nuclear, providing about 11,000 MWe or 0.1 percent of all power used by humanity. The nation of Iceland gets the majority of its power—over 500 MWe—from geothermal heat. A single geothermal power well on Earth will typically generate about 1 to 10 MWe—small by terrestrial power station standards, but large compared to Mars base requirements. On Earth, geothermal power stations of this size can be up and running within six months of the start of drilling, and have a record of being online 97 percent of the time, a figure which is exceeded only by hydroelectric power. Furthermore, in addition to supplying large amounts of power, a geothermal station could supply a Mars base with something else equally valuable, a copious supply of liquid water. On Earth, geothermal power suffers the disadvantage that power generating stations must be positioned wherever the Earth in its whimsy has chosen to place a geothermal heat source, and, since we have already chosen the locations of our cities, this frequently presents a problem. On Mars, on the other hand, the cities have yet to be built.

Given the value of a geothermal power/water supply, the discovery of such a source would probably dictate the location of the Mars base.

In short, geothermal power supplies would be enormously advantageous to Martian settlers. The question is, do they exist? Perhaps somewhat surprisingly, the answer is almost certainly yes.

Large-scale volcanic features exist on Mars, for example in Tharsis, that have been dated to less than 200 million years old. About 4 percent of the planet's surface (about 5 million square kilometers, mostly in the northern regions of Elysium, Arcadia, and Amazonia, as well as the equatorial Tharsis region) is classified by Mars geologists as "Upper Amazonian," which means that it has been resurfaced by either volcanic action or flooding sometime within the past 500 million years. Now ages of 200 to 500 million years may seem like ancient history, but given Mars' 4 billion year age, they actually qualify as "the present." From a geological point of view on Mars, 200 million years ago is "today." If volcanoes were active then, they are just as likely to be active now.

Furthermore, as we have seen, Mars possesses large supplies of water, with a liquid water table probably existing within a kilometer of the surface at least in some places. If an area was geothermally active in the recent past, this water could be hot enough to represent a practical power source.

If we consider only the upper Amazonian territories as viable candidates, and spread their formation equally in time across the 500 million years of that era, we find that 10 percent, or 0.5 million square kilometers, is probably less than 50 million years old; 1 percent, or 50,000 square kilometers, is probably less than 5 million years old; and 0.1 percent, or 5,000 square kilometers, has probably been active within the past 500,000 years.

You don't have to extract geothermal power from a region that is actually volcanically active now. The ground stays hot a long time after activity has subsided. In his seminal paper on Mars geothermal power, Fogg presented calculations of the temperature profiles of Martian land as a function of the time since the region was active. His results are summarized in Table 7.2.

As a point of reference, the current state of the art of terrestrial drill-

TABLE 7.2
Characteristics of Mars Geothermal Fields

Time since active (million years)	0.5	5	10	20	50	>150
Depth to reach 0°C (km)	0.29	0.65	0.91	1.29	2.04	3.53
Depth to reach 60°C (km)	0.62	1.38	1.95	2.76	4.35	7.53
Depth to reach 100°C (km)	0.84	1.87	2.64	3.73	5.88	~10
Depth to reach 200°C (km)	1.38	3.09	4.36	6.17	9.73	~17
Depth to reach 300°C (km)	1.92	4.30	6.09	8.61	~13	~24
Probable land available (1,000 km^2)	5	50	100	200	500	plenty

ing technology is to be able to drill down to about 10 kilometers. On Mars it should be easier to drill deeper, because the lower gravity will compact the soil less forcefully. It can be seen that the amount of territory that has had associated geothermal activity within the past five million years is quite large, and for these territories, wells just a few kilometers deep should be adequate to bring up very hot water. Once brought to the surface, the water would be flashed to steam and used to power a turbine to generate electric power. This will work even more efficiently on Mars than it does on Earth, because the low atmospheric pressure will allow the steam to be much more fully expanded before it is condensed. Some of the waste water generated by this process will be tapped off to supply the base with as much water as it needs. The rest will be channeled back down into the well to replenish the subsurface aquifer.

Geothermal energy cannot be generated on the Moon; it cannot be generated on asteroids. Of all extraterrestrial bodies in our solar system, only Mars has the potential to produce such a bountiful source of power to support human settlement.

The options for developing solar and wind power for outlying installation power, together with the use of geothermal energy for the main base load, indicate that once given a fair start with a nuclear reactor, a Mars base that has mastered an appropriate array of local resource utilization technologies can continue to expand its own power supply on the basis of its own efforts. The more power it has, the faster it will grow; the faster it grows, the more power it will have. Once it is possible to produce solar, wind, and especially geothermal power on Mars, the growth of the base will become exponential.

USE OF THE BASE TO SUPPORT LONG-RANGE MOBILITY ON MARS

While all this development is going on at the base, will our global exploration of Mars cease? Far from it. However well the base location is chosen, it is certain that some essential resources needed for its development will be available only at sites tens, hundreds, or thousands of kilometers distant. Global exploration for and transport of these resources will be an essential capability necessary for the growth of the base. In a symbiotic relationship, it will be the base itself that will create the capability for precisely such long-range mobility.

The situation is somewhat analogous to the development of human exploration of Antarctica. Prior to the International Geophysical Year in 1957, Antarctic exploration was conducted via a series of sorties, with each exploration party generally using its own ship as its base. Starting that year, however, a decision was made and implemented to build up a large permanently staffed base at McMurdo Sound. Today, this base provides facilities that allow the support of mechanized vehicles, helicopters, and aircraft that give Antarctic researchers access to every portion of the continent. By concentrating resources at a single point, a capability was created that affords much broader penetration than ever would have been possible by continuing a tradition of dog sled and ski sorties from individual exploration ships.

The terrain on Mars is far rougher than even Antarctica. To have true long-range mobility on Mars it is necessary to be able to fly. While balloons and perhaps subsonic aircraft can be used to loft small robotic packages across the windy Martian skies, the only systems trustworthy enough for human transport will be rocket-powered vehicles that can punch through any weather. These could be either purely ballistic vehicles that leap out of the Martian atmosphere to travel from one side of the planet to the other, or winged rocketplanes capable of supersonic flight. Both of these types of systems are by their nature propellant hogs, and the operation of either would be unthinkable unless the large amounts of propellant required are manufactured on Mars.

For example, consider a Mars piloted ballistic hopper with a mass of 10 tonnes powered by methane/oxygen rocket engines with a specific impulse of 380 seconds. Let's say we want to fly it 2,600 kilometers (which is 45 degrees of latitude or longitude across Mars), land, and then eventually return with no added cargo. In order to perform this maneuver, the vehicle will need a mass ratio of about 7 so the total propellant needed will be 60 tonnes. If we were to do it with a 15-tonne rocketplane (its wings will make the plane heavier) with a supersonic glide lift-to-drag (L/D) ratio of 4, the mass ratio would be about 5, so again 60 tonnes of propellant would be needed. It's clear that there is no way these sorts of vehicles are going to be doing much flying on Mars if their methane/oxygen propellant, or even the hydrogen feedstock required to manufacture it, must be transported from Earth.

The need to carry enough propellant for both the outbound and return legs of the exploratory sortie constrains the maximum range of chemical-propelled rockets on Mars to about 4,000 kilometers. These limits can be overcome if the vehicle can make its own propellant after it lands. Chemical bipropellants don't allow this because their manufacture requires too much power (about 5 kWh per kilogram), and therefore too large a power supply to make such a system flight-mobile. In the late 1980s, however, I came up with a vehicle concept I called a "NIMF" (for nuclear rocket using indigenous Martian fuel) that can overcome this problem.[41,42] In the case of the NIMF, raw carbon dioxide from the Martian air is used as a propellant, and is heated by an onboard nuclear thermal rocket (NTR) engine to create a hot rocket exhaust. Because NTRs don't convert their heat into electricity, all the power conversion gear that actually forms most of a nuclear power reactor's mass is eliminated, allowing these systems to be small and lightweight. Because the propellant is simply raw carbon dioxide, which can be acquired at low energy cost (less than 0.3 kWe-hrs/kg) through direct compression out of the atmosphere, not much onboard electric power is needed, and all the chemical synthesis gear is eliminated as well. Hot carbon dioxide is not a high-class rocket propellant, and specific impulses of about 260 seconds are all you can expect. But a prospector needs a mule that can eat mountain scrub; high-strung race horses that will eat only gourmet fodder are of little use in the hills. Essentially a far more powerful and potent evolution of the gashopper vehicles discussed in chapter 6, a NIMF is the ideal exploration craft because it can live on what it finds in the field. Rocket vehicles equipped with this type of propulsion system would give Mars explorers complete global mobility, allowing them to hop around the planet in a craft that can refuel itself each time it lands. NIMF ballistic hoppers and rocketplanes are depicted in the plates.

The advantage of the NIMF's mode of operation are manifold. Despite its lower specific impulse, the fact that the NIMF does not have to carry its return propellant gives the vehicle complete global reach, while even the best chemical systems are range limited. The NIMF offers another advantage: Because it makes its own propellant, it puts far less strain on the power resources of the base than the chemical systems. Producing the 60 tonnes of methane/oxygen required by the chemical rocket systems described at the beginning of this section would fully occupy a 100 kWe reactor stationed at the base for 123 days. A NIMF sortie would cost the base power supply nothing, nor would it tap into the base's hydrogen or water supply. Its only imposition on the base would be for crew supplies, maintenance, and repair. Another benefit of having a NIMF operating on Mars is its unique ability to engage in rapid global surface-to-surface shipment of large quantities of cargo. If you need 20 tonnes of copper sulfide ore, a 40-tonne cargo NIMF could fly to the other side of the planet and get it for you. No other system can offer this kind of performance.

You may recall (chapter 3) that in the period prior to the development of the Mars Direct mission architecture I had advocated a humans-to-Mars architecture based upon a single heavy-lift launch, nuclear thermal rocket (NTR) propulsion for trans-Mars injection, and the use of an NIMF vehicle to hop around Mars and then return. I dropped this in favor of Mars Direct because it became clear to me that the technology required for NTRs and NIMFs was too advanced to form the basis of initial Mars exploration missions. The missions they made possible were very attractive, but the schedule required for their development would postpone the first mission beyond the point of programmatic viability. That said, it remains the case that NIMF technology offers a set of extremely potent capabilities to support the development of the Mars base. Thus, in the context of an extended Mars program, it would be wise to spend a significant effort to bring NIMF vehicles into play. Then, a few years into the base-building phase, they will be ready, and the base will be able to access resources from all over the planet.

THE BEGINNINGS OF COLONIZATION

The first astronaut explorers on Mars will spend an eighteen-month tour on the Red Planet, waiting for the first good launch window to open for their return. But, as the base develops and living conditions improve, some future astronauts may choose to extend their surface stay beyond the basic one-and-a-half-year tour of duty, to four years, six years, and more. The base's sponsors will probably offer large financial bonuses to those who choose to do so. After all, most of the expense of the base actually lies in moving people back and forth. The longer the base operates, the more incentive there will be to develop new forms of interplanetary transportation that reduce logistics costs ever lower. As we will see, the government may do this, or perhaps it will be done by opening base cargo deliveries from Earth to private competition, but it will be done. It will become ever cheaper to go to Mars, and ever cheaper to maintain people once they are there. As more people steadily arrive and stay longer before they leave, the population of the base will come to resemble a town—and will actually grow into one.

The colonization of Mars will then begin.

8: THE COLONIZATION OF MARS

> This proposition being made publike and coming to the scanning of all, it raised many variable opinions amongst men, and caused many fears & doubts amongst themselves. Some, from their reasons & hops conceived, laboured to stirr up & incourage the rest to undertake and prosecute the same; others, againe, out of their fears, objected against it, & sought to divert from it, alledging many things, and those neither unreasonable nor unprobable; as that it was a great designe, and subjecte to many unconceivable perills & dangers . . .
>
> It was answered that all great & honourable actions are accompanied with great difficulties, and must be both enterprised and overcome with answerable courages.
>
> —Gov. William Bradford, *Of Plymouth Plantation,* 1621.

In the previous chapters we have looked at the process of opening Mars to human settlement largely from a technical point of view. We have seen that using twentieth-century technology, the first human explorers can reach Mars in about ten years for a cost well within the discretionary spending capabilities of the U.S. government. We have forseen that with a comparatively limited extension of this effort, a base can be built upon Mars capable of supporting dozens or even hundreds of people within a few decades of the first landing—people who will then proceed to master the techniques of local resource utilization that could someday make Mars the home for millions.

We thus come to the crux of the matter: the settlement phase. Can Mars really be colonized? From the technical point of view, there is little doubt that we can eventually do just about anything we want on Mars, including, as we shall see in the next chapter, terraforming Mars—transforming the planet from a frigid, arid world into a warm, wet planet once again. But how much can we afford? While the exploration and base-building phases can and probably must be carried out on the basis of government funding, during the settlement phase economics comes to the fore. While a Mars base of even a few hundred people can probably be supported out of pocket by governmental expenditures, a developing Martian society, one that may come to number in the hundreds of thousands, clearly cannot. To be viable, a real Martian civilization must be either completely autarkic (very unlikely until the far future) or be able to produce some kind of export that allows it to pay for the imports it requires.

Around this question will hang the future of Mars, and not just human civilization on Mars but the very nature of the planet itself. If a viable Martian civilization can be established, its population and powers to change the planet will continue to grow. Mars was once a temperate planet, and with enough work, it can be made so again. The advantages to Mars settlers of a terraformed world are so obvious, that put simply, if Mars is colonized, then it will also be terraformed. Therefore, ultimately, the feasibility or lack thereof of terraforming Mars is fundamentally a corollary to the economic viability of the Martian colonization effort.

Thus, the central objection raised against the human settlement and terraforming of Mars: *Such projects may be technologically feasible, but there is no possible way that they can be paid for.* On the surface, the arguments given supporting this position appear cogent, for Mars is a distant place, difficult to access, and possesses a hostile environment that holds no apparent resources of economic value. These arguments appear ironclad, yet it must be pointed out that they were also presented in the past as convincing reasons for the utter impracticality of the European settlement of North America and Australia. It is certainly true that the technological and economic problems facing Mars colonization in the twenty-first century are vastly different in detail than those that had to be overcome in the colonization of the New World. Nevertheless, it is my contention that these arguments are flawed by essentially the same false logic and lack of understanding that resulted in repeated misevaluations of the value of colonial settlements (as opposed to trading posts, plantations, and other extractive activities) on the part of numerous European governments during the four hundred years following Columbus.

During the period of their global ascendancy, the Spanish ignored North America; to them it was nothing but a vast amount of worthless wilderness. In 1781, while Cornwallis was being blockaded into submission at Yorktown, the British deployed their fleet into the Caribbean to seize a few high-income sugar plantation islands from the French. In 1803, Napoleon Bonaparte sold a third of what is now the United States for two million dollars. In 1867, the Czar sold off Alaska for a similar pittance. The existence of Australia was known to Europe for two hundred years before the first colony arrived, yet no European power even bothered to claim the continent until 1830. These pieces of shortsighted statecraft are legendary today. Yet their consistency shows a persistent blind spot among policy-making groups as to the true sources of wealth and power. I believe that two hundred years from now the current apathy of governments toward the value of extraterrestrial bodies, and Mars in particular, will be viewed in a similar light.

It is almost impossible to know what enterprises will be economically viable twenty years from now, let alone fifty or one hundred. Nevertheless, in this chapter I shall endeavor to show you how and why the economics of Mars colonization can be made to work, and why the success of this colonization effort will ultimately be the keystone to human expansion throughout our planetary system. While I shall return to historical analogies periodically, my arguments will not be primarily historical in nature. Rather, they are based on the concrete case of Mars itself, its unique characteristics, resources, technological requirements, and its relationships to the other important bodies within our solar system.

THE UNIQUENESS OF MARS

In proposing a new enterprise, for example in a business plan, it's generally necessary to assemble and list the advantages of your product or service. What have you got that the competition doesn't have? All right then, what's Mars got?

Among extraterrestrial bodies in our solar system, Mars is singular in that it possesses all the raw materials required to support not only life, but a new branch of human civilization. This uniqueness is illustrated most clearly if we contrast Mars with the Earth's Moon, the most frequently cited alternative location for extraterrestrial human colonization.

In contrast to the Moon, Mars is rich in carbon, nitrogen, hydrogen, and oxygen, all in biologically readily accessible forms such as carbon dioxide gas, nitrogen gas, and water ice and permafrost. Carbon and nitrogen are only present on the Moon in parts-per-million quantities. There is some water ice, but only in permanently shaded ultracold (–230°C) polar craters—locations so frigid as to make their contents virtually inaccessible outside of such environments. Oxygen is abundant, but only in tightly bound oxides such as silicon dioxide (SiO_2), ferrous oxide (Fe_2O_3), magnesium oxide (MgO), and alumina oxide (Al_2O_3), which require very high energy processes to reduce. Current knowledge indicates that if Mars were smooth and all its ice and permafrost melted into liquid water, the entire planet would be covered with an ocean over 100 meters deep. This contrasts strongly with the Moon, which is so dry that if concrete were found there, lunar colonists would mine it to get the water out. Thus, if plants could be grown in greenhouses on the Moon (an unlikely proposition, as we've seen) most of their biomass material would have to be imported.

The Moon is also deficient in about half the metals of interest to industrial society (copper, nickel, and zinc, for example), as well as many other elements of interest such as sulfur, fluorine, bromine, phosphorus, and chlorine. Mars has every required element in abundance. Moreover, on Mars, as on Earth, hydrologic and volcanic processes have occurred that are likely to have consolidated various elements into local concentrations of high-grade mineral ore. Indeed, the geologic history of Mars has been compared to that of Africa,[43] with very optimistic implications as to its mineral wealth as a corollary. In contrast, the Moon has had virtually no history of water or volcanic action, with the result that it is basically composed of trash rocks with very little differentiation into ores that represent useful concentrations of anything interesting.

You can generate power on either the Moon or Mars with solar panels, and here the advantages of the Moon's clearer skies and closer proximity to the Sun than Mars roughly balance the disadvantage of large energy storage requirements created by the Moon's 28-day light/dark cycle. But if you wish to *manufacture* solar panels, so as to create a self-expanding power base, Mars holds an enormous advantage, as only Mars possesses the large supplies of carbon and hydrogen needed to produce the pure silicon required for producing photovoltaic panels and other electronics. In addition, Mars has the potential for wind-generated power while the Moon clearly does not. But both solar and wind power offer relatively modest potential—tens or at most hundreds of kilowatts here or there. To create a vibrant civilization you need a richer power base, and this Mars has both in the short and medium term in the form of its geothermal power resources which offer potential for a large number of locally created electricity-generating stations in the 10 MWe (10,000 kilowatt) class. In the long term, Mars will enjoy a power-rich economy based upon exploitation of its large domestic resources of deuterium fuel for fusion reactors. Deuterium is five times more common on Mars than it is on Earth, and tens of thousands of times more common on Mars than on the Moon.

But, as we discussed in chapter 7, the biggest problem with the Moon, as with all other airless planetary bodies and proposed artificial free-space colonies, is that sunlight is not available in a form useful for growing crops. A single acre of plants on Earth requires 4 MW of sunlight power, a square kilometer needs 1,000 MW. The entire world put together does not produce enough electric power to illuminate the farms of the state of Rhode Island, that agricultural giant. Growing crops with electrically generated light is just economically hopeless. But you can't use natural sunlight on the Moon or any other airless body in space unless you put walls on the greenhouse thick enough to shield out solar flares, a requirement that enormously increases the expense of creating crop land. Even if you did that, it wouldn't do you any good on the Moon, because plants won't grow in a light/dark cycle lasting 28 days.

On Mars, there is an atmosphere thick enough to protect crops grown on the surface from solar flares. Therefore, thin-walled inflatable plastic greenhouses protected by unpressurized UV-resistant hard-plastic shield domes can be used to rapidly create crop land on the surface. Even without the problems of solar flares and a month-long diurnal cycle, such simple greenhouses would be impractical on the Moon because they would create unbearably high temperatures. On Mars, in contrast, the strong greenhouse effect created by such domes would be precisely what is necessary to produce a temperate climate inside. Such domes up to 50 meters in diameter are light enough to be transported from Earth initially, and later on they can be manufactured on Mars out of indigenous materials. Because all the resources to make plastics exist on Mars, networks of such 50-to-100 meter domes could rapidly be manufactured and deployed, opening up large areas of the surface to both shirtsleeve human habitation and agriculture. That's just the beginning, because, as we shall see in chapter 9, it will eventually be possible for humans to substantially thicken Mars' atmosphere by forcing the regolith to outgas its contents through a deliberate program of artificially induced global warming. Once that has been accomplished, the habitation domes could be virtually any size, as they would not have to sustain a pressure differential between their interior and exterior. In fact, once that has been done, it will be possible to raise specially bred crops outside the domes.

The point to be made is that unlike colonists on any other known extraterrestrial body, Martian colonists will be able to live on the surface, not in tunnels, and move about freely and grow crops in the light of day. Mars is a place where humans can live and multiply to large numbers, supporting themselves with products of every description made out of indigenous materials. Mars is thus a place where an actual civilization, not just a mining or scientific outpost, can be developed. And significantly for interplanetary commerce, *Mars and Earth are the only two locations in the solar system where humans will be able to grow crops for export.*

INTERPLANETARY COMMERCE

Mars is the best target for colonization in the solar system because it has by far the greatest potential for self-sufficiency. Nevertheless, even with optimistic development of robotic manufacturing techniques, Mars will not have the labor power required to make it fully self-sufficient until its population numbers in the millions. Thus, for centuries it will be necessary and forever it will be desirable, for Mars to be able to import specialized manufactured goods from Earth. These goods can be fairly limited in mass, as only small portions (by weight) of even very high-tech goods are actually complex. Nevertheless, these smaller sophisticated items will have to be paid for, and the high costs of Earth-launch and interplanetary transport will greatly increase their price. What can Mars possibly export to Earth in return?

It is this question that has caused many to deem Mars colonization intractable, or at least inferior in prospect to the Moon. For example, much has been made of the fact that the Moon has indigenous supplies of helium-3, an isotope not found on Earth and which could be of considerable value as a fuel for second-generation thermonuclear fusion reactors. Mars has no known helium-3 resources. On the other hand, because of its complex geologic history, Mars may contain concentrated mineral ores, with much greater concentrations of precious metal ores readily available than is currently the case on Earth—because the terrestrial ores have been heavily scavenged by humans for the past five thousand years. In a paper I coauthored with David Baker in 1990, I showed that if concentrated supplies of metals of equal or greater value than silver (such as silver, germanium, hafnium, lanthanum, cerium, rhenium, samarium, gallium, gadolinium, gold, palladium, iridium, rubidium, platinum, rhodium, europium, and a host of others) were available on Mars, they could potentially be transported back to Earth for a substantial profit.[44] Reusable Mars-surface–based single-stage-to-orbit vehicles such as NIMFs (discussed in chapter 7) could haul cargoes to Mars orbit for transportation to Earth via either cheap expendable chemical stages manufactured on Mars or reusable cycling solar or magnetic sail-powered interplanetary spacecraft. (These advanced propulsion systems are discussed in the focus section at the end of this chapter.) The existence of such precious metal ores, however, is still hypothetical.

But there is one commercial resource that is known to exist ubiquitously on Mars in large amounts—deuterium. Deuterium, the heavy isotope of hydrogen, occurs as 166 out of every million hydrogen atoms on Earth, but comprises 833 out of every million hydrogen atoms on Mars. Deuterium is the key fuel not only for both first- and second-generation fusion reactors, but it is also an essential material for the nuclear power industry today. If you have enough deuterium, you can moderate a nuclear fission reactor with "heavy water" instead of ordinary "light water," and such a heavy-water moderated reactor can run on natural uranium, with no enrichment required. Canadian-made nuclear power reactors known as CANDUs work on this principle today. The problem, however, is that you have to electrolyze 30 tonnes of ordinary "light" water to produce enough hydrogen to make one kilogram of deuterium, and unless you have a lot of very cheap hydroelectric power to burn, the process is prohibitively expensive. (This is why in World War II the German atomic bomb project had to conduct its heavy-water production near the large Norwegian hydroelectric dams at Vemork. When Norwegian resistance commandos and U.S. B-17s wrecked the place in a series of raids in 1943, Germany's nuclear program was effectively destroyed.) Even with cheap power, deuterium is very expensive; its current market value on Earth is about $10,000 per kilogram, roughly 12 times as valuable as silver (at $27 per ounce) or 25 percent as valuable as gold (at $1,200 per ounce). This is in today's prefusion economy. Once fusion reactors go into widespread use, deuterium prices will increase. As discussed in the previous chapters, the Mars base will be using most of its power in water electrolysis to drive its various life-support and chemical-synthesis processes. If a deuterium/hydrogen separation stage is applied to the hydrogen produced by the electrolysis operations prior to recirculating it back into the chemical reactors, then every 6 tonnes of Martian water electrolyzed can provide about one kilogram of deuterium as a by-product. Each person on Mars will require about 10 tonnes of water electrolysis per (Earth) year. If the amount of water electrolysis supporting the various materials-processing operations is twice this, a total of 6 million tonnes per year of water electrolysis will be required by a 200,000 person Mars colony. This will result in the production of 1,000 tonnes of deuterium per year, enough to produce 11 terrawatts (TW) of electricity, or about what the entire human race consumes today. At current deuterium prices this represents an annual export income potential of $10 billion—a figure comparable to a nation of much greater size on Earth. (For example, New Zealand booked $26 billion of gross exports in 2009, yet is a nation of 4.3 million.) At the current average rate of 7 cents/kilowatt-hour for electricity, the total value of the power produced on Earth as a result would total about $7 trillion per year.

Ideas may be another possible export for Martian colonists. Just as the labor shortage prevalent in colonial and nineteenth-century America drove the creation of "Yankee ingenuity," so the conditions of extreme labor shortage combined with a technological culture will tend to drive Martian ingenuity to produce wave after wave of invention in energy production, automation and robotics, biotechnology, and other areas. These inventions, licensed on Earth, could finance Mars even as they revolutionize and advance terrestrial living standards as forcefully as nineteenth-century American invention changed Europe and ultimately the rest of the world as well.

Inventions produced as a matter of necessity by a frontier culture can make Mars rich, but invention and direct export to Earth are not the only ways that Martians will be able to make a fortune. The other route is via trade in support of mining operations in the asteroid belt, the band of small, mineral-rich bodies lying between the orbits of Mars and Jupiter.

To understand this, it is necessary to consider the energy relationships between the Earth, Moon, Mars, and the main asteroid belt. The asteroid belt enters into the picture here because it is known to contain vast supplies of very high grade metal ore in a low-gravity environment that makes it potentially easy to export to Earth.[36] For example, John

Lewis of the University of Arizona has considered the case of a run-of-the-mill asteroid just one kilometer in diameter. This asteroid would have a mass of 2 billion tonnes, of which 200 million tonnes would be iron, 30 million tonnes would be high-quality nickel, 1.5 million tonnes would be the strategic metal cobalt, and 7,500 tonnes would be a mixture of platinum group metals whose average value at current prices would be in the neighborhood of $20,000 per kilogram. That adds up to $150 billion for the platinum alone. There is little doubt about this, for we have lots of samples of asteroids in the form of meteorites. As a rule, meteoritic iron contains between 6 and 30 percent nickel, between 0.5 and 1 percent cobalt, and platinum group metal concentrations at least 10 times the best terrestrial ore. Furthermore, since the asteroids also contain a good deal of carbon and oxygen, all of these materials can be separated from the asteroid and from each other using variations of the carbon-monoxide–based chemistry we discussed in chapter 7 for refining metals on Mars.

There are about 5,000 asteroids known today, of which about 98 percent are in the Main Belt between Mars and Jupiter, with an average distance from the Sun of about 2.7 astronomical units, or AU. (The Earth is 1.0 AU from the Sun.) This Main Belt group includes all the known asteroids residing within the orbit of Jupiter with diameters greater than 10 kilometers, hundreds larger than 100 kilometers, and one as large as 914 kilometers. Except for a few tiny objects that travel closer to the Sun than the Earth and a handful that have been spotted beyond Jupiter, the remaining 2 percent, all small, have orbits lying between Earth and Mars. The 2 percent figure, however, greatly overstates the proportion of such near-Earth asteroids to Main Belters, because their relative closeness to the Earth and Sun makes them much easier to see. A reasonable estimate would be that the Main Belt asteroids outnumber the near-Earth group at least a thousand to one. Of the near Earth asteroids, about 90 percent orbit closer to Mars than to the Earth.

As should be clear from Lewis's example, these asteroids collectively represent enormous economic potential. While much has been made recently of the importance of the near-Earth group (especially due to gradual realization that if we don't develop some serious space-faring capabilities, one of them is likely to wipe out the human race when it impacts our planet someday), the relative numbers of the two classes make it clear that for mining purposes, the real action is going to be in the Main Belt.

Miners operating among the asteroids will be unable to produce most of their necessary supplies locally. There will thus be a need to import food and other necessary goods either from Earth or Mars. As shown in the table below, Mars has an overwhelming positional advantage as a location from which to conduct such trade. This advantage results from the fact that the rocket propulsion ΔVs required to reach the asteroid belt from Mars are much less than those from Earth, and as a result, the mass ratio (a spacecraft's fully fueled mass divided by its dry mass) required of spacecraft leaving Mars is also much less.

In Table 8.1, Ceres is chosen as a typical Main Belt asteroid destination, as it is the largest asteroid and positioned right in the heart of the Belt. You'll notice, however, that I also give the Earth's Moon as a potential port of call. Despite the fact that it is much closer to Earth physically, we can see that from a propulsion point of view, it is much easier to reach the Moon from Mars than it is from the Earth! That is,

TABLE 8.1
Transportation in the Inner Solar System

	Earth		Mars	
	ΔV (km/s)	*Mass Ratio*	*ΔV (km/s)*	*Mass Ratio*
Surface to low orbit	9.0	11.40	4.0	2.90
Surface to escape	12.0	25.60	5.5	4.40
Low orbit to lunar surface	6.0	5.10	5.4	4.30
Surface to lunar surface	15.0	57.60	9.4	12.50
Low orbit to Ceres	9.6	13.40	4.9	3.80
Surface to Ceres	18.6	152.50	8.9	11.10
Ceres to planet	4.8	3.70	2.7	2.10
NEP round-trip low orbit to Ceres	40.0	2.30	15.0	1.35
Chemical to low orbit, NEP round-trip to Ceres	9/40	26.20	4/15	3.90

the required mass ratio is only 12.5 going from Mars to the Moon, while it is 57.6 from Earth. This would be even more forcefully the case for travel from either Earth or Mars to nearly any near-Earth asteroid as well.

All the entries presented in Table 8.1 except the last two are based upon a transportation system using methane/oxygen (CH_4/O_2) engines with a specific impulse (Isp) of 380 seconds and ΔVs appropriate for trajectories employing high-thrust chemical propulsion systems. These were chosen because methane/oxygen is the highest performing space-storable chemical propellant, and can be manufactured easily on Earth, on Mars, or on a carbonaceous asteroid. Hydrogen/oxygen propellants, while offering a higher Isp (450 seconds), are not storable for long periods in space. Moreover, it is an unsuitable propellant for a cheap reusable space transportation system, since its costs exceed methane/oxygen propellant by more than an order of magnitude and its bulk makes it very difficult to transport to orbit in any quantity using reusable single-stage-to-orbit (SSTO) vehicles (thus ruling it out for true cheap surface-to-orbit systems). The last two entries in the table are based upon nuclear electric propulsion (NEP) using argon propellant, available at either the Earth or Mars, with an Isp of 5,000 seconds for in-space propulsion, with methane/oxygen used to reach low orbit from the planet's surface. Such SSTO and NEP systems, while somewhat futuristic today, represent a conservative baseline for interplanetary transportation technology in the period we are discussing.

It can be seen that if chemical systems are used exclusively, then the mass ratio required to deliver dry mass to the asteroid belt from Earth is 14 times greater than from Mars. This implies a still (much) greater ratio of payload to takeoff mass ratio from Mars to Ceres than from Earth. In fact, looking at Table 8.1, we can safely say that useful trade between Earth and Ceres (or any other body in the main asteroid belt) using chemical propulsion is probably impossible, while from Mars it is relatively easy. We can also see that there is nearly a fivefold advantage in mass ratio delivering cargoes to the Earth's Moon from Mars over doing it from Earth.

If nuclear electric propulsion is introduced, the story changes, but not much. Mars still has a sevenfold advantage in mass ratio over Earth as a port of departure for the main belt, which translates into a payload-to-takeoff-weight ratio nearly two orders of magnitude higher for Mars departure than for Earth.

But those are just mass ratios, and as noted above, this understates Mars' advantage. Now let's compare some end-to-end missions that go from either Earth or Mars to Ceres. This is shown in Table 8.2, with both all-chemical and combined chemical/NEP transportation systems considered. Both missions deliver 50 tonnes of cargo. In addition, both NEP and chemical systems have to carry fuel tanks whose mass I estimated at 7 percent of the propellant they carry. For surface-to-orbit vehicles, I employed methane/oxygen SSTO rockets, and assumed that the vehicles need to have a dry mass (for thermal protection, engines, landing gear, etc.) excluding tankage that is equal to their payload, or 50 tonnes. Chemical interplanetary transportation systems can be built flimsier, so I assigned them a dry inert mass excluding tankage equal to 20 percent of the payload. The NEP engines in Table 8.2 require 10 megawatts of electricity (MWe) of power for delivery to Ceres from Mars and 30 MWe for delivery from Earth, with each NEP system massing 5 tonnes/MWe. (This is a much lighter mass/power ratio than the 40 tonnes/MWe projected for the 100 kWe reactor to be used by the Mars Direct mission, but given the much larger size unit and more futuristic context, it may be considered reasonable.) The different power ratings give both systems about equal power/mass ratios. Nevertheless, the NEP system leaving Earth still has to burn its engine 2.4 times as long. If you wanted to increase the power rating of the Earth-based NEP vessel so that its burn time were the same as the Mars-based system, the mass of the Earth-based mission would go to infinity. In Table 8.2, the mass numbers are for the total mission. It is clear that the total launch requirement would probably be divided up into many launch vehicles.

You can see that the launch burden for sending the cargo to Ceres is about *50 times* less for missions starting from Mars than those departing from Earth, regardless of whether the technology employed is all chemical propulsion or chemical launch vehicles combined with nuclear electric propulsion for interplanetary transfer. If the launch vehicle used has a lift-off mass of 1,000 tonnes, it would require *107 launches* to assemble a methane/oxygen freighter mission launched from Earth, but only

TABLE 8.2
Mass of Freighter Missions to the Main Asteroid Belt (tonnes)

	Departure from Earth		Departure from Mars	
	CH_4/O_2	*Chem/NEP*	*CH_4/O_2*	*Chem/NEP*
Propulsion system				
Payload	50	50	50	50
Interplanetary spacecraft	10	150	10	50
Interplanetary tankage	85	19	15	3
Interplanetary propellant	1,220	268	205	37
Total mass in low orbit	1,365	487	280	140
Launch vehicle inert mass	1,365	337	280	90
Launch vehicle tankage	6,790	1,758	88	28
Launch vehicle propellant	97,000	25,127	1,250	401
Total ground lift-off mass	106,520	27,559	1,898	609

two launches for a departure from Mars. Even if propellant and other launch costs were 10 times greater on Mars than on Earth, it would still be enormously advantageous to launch from Mars. Furthermore, all this analysis assumes that the ships return from the asteroid belt without cargo. If the added burden of hauling enough propellant all the way out from Earth to return substantial amounts of asteroidal metal cargo without refueling at Mars is thrown into the mission requirement, then the Earth-based mission becomes even more hopeless.

The result that follows is simply this: Anything that needs to be sent to the asteroid belt that can be produced on Mars will be produced on Mars.

The outline of future interplanetary commerce thus becomes clear. There will be a "triangle trade," with Earth supplying high-technology manufactured goods to Mars, Mars supplying low-technology manufactured goods and food staples to the asteroid belt and possibly to the Moon as well, and the asteroids sending metals (and perhaps the Moon sending helium-3) back to Earth. This triangle trade is directly analogous to the triangle trade of Britain, her North American colonies, and the West Indies during the colonial period. Britain would send manufactured goods to North America, the American colonies would send food staples and needed craft products to the West Indies, and the West Indies would send cash crops such as sugar to Britain. A similar triangle trade involving Britain, Australia, and the Spice Islands also supported British trade in the East Indies during the nineteenth century.

POPULATING MARS

The difficulty of interplanetary travel may make Mars colonization seem visionary. However, colonization is, by definition, a one-way trip, and it is this fact that makes it possible to transport the large numbers of people that a colony in a new world needs to succeed.

Let us consider two models of how humans might emigrate to Mars: a government-sponsored model and a privately sponsored model.

If government sponsorship is available, the technological means required for immigration on a significant scale are essentially available today. In Figure 8.1, we see one version of such a concept that could be used to transport immigrants to Mars. A shuttle-derived heavy-lift launch vehicle lifts 145 tonnes (a Saturn V had about this same capacity) to low Earth orbit, then a nuclear thermal rocket (NTR, such as was demonstrated in the United States in the 1960s NERVA program) with an Isp of 900 seconds hurls a 70-tonne habcraft onto a seven-month trajectory to Mars. Arriving at Mars, the habcraft uses its conical aeroshell to aerobrake, and then parachutes and lands on its own sets of methane/oxygen engines.

The habcraft is 8 meters in diameter and includes four complete habitation decks, for a total living area of 200 square meters, allowing it to house 24 people adequately in space and on Mars. Expansion area is available in the fifth (uppermost) deck after the cargo it contains is unloaded upon arrival. Thus, in a single booster launch, 24 people, complete with their housing and tools, can be transported one way from Earth to Mars.

Now, let us assume for purposes of example that starting in 2030, an average of four such boosters are launched every year from Earth. If we then make various reasonable demographic assumptions, the pop-

FIGURE 8.1
An NTR augmented heavy-lift launch vehicle, capable of transporting 24 colonists one-way to the Red Planet.

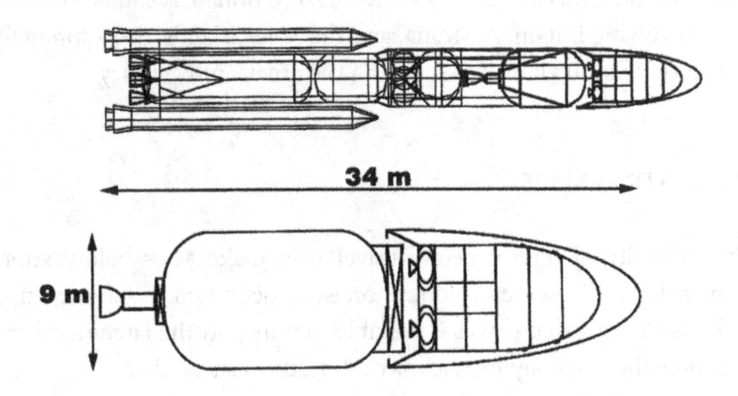

ulation curve for Mars can be computed. The results are shown in Figure 8.2. Examining the graph, we see that with this level of effort (and the technology frozen at early twenty-first-century levels), the rate of human population growth of Mars in the coming century would be about one-fifth that experienced by colonial America in the seventeenth and eighteenth centuries.

This in itself is a very significant result. What it means is that the distance to Mars and the transportation challenge that it implies is not a major obstacle to the initiation of a human civilization on the Red Planet. Rather, the key questions, as discussed in chapter 7, are those of resource utilization, growing food, building housing, and manufacturing all sorts of useful goods on the surface of Mars. Moreover the projected population growth rate, one-fifth that of colonial America, while a bit slow, is significant on a historical scale, and assuming a cost of $1 billion per launch, the $4 billion-per-year program cost could be sustained for some time by any major power on Earth that cared to plant the seeds of its posterity on Mars.

However, with a cost per launch of about $1 billion, the cost per immigrant would be $40 million. Such a price might be affordable to governments (for a time), but not to individuals or private groups. If Mars is ever to benefit from the dynamic energy of large numbers of immigrants motivated by personal choice seeking to make their mark

FIGURE 8.2
Colonization of Mars compared to North America. Analysis assumes 100 immigrants per year starting in 2030, increasing at a 2 percent annual rate, 50/50 male/female. All immigrants are between ages 20 and 40. Average of 3.5 children to an ideal Martian family. Mortality rates are 0.1 percent per year between ages 0 and 59, 1 percent between ages 60 and 79, 10 percent per year for those over 80.

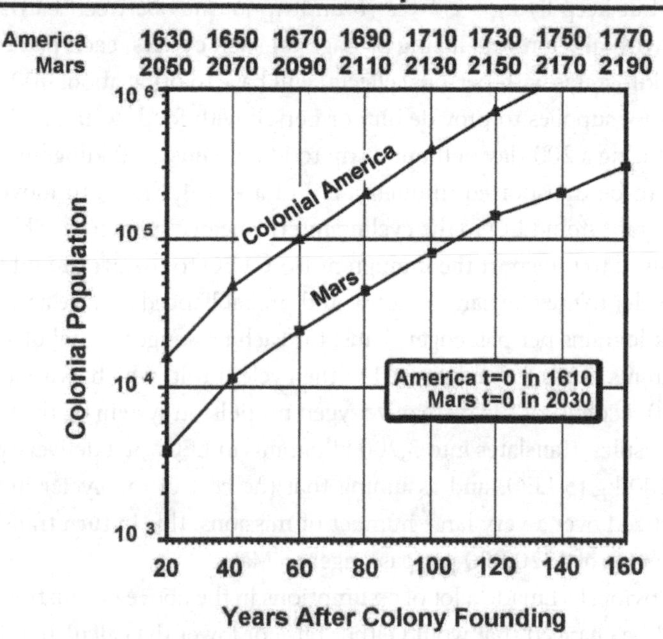

in a new world, the transportation fee will have to drop a lot lower than this. Let us therefore examine an alternative model to see how low it is likely to drop.

Consider once again our methane/oxygen SSTO vehicles used to transport payloads from the surface of the Earth to low-Earth orbit (LEO). For every kilogram of payload delivered to orbit, about 70 kilograms of propellant are required. Methane/oxygen bipropellant costs about $0.20/kilogram, so $14 of propellant costs will be incurred for every kilogram lifted to orbit. If we then assume total system operation cost is seven times propellant costs (roughly double the total cost/fuel cost ratio of airlines), then the cost of delivery to LEO could be about

$100 per kilogram. Let's assume that there is operating between Earth and Mars a spacecraft in a permanent cycling orbit which has the ability to recycle water and oxygen with 95 percent efficiency. Such interplanetary "cyclers," proposed by Apollo 11 pilot Buzz Aldrin as the basis of a permanent Earth-Mars transportation system, could efficiently provide ample quarters for many immigrants, as they only have to be launched once but keep flying a 2.2-year roundtrip journey between Earth and Mars virtually forever. Taking passage on such cyclers, each passenger (100 kilograms with personal effects) will have to bring about 400 kilograms of supplies to provide him or herself with food, water, and oxygen during a 200-day outbound trip to Mars. Thus, 500 kilograms will need to be transported through a ΔV of about 4.3 km/s to move the immigrant from LEO to the cycling interplanetary spacecraft. The capsule used to transport the immigrant from LEO to the cycler and from the cycler to the Martian surface would probably need to have a mass of 500 kilograms per passenger. Thus, for each passenger a total of 1,000 kilograms needs to be delivered to the cycler orbit, which, with an Isp of 380 seconds for the methane/oxygen propulsion system on the transfer capsules, translates into 3,200 kilograms in LEO. At a delivery price of $100/kg to LEO, and assuming that the cost of the cycler itself is amortized over a very large number of missions, this in turn translates into a cost of $320,000 per passenger to Mars.

Obviously, I made a lot of assumptions in the above calculation that could be changed that would either raise or lower the calculated ticket price significantly. For example, use of air-breathing supersonic ramjet ("scramjet") propulsion to perform a significant part of the Earth-to-orbit ΔV could cut orbit delivery costs by as much as a factor of three. An electric propulsion ferry could be used to lift the transfer capsule nearly to Earth escape, after which the capsule would be dropped to execute a powered flyby close to the Earth using high-thrust chemical stage. This would allow it to kick out of orbit and reach the cycler with a chemical ΔV of only 1.3 km/s, thereby doubling payload and reducing costs yet again. If the cycler employs a magnetic sail (see focus section) instead of simply using natural interplanetary orbits (with gravity assists), the hyperbolic velocity departing Earth required for cycler rendezvous can be essentially zero, thereby allowing the entire

TABLE 8.3
Possible Cost Reductions of Earth-to-Mars Transportation System

	Baseline	Advanced	Reduction Factor	Fare to Mars (1996 dollars)
Baseline mission	—	—	1.0	$320,000
Earth to orbit	Rockets	Scramjets	0.3	$ 96,000
Life-support closure	95%	99%	0.7	$ 67,000
LEO escape propulsion	CH_4/O_2	NEP	0.6	$ 40,000
Cycler propulsion	Natural	Magsail	0.7	$ 28,000

LEO-to-cycler delivery to be done by electric propulsion, or conceivably even solar or magnetic sails. Increasing the degree of closure of the life-support system on the cycler from our baseline of 95 percent of water and oxygen recycled to 99 percent would reduce the consumable delivery requirement for each passenger, thereby reducing passage costs still more. Thus, eventually Earth-to-Mars transportation costs could be expected to drop another order of magnitude, to $30,000 per passenger or so. The cost impacts as each of these innovations is progressively introduced are displayed in Table 8.3.

Nevertheless, the $320,000 fare cited for early immigrants is interesting. It's not a sum of money that anyone would spend lightly, but it is a sum of money—roughly the cost of an upper-middle-class house in many parts of suburban America, or put another way, roughly the life's savings of a successful middle-class family—that a large number of people could finance if they really wanted to do so. Why would they want to do so? Simply this: because of the small size of the Martian population and the large transport cost itself, it is certain that the cost of labor on Mars will be much greater than on Earth. Therefore, wages may be much higher on Mars than on Earth. While $320,000 might be six years' salary to an engineer on Earth, it would likely represent only one or two years' salary on Mars. This wage differential, precisely analogous to the wage differential between Europe and America during most of the past four centuries, can make emigration to Mars both desirable and possible for the individual. From the seventeenth through nineteenth centuries, many European families followed a classic pattern of pooling their resources to enable one family member to emigrate to America. That emigrant, in turn, would proceed to earn enough money to bring the rest of the family over. Today, the same method of obtaining passage is used by Third World immigrants whose salaries in their native lands are dwarfed by current airfares. Because the necessary income will be there to pay for the trip after it has been made, loans can even be taken out to finance the journey. It's been done in the past, and can be done in the future.

As mentioned before, the labor shortage that will prevail on Mars will drive Martian civilization toward both technological and social advances. If you're paying five times the terrestrial wage rate, you're not going to want to waste any of your worker's time with stoop labor or filling out forms, and you will not seek to exclude someone who can perform some desperately needed profession from doing so just because he or she has not taken the trouble to run some institutional obstacle course. In short, Martian civilization will be practical because

FIGURE 8.3
Over time, the Mars base will grow into a true settlement, the beginning of a new branch of human civilization. (Artwork by Robert Murray, Mars Society)

it will have to be, just as nineteenth-century American civilization was. This forced pragmatism will give Mars an enormous advantage in competing with the less-stressed and therefore more tradition-bound society remaining behind on Earth. If necessity is the mother of invention, Mars will provide the cradle. A frontier society based on technological excellence and pragmatism, and populated by people self-selected for personal drive, will perforce be a hotbed of invention, and these inventions will not only serve the needs of Mars but of the terrestrial population as well. Therefore, they will bring income to Mars (via terrestrial licensing) while at the same time they disrupt the labor-rich terrestrial society's inherent tendency toward stagnation. This process of rejuvenation, as we'll discuss in later chapters, will ultimately be the greatest benefit that the colonization of Mars will offer Earth. And it will be those terrestrial societies who have the closest social, cultural, linguistic, and economic links with the Martians who will benefit the most.

SELLING MARTIAN REAL ESTATE

Martian real estate can be broken down into two categories: habitable and open. By habitable real estate I mean that which is under a dome, allowing human settlers to live there in a relatively conventional shirt-sleeve environment. Open real estate, on the other hand, is that which lies outside the domes. It is obvious that habitable real estate is far more valuable than open. Nevertheless, both of these can be bought and sold, and as transportation costs drop, both forms of Martian real estate will rise in value.

The only kind of land that exists on Mars right now is open. There is an immense amount of it—144 million square kilometers—but it might seem that it is all completely worthless because it cannot currently be exploited. Not so. Enormous tracts of land were bought and sold in Kentucky for large sums of money a hundred years before settlers arrived. For purposes of development, trans-Appalachian America in the 1600s might as well have been Mars. Two things, though, made these distant lands valuable and salable. For one, at least a few people believed that the land would be exploitable someday, and a juridical arrangement existed in the form of British Crown land patents that allowed trans-Appalachian land to be privately owned. In fact, if a mechanism were put in place that could enforce private property rights on Mars, land there could probably be bought and sold now. Such a mechanism would not need to employ enforcers on the surface of Mars (no space patrol required). Instead, the patent or property registry of a sufficiently powerful nation, such as the United States, would be entirely adequate. For example, if the United States chose to grant a mining patent to any private group that surveyed a piece of Martian real estate to some specified degree of fidelity, such claims would be tradable today on the basis of their future speculative worth (and could probably be used to privately finance robotic mining survey probes in the near future). Furthermore, such claims would be enforceable internationally and throughout the solar system simply by having the U.S. Customs Office penalize with a punitive tariff any U.S. import made anywhere, directly or indirectly, with material that was extracted in defiance of the claim. This sort of mechanism would not necessarily imply U.S. sovereignty over Mars, any more then the current U.S. Patent and Copyright Office's coining of ideas into intellectual property implies U.S. government sovereignty over the universe of ideas. But, whether it's the United States, NATO, the United Nations, or the Martian Republic, some government's agreement is needed to give worthless terrain real estate property value.

Once that is in place, however, even the undeveloped open real estate on Mars represents a tremendous source of capital to finance the initial development of Martian settlements. Sold at an average value of $20 per acre, Mars could be worth $700 billion. Should Mars be terraformed, these open land prices could be expected to grow a hundredfold, with a rough planetary land value of $70 trillion implied. Assuming, as appears to be the case, that a method of terraforming Mars could be found with a total cost much less than this, those who own Mars would have every reason to seek to develop their property via planetary engineering.

Of course, all open real estate on Mars will not be of equal value. Those sections known to contain valuable minerals, water, geothermal power potential, or other resources, or which are located closer to the habitable areas, will be worth much more. For these reasons, as with land speculators on Earth in the past, the owners of open unexplored real estate on Mars will exercise all their influence to further the exploration of, and encourage the settlement of, land under their control.

Far more valuable than the open real estate will be habitable real estate beneath the domes. Each 100-meter–diameter dome, massing about 80 tonnes, would enclose an area of about two acres. Assuming that dwelling units for 20 families are erected within, and each family is willing to pay $50,000 for their habitation land (a plot 20 meters on a side), then the total real estate value enclosed by a single dome would be $1 million. At this rate, the creation of habitable land by the mass production and erection of large numbers of domes to house the waves of immigrants should prove to be one of the biggest businesses on Mars and a major source of income for the colony.

In the twenty-first century, Earth's population growth will make real estate here ever more expensive, making it harder for people to own their own homes. At the same time, the ongoing bureaucratization of daily life will make it ever harder for strong spirits to find adequate means for expressing their creative drive and initiative on Earth. Regulation to protect what is will become ever more burdensome to those who would create what is not. A confined world will limit opportunity for all and seek to enforce behavioral and cultural norms that will be unacceptable to many. For example, consider the offense to human nature imposed by China's current "one child per family" policy. When the frictions caused by such oppression turn into inevitable revolts and wars, there will be losers. Looking around the world today, it is not difficult to pick out dozens of small nations in Asia, Africa, the Middle East, the former Soviet Union, and Europe that adjoin larger nations that now or in the past have demonstrated the desire to conquer their neighbors. Again, there will be wars, and losers, and millions of emigrants willing to take on the hard challenges of making a new life on the frontier rather than accept subjugation. A planet of refuge will be needed, and Mars will be there.

HISTORICAL ANALOGIES

The primary analogy I wish to draw is that Mars is to the new age of exploration what North America was to the last. The Earth's Moon, close to the metropolitan planet but impoverished in resources, compares to Greenland. Other destinations, such as the Main Belt asteroids, may be rich in potential future exports to Earth, but lack the preconditions for the creation of a fully developed indigenous society—these compare to the West Indies. Only Mars has the full set of resources required to develop a native civilization, and only Mars is a viable target for true colonization. Like America in its relationship to Britain and the West Indies, Mars has a positional advantage with respect to the asteroids that will allow it to participate in a useful way to support extractive activities on behalf of Earth. But despite the shortsighted calculations of eighteenth-century European statesmen and financiers, the true value of America never was as a logistical support base for the West Indies' sugar-and-spice trade, or inland fur trade, or as a potential market for manufactured goods. The true value of America was as the future home for a new branch of human civilization, one that combined its humanistic antecedents and its frontier conditions to develop into the most powerful engine for human progress and economic growth the world has ever seen. The wealth of America was in the fact that she could support people, and that the right kind of people chose to go to her. Every feature of frontier American life that acted to create a practical can-do culture of innovative people will apply to Mars one hundredfold.

Mars is a harsher place than any on Earth. But provided one can survive the regimen, it is the toughest schools that are the best. The Martians shall do well.

FOCUS SECTION—ADVANCED INTERPLANETARY TRANSPORTATION

Destination drives transportation. Just as the opening of the New World drove a revolution in European naval architecture, so will the establishment of a Mars base summon new types of space propulsion systems that will make the colonization of Mars commercially feasible. These new systems, vastly more capable than anything we have today, have been on the drawing boards for some time, waiting for the spur of necessity to call them into being. Let's look and see what the future might hold.

AIR-BREATHING LAUNCH SYSTEMS

Current rocket-based launch systems are only about 2 percent as efficient in hauling cargo as jet aircraft. The reason for this difference is simple—rockets haul their own oxidizer while jets get theirs from the air. Since the oxidizer makes up about 75 percent of the total propellant weight, this enormously compromises a rocket vehicle's performance. Launch vehicles attempting to reach orbit are flying through an ocean of oxidizer. Why don't they try to use any of it?

Unfortunately, technical difficulties and lack of will have intersected to stall the development of hypersonic air-breathing propulsion. Current ramjet engines used on some missiles can make it to Mach 5.5, but beyond this speed it becomes impossible to slow the air that enters the jet engine to subsonic speeds without heating the air too much in the process. Thus, the combustion inside the engine must take place in a supersonic flow. An engine that can do this is a new type of animal, a "scramjet," and is in a sense as much of an advance over existing jet engines as jets were over propellers. The National Aerospace Plane (NASP) program—canceled in 1993 due to lack of perceived necessity—conducted extensive computer calculations showing that scramjets will work. A somewhat less technologically challenging approach that can obtain much of the scramjet's benefits is the air-augmented rocket: a rocket that obtains part of its needed oxidizer from the atmosphere during its upward flight. Air-augmented rockets that could get a specific impulse over 1,000 seconds were demonstrated on the test stand at The Marquardt Company in 1966. Unfortunately, a change in governmental bureaucratic whims canceled the program before the engines could be flight tested.

The use of scramjets or air-augmented rockets on even part of the launch trajectory of a single-stage-to-orbit (SSTO) vehicle would greatly increase its payload. This is exactly what is needed to meet the logistics demands of a developing Mars settlement, which will call for the cheap delivery of large amounts of cargo to orbit, and beyond. The colonization of Mars is thus central to the development of the technologies that will give us cheap access to space.

ELECTRIC PROPULSION

The key metric of a rocket's performance is its specific impulse, the number of seconds it can use a pound of propellant to make a pound of thrust. The best chemical rockets available today have a specific impulse of about 450 seconds, while nuclear thermal rockets can get about 900 seconds.

FIGURE 8.4
Nuclear Electric Propulsion spacecraft will require very large reactor systems. Current hype proclaiming them the key to quick trips to Mars is false, as they accelerate very slowly. However, because they are very efficient in the use of propellant, they could someday be used to significantly reduce the cost of transporting cargo to Mars. (Artwork courtesy NASA)

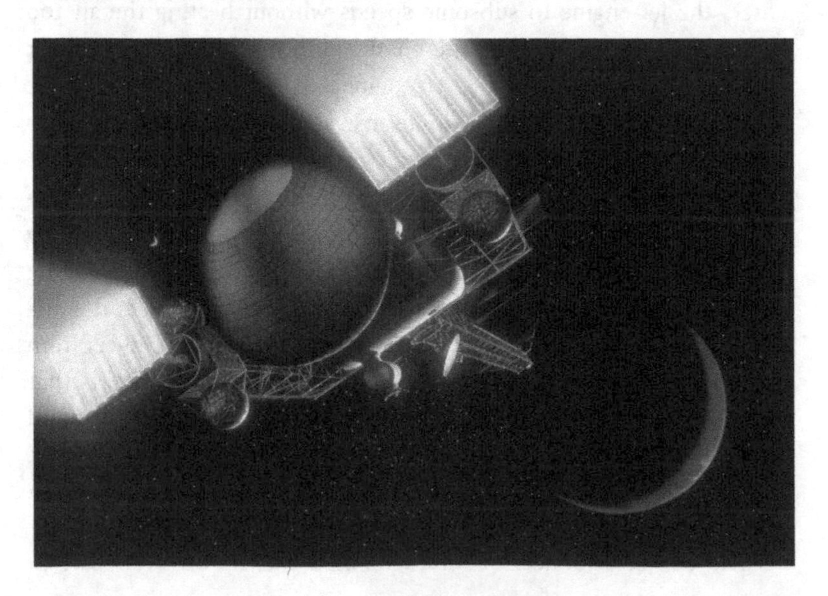

But there is another way to make a high specific impulse. That is to ionize a gas by stripping some of the electrons off of its atoms and then accelerate it by means of the attractive and repulsive forces of an electrostatic grid. This technique is known as electric propulsion, or "ion drive." A related concept converts the gas to a plasma, which is then ejected from a magnetic nozzle to produce thrust. Either way, using electric propulsion you can generate specific impulses of many thousands of seconds, without ever heating the exhaust gas to very high temperatures. This is not just a theory, but a fact—ion drives are used for station-keeping propulsion on many satellites today. But you need a lot of electric power if you are going to create much thrust. For example, a 120-tonne spacecraft would need a power of 5 megawatts (about 70 times that planned for the international space station) to generate a thrust of 280 newtons (about 60 lbs) with a specific impulse of 5,000 seconds. Assuming it had this much power, however, it could generate the 30 km/s ΔV it would need to travel from low-Earth orbit to Mars and back in about one year of continuous thrusting. Such a nuclear electric propulsion (NEP) spaceship could accomplish this incredible ΔV with a mass ratio of only about 1.82. The trajectories that electric propulsion vehicles must follow generally require rather more ΔV (typically a factor of two) to get from one point in the solar system to another than chemical propulsion systems do, but since the Isp is about 10 times higher, you can still come out way ahead, *provided that you don't let the mass of the nuclear electric propulsion system itself become excessive.*

Electric ion thrusters already exist in kilowatt-sized units, and bulking them up to the several megawatt sizes needed for NEP transportation systems offers no fundamental challenges. The real problem in enabling NEP propulsion systems to date has been to obtain the funds and sustained commitment required to develop a multi-megawatt space nuclear reactor required to power them. In this context it should be noted that claims by some ardent electric propulsion advocates, such as the VASIMR group led by former astronaut Franklin Chang-Diaz, that their plasma drive technology could enable quick (~40-day) trips to Mars if only they were provided with 200 megawatt nuclear reactors are risible. Even assuming our optimistic forecast of an eightfold decrease in the mass/power ratio for late twenty-first century space reactor systems compared to those projected based on current technology (from today's' 40 tonnes per MWe to a future performance of 5 tonnes per MWe), a 200 MWe reactor would have a mass of 1,000 tonnes, and outweigh its payload by an order of magnitude. But since the reactor needs to push not only its relatively small payload, but itself as well, no matter how big it is it would never be able to accelerate the spacecraft at the rate required to achieve a fast transit. Thus, the VASIMR group's claims of possessing a breakthrough fast propulsion system are unfounded, and rather unfortunate, as they have been embraced by those opposed to sending humans to Mars in our time as a reason to wait until such fantastical space drives become available.

However, provided that the illusory goal of using electric propulsion to achieve a fast transit is discarded, the size of the nuclear reactor system can be kept small relative to the cargo, allowing the technology to potentially play a useful role in reducing the launch-mass, and thus cost, required to conduct future interplanetary commerce.

SOLAR SAILS

Ships and sails proper for heavenly breezes should be fashioned . . .

—Johannes Kepler, 1609

Nearly four hundred years ago our old friend Kepler observed that regardless of whether a comet is moving toward or away from the Sun, its tail always points away from the Sun. This caused him to guess that light emanating from the Sun exerts a force that pushes the comet's tail away. He was right, although the fact that light exerts force had to wait till 1901 to be proven.

Well, if sunlight can push comet tails around, why can't we use it to move spaceships around? Why can't we just deploy big mirrors on our spacecraft, solar sails if you will, and have sunlight push on them to create propulsive force? The answer is that we can, but it takes an awful lot of sunlight to exert any significant amount of push. For example, at 1 AU, the Earth's distance from the Sun, a solar sail the size of a square kilometer would receive a total force of 10 newtons, about 2.2 pounds, pushing on it from the Sun. So, in order to make a solar sail into a practical propulsion system, you need to make it out of very thin material and cover a huge area. Let's say we made a 1-square-kilometer sail 0.01 mm or 10 microns thick, about one fourth the thickness of a kitchen trash bag. In that case, the sail would weigh 10 tonnes, and it could accelerate itself 32 km/s in just about a year. Of course, if the sail were hauling a payload equal to its own weight, that would slow it down by a factor of two. Still, a 10-micron-thick solar sail would be in the ballpark for an effective propulsion device supporting Earth-Mars transportation. And if a 1-micron sail could be made, then we'd really be flying . . .

No one has ever flown a solar-sail–propelled mission, but there was a very serious study done at NASA's Jet Propulsion Laboratory in the 1970s to use one to drive a probe to Halley's comet during its 1986 appearance. Unfortunately, the proposal went down the drain when Congress failed to fund the mission. Amateur groups, such as Robert Staehle's World Space Foundation and the French Union pour la Promotion de la Propulsion Photonique, have built solar sails. They had hoped to fly a solar sail regatta race to the Moon during the Columbus anniversary year 1992, but failed to hitchhike rides on launch vehicles that would allow them to get their craft into space.

Solar sails do have some real technical problems that bear on packing, unpacking, deploying without damage, and controlling huge space structures made of very thin materials. Still, it has to be said that the main impediment to the demonstration of solar sails has not been the technical obstacles, but the refusal of the world's space agencies to allocate any significant funds to the their development and testing. Let's hope that the Martians will do better.

MAGNETIC SAILS

Sunlight is not the only forceful breeze that emanates from the Sun. There is another, known as the solar wind.

The solar wind is a flood of plasma, protons, and electrons, that streams out constantly from the Sun in all directions at a velocity of about 500 km/s. We never encounter it here on Earth, because we are protected from it by the Earth's magnetosphere.

If the Earth's magnetosphere blocks the solar wind, it must be creating drag, and therefore feel a force as a result. Why not create an artificial magnetosphere on a spacecraft and use the same effect for propulsion? This was an idea that Boeing engineer Dana Andrews and I hit on in 1988. The idea was timely. In 1987, high-temperature superconductors had been discovered. These are essential to making a magnetic propulsion device practical, as low-temperature superconductors require too much heavy cooling equipment and ordinary conductors require far too much power. The amount of force per square kilometer of solar wind is much less even than that created by sunlight, but the area blocked off by a magnetic field could be made much larger than any practical solid solar sail. Working in collaboration, Dana and I derived equations and ran computer simulations of the solar wind impacting a spacecraft generating a large magnetic field. Our results: If practical high-temperature superconducting cable can be made that can conduct electrical current with the same density as the state-of-the-art low-temperature superconductors such as niobium titanium (NbTi)—about 1 million amps per square centimeter—then magnetic sails or "magsails" can be made that will have thrust-to-weight ratios *a hundred times better* than that of a 10-micron-thick solar sail.[45] Furthermore, unlike an ultrathin solar sail, the magsail would not be difficult to deploy. Instead of being made of thin plastic film, it would be made of rugged cable, which due to magnetic forces would automatically "inflate" itself into a stiff hoop shape as soon as electrical current was put in it. It would take power to get current flowing through the cable, but because superconducting wire has no electrical resistance, once the current was in the cable, no further power would be needed to keep it going. In addition, the magsail would shield the ship completely against solar flares.

A magsail can exert enough force in the direction away from the Sun to completely negate the Sun's gravitational attraction, or it can have its current turned down so as to negate whatever portion of the Sun's gravity is desired. Without going into details here, this capability would allow a ship co-orbiting the Sun with the Earth to shift itself into orbits that take it to any planet in the solar system, just by turning the magsail power up and down. And it all can be done without an ounce of propellant.

Magsails are not practical now, because the high-temperature superconducting cable they need does not exist. However, research in this field is proceeding fast. I think it's a pretty good bet that ten or twenty years from now the type of cable required to make an excellent magsail will be widely available.

FUSION

Thermonuclear fusion reactors work by using magnetic fields to confine a plasma consisting of certain species of ultrahot charged particles within a vacuum chamber where they can collide and react. Since high-energy particles have the ability gradually to fight their way out of the magnetic trap, the reactor chamber must be of a certain minimum size to stall the particles' escape long enough for a reaction to occur. This minimum size requirement tends to make fusion power plants unattractive for low-power applications, but in the world of the future where human energy needs will be on a scale tens or hundreds of times greater than today, fusion will be far and away the cheapest game in town.

In addition to providing the power base needed for continued societal growth, fusion reactors can provide a very advanced spacecraft propulsion system, especially since in space the vacuum the reaction requires can be had for free in any size desired. The deuterium/helium-3 (D/He3) reaction provides the best performance, because the fuel has the highest energy-to-mass ratio of any substance found in nature, but the much cheaper pure deuterium fueled reaction (D-D) is about 60 percent as good. A rocket engine based upon controlled fusion could work simply by allowing the plasma to leak out of one end of a magnetic trap, adding ordinary hydrogen to the leaked plasma, and then directing the exhaust mixture away from the ship with a magnetic nozzle. The more hydrogen added, the higher the thrust, but the lower the exhaust velocity. For travel to Mars or the outer solar system, the exhaust would be about 99 percent ordinary hydrogen, and the exhaust velocity would be over 100 km/s (10,000 seconds Isp). If no hydrogen is added, a fusion configuration could theoretically yield exhaust velocities as high as 18,000 km/s (1.8 million seconds Isp), or 6 percent the speed of light using deuterium/helium-3, or 4 percent the speed of light using pure deuterium! Although the thrust level of such pure D/He3 or D-D rockets would be too low for in-solar-system travel, the terrific exhaust velocity would make possible voyages to *nearby stars* with trip times of less than a century. Such a fusion-powered starship would only need to burn fuel to accelerate, since stopping could be accomplished by deploying a magsail to create drag with the interstellar plasma.

Fusion propulsion could ultimately make travel to Mars possible on a time-scale of weeks instead of months, travel to Jupiter and Saturn possible in months instead of years, and travel to other solar systems on time scales of decades instead of millennia. It may be that fusion spacecraft propulsion will evolve as an outgrowth of terrestrial power plants, but the reverse is at least equally likely. Recall that the first really efficient steam engines were built to power steamships, and the first practical nuclear power plants were on nuclear submarines. There is a reason why this happened. Mobile systems constantly demand higher technology, whereas static systems do not. To a consumer, a kilowatt is a kilowatt, whether produced by thermonuclear fusion or burning coal. But a fusion-powered spacecraft offers totally new and dramatically superior possibilities over any lower technology. Thus, the most forceful initial driver for the introduction of fusion may well be space propulsion, servicing the demands of business people engaged in Earth-Mars trade for increasingly fast means of transportation.

Currently, the world's fusion research programs are proceeding at a snail's pace, devastated by budget cuts from shortsighted politicians who have neither the capacity nor the inclination to address future necessities.

By forcing us to tackle the problems of fusion technology development, the growth of Martian civilization may well provide the basis for the survival of technological society.

9: TERRAFORMING MARS

God made the world, but the Dutch made Holland.

—Traditional saying in the Netherlands

Thus far in this book we have discussed the prospects for relatively near-term exploration and settlement of Mars. Now we'll address the ultimate challenge that the Red Planet presents to humanity—terraforming.[46,47] Can we transform Mars to make it fully habitable?

On the surface the idea appears to be utterly fantastical, simply science fiction. Yet, not so long ago, the subject of human voyages to the Moon was the domain of science fiction. Today, lunar expeditions are a subject for *historians*, and manned Mars exploration the province of working engineers. Most people may *believe* that the prospect of drastically changing the Red Planet's temperature and atmosphere to create more Earthlike conditions—"terraforming" Mars—is either sheer fantasy or at best a technological challenge for the far distant future. However, unlike some other extreme engineering concepts—faster-than-light travel or nanotechnology, say—terraforming has some history to it, about four billion years' worth.

The history of life on Earth is one of terraforming—that's why our beautiful blue planet exists as it does. When the Earth was born, it had no oxygen in its atmosphere, only carbon dioxide and nitrogen, and the land was composed of barren rock. It was fortunate that the Sun was only about 70 percent as bright then as it is now, because if the present-day Sun had shined down on that Earth, the thick layer of carbon dioxide in the atmosphere would have provided enough of a greenhouse effect to turn the planet into a boiling Venus-like hell. Fortunately, however, photosynthetic organisms evolved that transformed the carbon dioxide in Earth's atmosphere into oxygen, in the process completely changing the surface chemistry of the planet. As a result of this activity, not only was a runaway greenhouse effect on Earth avoided, but the evolution of aerobic organisms, those using oxygen-based respiration, was begun. These animals and plants then proceeded to modify the Earth still more, colonizing the land, creating soil, and drastically modifying global climate. Life is selfish, so it's not surprising that all of the modifications that life has made to the Earth have contributed to enhancing its prospects, expanding the biosphere, and accelerating its rate of development for new capabilities to improve the Earth as a home for life still more.

Humans are only the most recent practitioners of this art. Starting with our earliest civilizations, we used irrigation, crop seedings, weeding, domestication of animals, and protection of their herds to enhance the activity of those parts of the Earth most efficient in supporting human life. In so doing, we have expanded the biospheric basis for human population, which has expanded our numbers and thereby our power to change our environment to support a continued cycle of exponential growth. As a result, we have literally remade the Earth into a place that can support billions of people, a substantial fraction of whom have been sufficiently liberated from the need to toil for daily survival—such that some can now look out into the night sky for new worlds to conquer.

Some people consider the idea of terraforming Mars heretical—humanity playing God. Yet others would see in such an accomplishment the most profound vindication of the divine nature of the human spirit, exercised in its highest form to bring a dead world to life. My own sympathies are with the latter group. Indeed, I would go farther.

I would say that failure to terraform Mars constitutes failure to live up to our human nature and a betrayal of our responsibility as members of the community of life itself. Today, the living biosphere has the potential to expand its reach to encompass a whole new world. Humans, with their intelligence and technology, are the unique means that the biosphere has evolved to allow it to make that land grab, the first among many. Countless beings have lived and died to transform the Earth into a place that could create and allow human existence. Now it's our turn to do our part.

So let's pose the question once again: Can we *transform* Mars to make it fully habitable?

Let's consider the matter. Despite the fact that Mars today is a cold, dry, and possibly lifeless planet, it has all the elements required to support life: water, carbon, and oxygen (as carbon dioxide), and nitrogen. The physical aspects of Mars, its gravity, rotation rate, axial tilt, and distance from the Sun, are close enough to those of Earth to be acceptable. Mars does want in one area, though: It does not have much of an atmosphere.

Earth's atmospheric pressure at sea level is 14.7 pounds per square inch, or approximately 1 bar. (A "bar" is simply a unit for measuring pressure. "Bar" and "millibar," one-thousandth of a bar, are commonly used in meteorology and will be used in our discussion of terraforming.) Mars' current carbon dioxide atmosphere is less than 1 percent that of Earth's at sea level, varying between 6 and 10 millibars (mbar or mb). However, we know that Mars' atmosphere was once much thicker than it is now. The channels that snake along the Martian surface provide evidence that liquid water once coursed over the planet, and liquid water can exist only within a certain temperature and pressure range. At sea level on Earth, the temperature range is between 0° centigrade—the freezing point of water—and 100° centigrade—the boiling point of water. For water to run across the surface of Mars, the atmospheric pressure and temperature had to be greater than they are today.

Though Mars' atmosphere currently is rather thin, most researchers believe that there are enough reserves of carbon dioxide on the planet to substantially thicken it. Some of this carbon dioxide exists in frozen form as dry ice and makes up the south polar cap. Additional reserves are trapped within its regolith, the loose surface material that overlays the planet. ("Regolith" is an astrogeologist's term for dirt, and applies to any planetary body. "Soil" refers to Earth's regolith.) Releasing all of this carbon dioxide would greatly thicken the atmosphere, possibly to the point where its pressure would be about 30 percent that of Earth, or 300 mbar (almost one-third of a bar). Heating the planet will cause these vast reservoirs of trapped carbon dioxide to emerge. This is not just theory: We know for a fact that Mars' temperature and atmospheric pressure vary as the planet cycles between its nearest and furthest positions from the Sun over the course of a Martian year. In fact, as Mars warms and cools over the course of a year, its atmospheric pressure varies by plus or minus 20 percent compared to the average value on a seasonal basis.

We cannot, of course, move Mars to a warmer orbit. But we do know another way to heat a planet, one that we apparently have unwittingly practiced on Earth for perhaps the past century. That is to release or produce gases that will trap infrared radiation—the Sun's heat—and thereby warm the planet. On Earth it's called the "greenhouse effect" and has resulted from the introduction of carbon dioxide released from fossil fuel burning, as well as other greenhouse gases produced by industry. Call it terraforming or call it greenhousing, but the same can happen on Mars. An atmospheric greenhouse could be created on Mars in at least three different ways: by warming selected areas of the planet to release large reservoirs of the native greenhouse gas, carbon dioxide; by establishing factories on Mars to produce very powerful artificial greenhouse gases such as halocarbons ("CFCs"); or by releasing bacteria that could produce natural greenhouse gases more powerful than carbon dioxide (but much less so than halocarbons) such as ammonia or methane, once acceptable living conditions for bacteria were produced on Mars by one of the other methods.

While the concept of terraforming Mars may seem fantastic, the concepts supporting the notion are straightforward. Chief among them is that of *positive feedback*, a phenomenon that occurs when the output of a system enhances what is input to the system. For a Mars greenhouse effect, we find a positive feedback system in the relationship between atmospheric pressure—its thickness—and atmospheric temperature. Heating Mars will release carbon dioxide from the polar caps and from Martian regolith. The liberated carbon dioxide thickens the atmosphere and boosts its ability to trap heat. Trapping heat increases the surface temperature and, therefore, the amount of carbon dioxide that can be liberated from the ice caps and Martian regolith. And that is the key to terraforming Mars—the warmer it gets, the thicker the atmosphere becomes; and the thicker the atmosphere becomes, the warmer it gets.

In the sections below, we'll see how this system can be modeled, and present the results of calculations using such a model. These results give very strong support to the belief that humans will be able to effect radical improvements in the habitability of the Martian environment in the course of the twenty-first century We can, indeed, transform Mars.

TERRAFORMING CALCULATIONS

As I've noted, Mars is awash with carbon dioxide, a premier greenhouse gas, but much of that is trapped at the poles in frozen form, or locked in the planet's regolith. Both sources of carbon dioxide will help greenhouse Mars, but it's the frozen carbon dioxide at the poles that will initiate the process.

In computational studies, Chris McKay and I utilized Mars climate models to reveal that a small but sustained change in temperature at the Martian south pole—just 4°C—can initiate a runaway greenhouse effect in the polar region that will result in the evaporation of the polar cap. (For those wanting to delve into the fine points, I've included a technical note at the end of the chapter that details the model we used as a basis for this discussion of terraforming.) As the cap evaporates, global atmospheric temperature and pressure will rise, and that, in turn, will initiate the liberation of huge quantities of carbon dioxide locked in the regolith. In short, a modest 4°C rise in temperature at the south pole can globally raise temperatures by tens of degrees and transform a 6 millibar atmosphere into one measured in hundreds of millibars.

Raising the temperature of the south pole by just 4°C hardly seems enough of a change to trigger such a planetary transformation, but it's akin to removing just one apple from the bottom of a carefully stacked pyramid at the grocery store. Someone worked long and hard to arrange those apples in a state of delicate balance, of equilibrium. It doesn't take much to knock that equilibrium askew. So it is with Mars' south polar cap. Frozen carbon dioxide—dry ice—forms the cap. Carbon dioxide can be characterized by a measure known as "vapor pressure," which is a measure of the tendency of a substance to change into a gaseous or vaporous state. Temperature alone affects the vapor pressure of a substance, and as you turn up the heat on a substance, you raise its vapor pressure—it will change into a vapor or gas more vigorously. The vapor pressure of carbon dioxide at 147° Kelvin is 6 millibars—the current conditions found at Mars' south pole. (Kelvin degrees are chilly; if you have the temperature in degrees Kelvin, you need to subtract 273 to translate it into degrees centigrade. Therefore, 273°K equals 0°C, or 32°F. The temperature at Mars' south polar cap, 147°K, is –126°C or –195°F.) This is a state of equilibrium for the polar cap. So long as the pole remains at that temperature, it's difficult for the carbon dioxide pressure to get much above 6 millibar because excess carbon dioxide will simply condense out of the atmosphere and return to its frozen, dry ice form.

Now, what if we increased the temperature at the pole artificially? Later, I will detail how this could be accomplished using large, orbital mirrors to concentrate sunlight on the pole, but, for the moment, let's say that we have started to artificially heat up the pole. As a consequence of raising the temperature, the vapor pressure of carbon dioxide will increase, and therefore more carbon dioxide will evaporate from the cap into the atmosphere. Vapor pressure—the tendency for a substance to transform into a gas or vapor—and atmospheric pressure—the actual weight of an atmosphere over the surface—are two very different concepts, but we can say that as the vapor pressure of carbon dioxide rises at the pole, Mars' global atmospheric pressure will rise as a consequence of the carbon dioxide being pumped into the atmosphere as the cap evaporates. The vapor pressure of carbon dioxide at any temperature is a known piece of scientific information—you can look it up in a chemical handbook—and what holds for carbon dioxide on Earth will hold for carbon dioxide on Mars. The greenhousing capability of a layer of carbon dioxide gas in a planetary atmosphere is also known, albeit with somewhat less precision, so it is possible to estimate with reasonable accuracy how much the temperature on Mars will increase as a result of the thickening of its atmosphere. With this basic understanding of the conditions at the pole, of the vapor pressure and its relation to temperature, we can now venture forward into the nitty-gritty calculations that reveal how we can jump start the terraforming of Mars.

To start, take a look at Figure 9.1. In this figure, you can see the results of a model developed by McKay and me when applied to the situation at Mars' south polar cap, where we believe there may exist enough frozen carbon dioxide to give Mars an atmosphere on the order of 50 to 100 mbar. I have plotted the polar temperature as a function of the atmospheric pressure, and the vapor pressure as a function of the polar temperature. Note the two points, A and B, where the

FIGURE 9.1
Mars polar cap/atmosphere dynamics. Current equilibrium is at point A. Raising polar temperatures by 4°K would drive equilibria A and B together, causing runaway heating that would lead to the elimination of the cap.

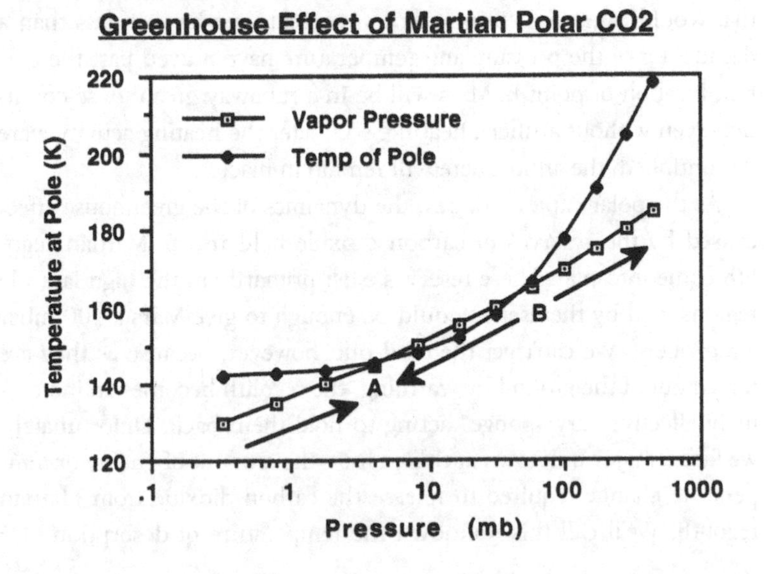

curves cross. Each is an equilibrium point where Mars' mean atmospheric pressure (P—the atmospheric pressure at Mars' mean surface elevation, in millibars) and the polar temperature (T—in degrees Kelvin) given by these two curves are mutually consistent. However, A is a stable equilibrium, while B is unstable. This can be seen by examining the dynamics of the system wherever the two curves do not coincide. Whenever the temperature curve lies above the vapor pressure curve, the system will move to the right, or toward increased temperature and pressure; this would represent a runaway greenhouse effect. Whenever the temperature curve lies below the pressure curve, the system will move to the left, or toward decreased temperatures and pressure; this would represent a runaway "icebox effect." Mars today is at point A, with 6 millibar of pressure and a temperature of about 147° Kelvin at the pole.

Now, consider what would happen if someone artificially increased the temperature of the Martian pole by several degrees Kelvin. As the temperature is increased, the whole temperature curve would move upward, causing points A and B to move toward each other until they met. If the temperature increase were 4°K, the temperature curve would move far enough up the graph to lie above the vapor pressure curve everywhere. The result would be a runaway greenhouse effect that would cause the entire pole to evaporate, perhaps in less than a decade. Once the pressure and temperature have moved past the current location of point B, Mars will be in a runaway greenhouse condition even without artificial heating, so if later the heating activity were discontinued, the atmosphere will remain in place.

As the polar cap evaporates, the dynamics of the greenhouse effect caused by the reserves of carbon dioxide held in the Martian regolith come into play. These reserves exist primarily in the high latitude regions, and by themselves could be enough to give Mars a 400 mbar atmosphere. We can't get them all out, however, because as they are forced out of the ground by warming, the regolith becomes an increasingly effective "dry sponge" acting to hold them back. Unfortunately, we face a major unknown at this point—the amount of energy or temperature change required to release the carbon dioxide from Martian regolith. We'll call this unknown the temperature of desorption (T_d)

FIGURE 9.2

Mars regolith/atmosphere dynamics under conditions of $T_d = 20$ with a volatile inventory of 500 mb of CO_2.

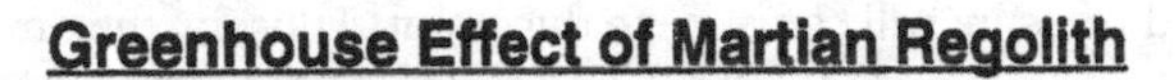

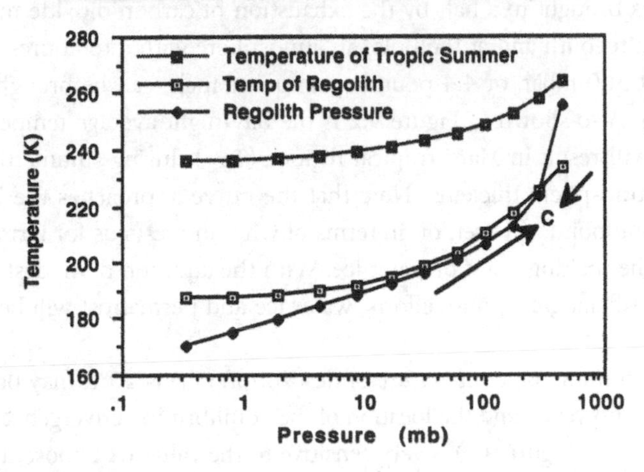

and estimate it at 20° Kelvin, though later we will vary its value to see how our model plays out. The dynamics of the atmosphere and regolith are shown in Figure 9.2. This figure displays the regolith-created atmospheric pressure on Mars (denoted "regolith pressure" in the figure) as a function of the regolith temperature, T_{reg}. (T_{reg} is the average of the planet's regolith temperature with different regions weighted in accordance with how much adsorbed gas they can hold at their own local temperature. Because colder soil holds more CO_2, T_{reg} tends to be representative of the temperatures in Mars' near-Arctic or near-Antarctic regions.) The figure also displays a plot of the regolith temperature as a function of the carbon dioxide pressure in the atmosphere. To arrive at these plots, I have assumed that releasing all available current polar carbon dioxide reserves would boost atmospheric pressure 100 millibars, and releasing all of the regolith carbon dioxide reserves would boost atmospheric pressure 394 millibars. Thus, together with the 6 millibars already in the atmosphere, Mars in this example is assumed to have a total carbon dioxide inventory of 500 millibars.

It can be seen in Figure 9.2 that the atmosphere/regolith system under the chosen assumption of the temperature of desorption (T_d) equaling 20°K has only one equilibrium point (where the two curves cross) that is stable. Once the polar cap is gone, Mars' global temperature and pressure will converge to this point. Thus, by the time the process is brought to a halt by the exhaustion of carbon dioxide reserves in the regolith and at the pole, an atmosphere with a total pressure of about 300 mbar, or 4.4 pounds per square inch, can be brought into being. Also shown in Figure 9.2 is the day-night average temperature that will result in Mars' tropical regions (T_{max}) during summertime as the atmosphere thickens. Note that the curve approaches the 273°K freezing point of water, or, in terms of what interests us for terraforming, the melting point of water ice. With the addition of modest ongoing artificial greenhouse efforts, water ice and permafrost will begin to melt.

Assuming the temperature of desorption (T_d) is 20°K may be optimistic, however, and the location of the equilibrium convergence point (point C in Figure 9.2) is very sensitive to the value we choose. In Figure 9.3, we show what happens if we instead assume values of 25° Kelvin and 30° Kelvin for the temperature required to drive carbon dioxide out of the regolith. In these cases, the convergence point shifts fairly drastically from 300 millibar when assuming T_d equals 20° Kelvin to 31 millibar when assuming T_d equals 25° Kelvin, and 16 millibar when assuming T_d equals 30° Kelvin. Such extraordinary sensitivity of the final condition to the unknown value of T_d may appear at first to put the entire viability of the terraforming concept at risk. However, in Figure 9.3 we also show (dotted line) what happens if we use artificial greenhouse methods to maintain the regolith temperature (T_{reg}) 10° Kelvin above those produced by the carbon dioxide outgassing itself. As mentioned previously, this could be done by pumping factory-produced CFCs into the atmosphere. You'll see that this drastically improves the final global temperature and atmospheric pressure values when assuming the temperature of desorption (T_d) to be 25° Kelvin or 30° Kelvin. In addition, we see that all three cases (T_d equals 20° Kelvin, 25° Kelvin, or 30° Kelvin) converge upon final states with Mars possessing atmospheres with several hundred millibars pressure.

There's one further unknown in the model that we should investi-

FIGURE 9.3
An induced 10°K rise in regolith temperature can counter effect of T_d variations. Data shown assume a planetary volatile inventory of 500 mb CO_2.

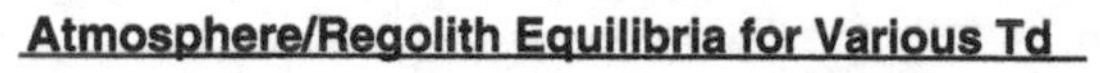

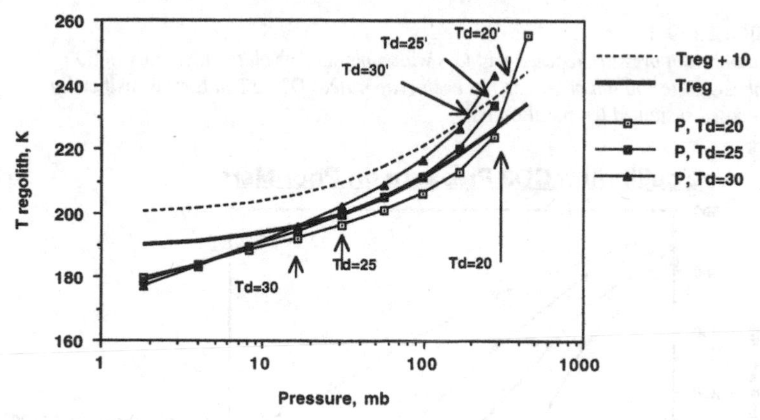

gate, though it's not quite as unknown as the temperature of desorption. That is the actual quantity of available carbon dioxide reserves we could find on Mars. The more reserves, the more carbon dioxide we can drive out of the regolith, and, hence, the thicker the atmosphere we can create. So we need to ask the questions, is Mars rich or poor in carbon dioxide reserves, and how does the answer play out in our model? At the moment, the best we can do is assume either condition, rich or poor, and run it through our model to see what happens.

To understand just how the abundance of carbon dioxide could affect our terraforming efforts, and how the value of T_d would interact with the quantities of carbon dioxide available, refer to Figures 9.4, 9.5, 9.6, and 9.7. In these figures, we can see the final atmospheric pressure and maximum seasonal average temperature equilibrium points in the Martian tropics based on assumptions of a poor Mars, possessing a total supply of about 500 millibar of carbon dioxide (50 mbar of carbon dioxide in the polar cap and 444 millibar in the regolith), or a rich Mars, possessing about 1,000 millibar of carbon dioxide (100 millibar in the polar cap and 894 millibar in the regolith). Recall that boosting the temperature of the regolith via artificial greenhousing methods made a significant difference in the final state of the atmosphere under differing values for the temperature of desorption. The same holds here, where different curves are shown under the assumptions that either no sus-

FIGURE 9.4

Equilibrium pressure reached on Mars with a planetary volatile inventory of 500 mb CO_2 after 50 mb polar cap has been evaporated. DT (ΔT in text) is artificially imposed sustained temperature rise.

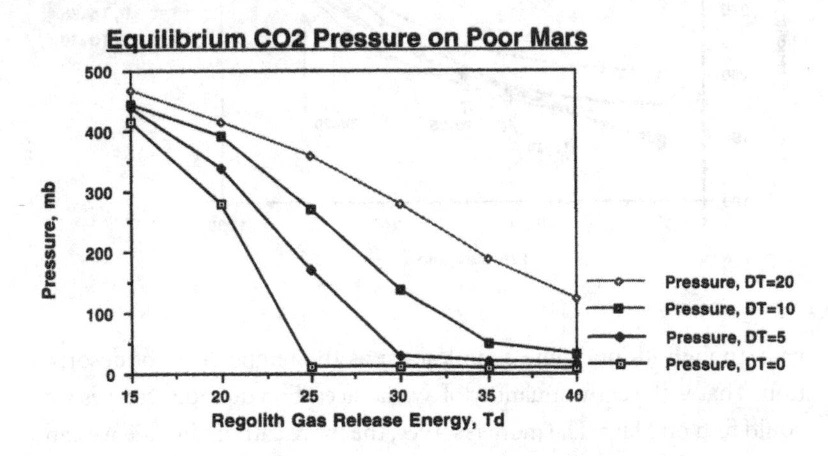

FIGURE 9.5

Equilibrium maximum seasonal (diurnal average) temperature reached on Mars with a planetary volatile inventory of 500 mb CO_2 after 50 mb polar cap has been evaporated.

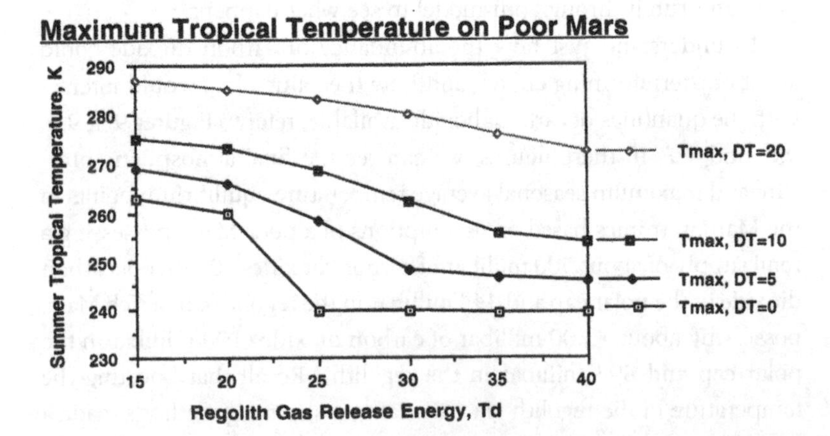

tained artificial greenhouse effort is mounted after the initial polar cap release, or that continued efforts are employed to maintain the planet's mean temperature at a level 5°, 10°, or 20° Kelvin above the value pro-

FIGURE 9.6
Equilibrium pressure reached on Mars with a planetary volatile inventory of 1,000 mb CO_2 after 100 mb polar cap has been evaporated.

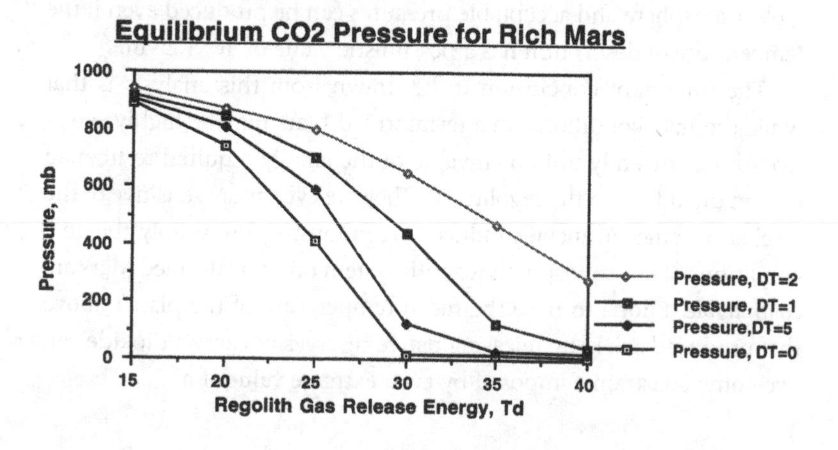

FIGURE 9.7
Equilibrium maximum seasonal temperature (diurnal average) reached on Mars with a planetary volatile inventory of 1,000 mb CO_2 after 100 mb polar cap has been evaporated.

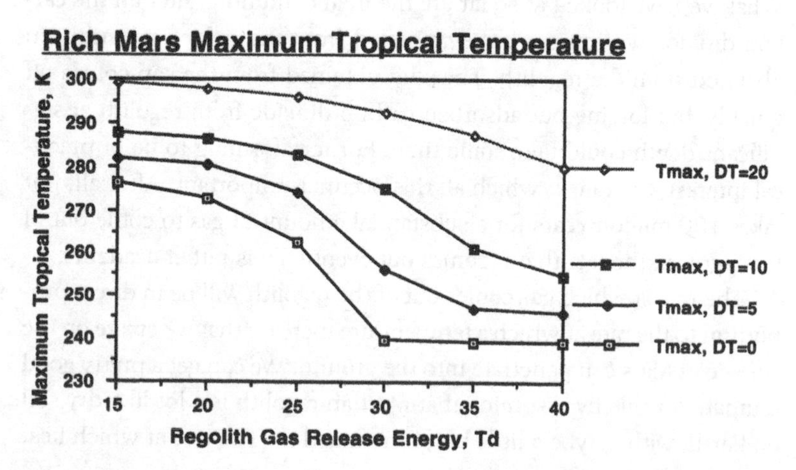

duced by the carbon dioxide atmosphere alone. For instance, you'll see in Figure 9.5 that even under an assumption of 40° Kelvin for the temperature of desorption, artificially maintaining atmospheric temperature at 20° Kelvin results in an overall temperature boost of more than 40° Kelvin. More important, though, we can see that if a sustained effort is mounted to maintain the planet's mean temperature 20° Kelvin above what native carbon dioxide reserves would produce, then a tangible atmosphere and acceptable pressures can be produced even if the temperature of desorption has a pessimistic value of 40° Kelvin.

The important conclusion to be drawn from this analysis is that while the final conditions on a terraformed Mars may be highly sensitive to the currently unknown value of the energy required to liberate carbon dioxide from the regolith, T_d, they are even more sensitive to the level of sustained artificially induced greenhousing. Put simply, the final conditions of the atmosphere/regolith system on a terraformed Mars are *controllable.* Efforts to raise the mean temperature of the planet above that produced solely by releasing native reserves of carbon dioxide can overcome constraints imposed by even extreme values for T_d.

HOW FAST WOULD THE ATMOSPHERE COME OUT OF THE REGOLITH?

What we have looked at so far are the final conditions after all the carbon dioxide we can get has evaporated from the polar cap and been liberated from the regolith. The stuff obtained from the cap comes off quickly, but forcing out adsorbed carbon dioxide from regolith at significant depth could take some time. For terraforming to be of practical interest, the rate at which all this occurs is important. After all, if it takes 100 million years for a substantial amount of gas to come out of the regolith, the fact that it comes out eventually is rather academic.

The rate at which gas comes out of the regolith will be in direct proportion to the rate at which a temperature increase that we create on the surface of Mars can penetrate into the ground. We can get a pretty good estimate for this by assuming that Martian regolith is a lot like dry soil on Earth, with maybe a little bit of ice mixed in. The rate at which heat will spread through such a medium will be governed by the process of thermal conduction. The equations of thermal conduction predict that the time a temperature rise needs to travel a given distance through a medium is proportional to the square of the distance. Based upon dry terrestrial soils, a reasonable estimate for this rate on Mars would be about 16 square meters per year. We also need to estimate how much gas is in the regolith. If you take zeolite minerals down to Martian temperatures and expose them to carbon dioxide, they will adsorb enough carbon dioxide to comprise 20 percent of their solid weight. Martian regolith is not made of zeolite, but probably includes a lot of clay-like minerals, which are not too dissimilar. As a rough guess, let's say that Martian regolith is saturated with about 5 percent carbon dioxide, and that the loose material has an average density of 2.5 tonnes per cubic meter. If that were the case, you would have to force out carbon dioxide held in regolith—outgas it—down to a depth of 200 meters to produce a 1,000-millibar (1 bar, Earth sea-level) pressure on Mars. Let's say we induced a sustained artificial temperature rise at the surface of 10° Kelvin, good enough to outgas a significant fraction of what is in the regolith. This temperature rise would then travel down into the ground. The rate at which this would occur is shown in Table 9.1.

It can be seen that while it takes a very long time to reach significant depths, modest depths can be reached rather quickly. So, while it might take thousands of years to penetrate 200 meters to get a full 1,000 millibars out of the regolith, the first 100 millibars can be gotten out in just a few decades.

Once significant regions of Mars rise above the freezing point of water on at least a seasonal basis, the large amounts of water frozen into the regolith as permafrost would begin to melt, and eventually flow out into the dry riverbeds of Mars. Water vapor is also a very effective greenhouse gas, and since the vapor pressure of water on Mars would rise enormously under such circumstances, the reappearance of liquid water on the Martian surface would add to the avalanche of self-accelerating effects all contributing toward the rapid warming of the planet. The seasonal availability of liquid water is also the key factor in allowing the establishment of natural ecosystems on the surface of Mars.

TABLE 9.1
Rate of Outgassing of Atmosphere from the Martian Regolith

Time (Earth years)	Depth Penetrated (meters)	Atmosphere Created (millibars)
1	4	20
4	8	40
9	12	60
16	16	80
25	20	100
36	24	120
49	28	140
64	32	160
81	36	180
100	40	200
144	48	240
196	56	280
256	64	320
324	72	360
400	80	400
900	120	600
1,600	160	800
2,500	200	1,000

The dynamics of the regolith gas-release process are only approximately understood, and the total available reserves of carbon dioxide won't be known until human explorers journey to Mars to make a detailed assessment, so these results must be regarded as approximate and uncertain. Nevertheless, it is clear that the positive feedback generated by the Martian carbon dioxide greenhouse system greatly reduces the amount of engineering effort that would otherwise be required to transform the Red Planet. In fact, since the amount of a greenhouse gas needed to heat a planet is roughly proportional to the square of the temperature change required, driving Mars into a runaway greenhouse with an artificial 10° Kelvin temperature rise only requires about

4 percent of the engineering effort that would be needed if the entire 50° Kelvin rise needed to raise the Martian tropics above the freezing point of water had to be engineered by brute force. The question we shall now examine is how such a 10° Kelvin global temperature rise could be induced.

METHODS OF ACCOMPLISHING GLOBAL WARMING ON MARS

The three most promising options for inducing the required temperature rise to produce a runaway greenhouse on Mars appear to be the use of orbital mirrors to change the heat balance of the south polar cap (thereby causing its carbon dioxide reservoir to vaporize); the mass production of artificial halocarbon (CFC) gases in industrial facilities on the Martian surface; and the creation of widespread bacterial ecosystems capable of warming the planet through emission of large amounts of strong natural greenhouse gases such as ammonia and methane. We'll look at each of these in turn. It may be, however, that the synergistic combination of several such methods may yield better results than any one of them used alone.[47]

Orbiting Mirrors

While the production of a space-based mirror capable of warming the entire surface of Mars to terrestrial temperatures is theoretically possible, the engineering challenges involved in such a task place such a project well outside the technological horizon of this book. A much more practical idea would be to construct a more modest mirror capable of warming a limited area of Mars by a few degrees. As shown by the data in Figure 9.1, a 4° Kelvin temperature rise imposed at the pole should be sufficient to cause the evaporation of the carbon dioxide reservoir in the south polar cap. Based upon the total amount of solar energy required to raise the temperature of a given area a certain number of degrees above the polar value of 150° Kelvin, it turns out that a space-based mirror with a radius of 125 kilometers could reflect enough sunlight to raise the entire area south of 70° south latitude by 5° Kelvin—more than enough. If made of solar sail-type aluminized mylar material with a density of 4 tonnes per square kilometer (about 4 microns thick), such a sail would have a mass of 200,000 tonnes. Many ships of this size are currently sailing the Earth's oceans. Thus, while this is too large to consider launching from Earth, if space-based manufacturing techniques are available, its construction in space out of asteroidal or Martian moon material is a serious option. The total amount of energy required to process the materials for such a reflector would be about 120 MWe-years, which could be readily provided by a set of 5 MWe nuclear reactors such as might be used in piloted nuclear electric propulsion (NEP) spacecraft. Interestingly, if stationed near Mars, such a device would not have to orbit the planet. Rather, solar light pressure could be made to balance the planet's gravity, allowing the mirror to hover as a "statite" with its power output trained constantly at the polar region.[48] For the sail density assumed, the required operating altitude would be 214,000 kilometers. The statite reflector concept and the required mirror size to produce a given polar temperature rise are shown in Figures 9.8 and 9.9.

If the value of T_d is lower than 20° Kelvin, then the release of the polar carbon dioxide reserves by themselves could be enough to trigger the release of the regolith's reserves in a runaway greenhouse effect. If, however, as seems probable, T_d is greater than 20°K, then the addition of strong greenhouse gases to the atmosphere will be required to force a global temperature rise sufficient to create a tangible atmospheric pressure on Mars.

FIGURE 9.8
Solar sails of 4 tonnes/km² density can be held stationary above Mars by light pressure at an altitude of 214,000 km. Wasting a small amount of light allows shadowing to be avoided.

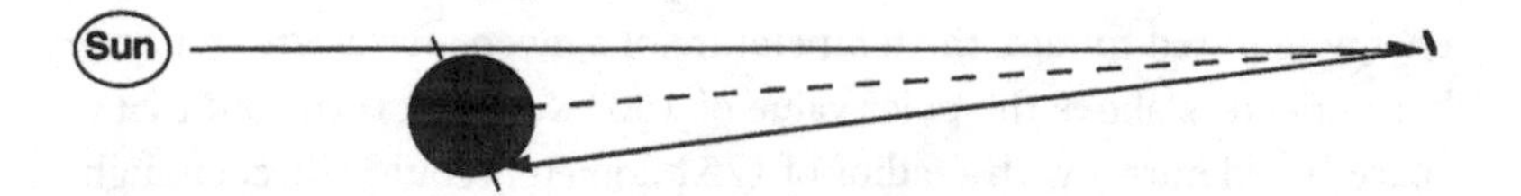

FIGURE 9.9
Solar sail mirrors with radii on the order of 100 km and masses of 200,000 tonnes can produce the 5°K temperature rise required to vaporize the CO_2 in Mars' south polar cap. It may he possible to construct such mirrors in space.

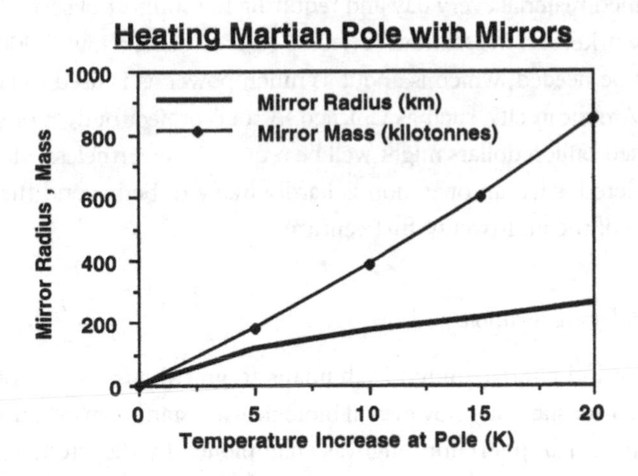

Producing Halocarbons on Mars

The most obvious way to increase the temperature on Mars is simply to set up factories there to produce the strongest greenhouse gases known to man, the halocarbons or CFCs that many consider to be currently threatening the Earth with a dangerous greenhouse effect, and then release them into the atmosphere. Here on Earth, CFCs are also blamed for the destruction of the ozone layer. However, if we choose our halocarbon greenhouse gases carefully and employ varieties lacking in chlorine, we can actually build up an ultraviolet-shielding ozone layer in the Martian atmosphere. One good candidate for such a gas would be perfluoromethane, CF_4, which also has the desirable feature of being very long lived (stable for more than 10,000 years) in an upper atmosphere. In Table 9.2, we show the amount of halocarbon gases needed in Mars' atmosphere to create a given temperature rise, and the power that would be needed on the Martian surface to produce the required CFCs over a period of twenty years. If the gases have an atmospheric lifetime of one hundred years, then approximately one-fifth the power levels shown in the table will be needed to maintain the CFC concentration after it has been built up. The industrial effort associated with such a power level would be substantial, producing about a trainload of refined material every day and requiring the support of several thousand workers on the Martian surface. Power levels of about 5,000 MWe might be needed, which is about as much power as is used today by a large American city, such as Chicago. A total project budget of several hundred billion dollars might well be required. Nevertheless, all things considered, such an operation is hardly likely to be beyond the capabilities of the mid-twenty-first century.

The Biological Solution

The level of effort required by humans to greenhouse Mars could be reduced substantially, however, if biological assistants can be employed. This approach to terraforming was championed by the late Carl Sagan starting in the 1960s, when he initiated the field of scientific terraforming speculation by suggesting that Venus might be made more habitable by seeding its atmosphere with algae that would consume the carbon dioxide in its atmosphere, thereby diminishing that planet's hellish greenhouse effect.[49] That idea probably will never work, but in his later Mars studies, Sagan and his collaborator James Pollack pointed out that bacteria exist that can metabolize nitrogen and water to produce ammonia.[50] In addition to its minority presence in the atmosphere, nitrogen can likely be found on Mars in substantial amounts in regolith nitrate beds. Other bacteria can synthesize water and carbon dioxide into methane. Now, while not as good as halocarbons, both ammonia and methane are excellent greenhouse gases, thousands of times more powerful on a molecule-for-molecule basis than carbon dioxide. If an initial greenhouse condition were to be created by polar mirrors or CFC manufacture, thereby putting some liquid water into circulation, it may be possible that a bacterial ecology could be set up on the planet's surface that would accelerate the process by producing large amounts of ammonia and methane. In fact, if 1 percent of the planet's surface were to be covered by such bacteria, and we assume that they operate at about 0.1 percent efficiency in converting the energy of sunlight into

TABLE 9.2
Greenhousing Mars with CFCs

Induced Heating (degrees K)	CFC Pressure (micro-bar)	CFC Production (tonnes/hour)	Power Required (MW_e)
5	0.012	260	1,310
10	0.04	880	4,490
20	0.11	2,410	12,070
30	0.22	4,830	24,150
40	0.39	8,587	42,933

chemical compounds, then around one billion tonnes of methane and ammonia would be produced each year. This is enough to warm the planet 10° Kelvin in about thirty years.

As an added benefit, ammonia and methane will shield the planet's surface against solar ultraviolet radiation. In the process, though, the ammonia and methane will be continuously destroyed, with a typical molecule having an atmospheric lifetime of several decades. The bacteria will constantly replace them, however. Also, as the planet warms and carbon dioxide outgasses from the regolith, Mars' ozone layer will thicken, providing extra UV shielding to both the surface and the ammonia-methane greenhouse gases in the atmosphere. (Carbon dioxide contributes to ozone formation. In fact, Mars currently has an ozone layer about 1/60th as thick as Earth's, which is pretty good when you consider that its atmosphere is only 1/120th as thick.)

In a matter of several decades, using a combination of these approaches Mars could be transformed from its current dry and frozen state into a relatively warm and slightly moist planet capable of supporting life. Humans could not breathe the air of the transformed Mars, but they would no longer require space suits and instead could travel freely in the open wearing ordinary clothes and a simple scuba-type breathing gear. In addition, because the outside atmospheric pressure will have been raised to human tolerable levels, it will be possible to have enormous habitable areas for humans under huge dome-like inflatable tents containing breathable air. (The domes could be of unlimited size because, unlike the pressurized domes employed during the base-building phase, there would be no pressure differential between their interior and the outside environment.) On the other hand, simple hardy plants could thrive in the carbon-dioxide–rich outside environment and spread rapidly across the planet's surface. In the course of centuries, these plants would introduce oxygen into Mars' atmosphere in increasingly breathable quantities, opening up the surface to advanced plants and growing numbers of animal types. As this occurred, the carbon dioxide content of the atmosphere would be reduced, which would cause the planet to cool unless greenhouse gases were introduced capable of blocking off those sections of the infrared spectrum previously protected by carbon dioxide. Providing these matters are attended to, however, the day would eventually come when the domed tents would no longer be necessary.

ACTIVATING THE HYDROSPHERE

The first steps required in the terraforming of Mars, warming the planet and thickening its atmosphere, can be accomplished with surprisingly modest means using in-situ production of halocarbon gases supplemented by helpful bacteria. The oxygen and nitrogen levels in the atmosphere, however, would be too low for many plants, and, if left in this condition, the planet would remain relatively dry, as the warmer temperatures would take centuries to melt Mars' ice and deeply buried permafrost. It is in this second phase of terraforming Mars when the hydrosphere is activated, the atmosphere is made breathable for advanced plants and primitive animals, and the temperature is increased further, that space-based manufacturing of large solar concentrators is likely to assume an increasingly important role.

The use of orbiting mirrors provides a potentially rapid method for hydrosphere activation. For example, if the 125-kilometer radius reflector discussed earlier for use in vaporizing the pole were to concentrate its power on a smaller region, 27 terrawatts would be available to melt lakes (one terrawatt, or TW equals one million megawatts). This is enough to melt 2 trillion tonnes of water per year (a lake 200 kilometers on a side and 50 meters deep). A single such mirror could also drive vast amounts of water out of the permafrost and into the nascent Martian ecosystem very quickly. The more rapidly water gets into circulation, the more action of denitrifying bacteria in breaking down nitrate beds to increase the atmospheric nitrogen supply and the spread of plants to produce oxygen will be accelerated. Activating the hydrosphere will also serve to destroy the oxidizing chemicals in the Martian regolith (which *Viking* showed are unstable in the presence of water), thereby releasing some additional oxygen into the atmosphere in the process. Thus, while the engineering of such mirrors maybe somewhat grandiose, the benefits to terraforming of being able to wield tens of terawatts of power in a controllable way can hardly be overstated.

OXYGENATING THE PLANET

The most technologically challenging aspect of terraforming Mars will be the creation of sufficient oxygen in the planet's atmosphere to support animal life. While bacteria and primitive plants can survive in an atmosphere without oxygen, advanced plants require at least 1 mbar and humans need 120 mbar. While Mars may have super-oxides in its regolith or nitrates that can be heated to release oxygen and nitrogen gas, the process would require enormous amounts of energy, about 2,200 TW-years for every millibar produced. Similar amounts of energy are required for plants to release oxygen from carbon dioxide. Plants, however, offer the advantage that once established they can propagate themselves. The production of an oxygen atmosphere on Mars thus breaks down into two phases. In the first phase, brute-force engineering techniques supplemented by pioneering cyanobacteria and primitive plants are employed to produce sufficient oxygen (about 1 millibar) to allow advanced plants to propagate across Mars. Assuming three 125-kilometer radius space mirrors active in supporting such a program and sufficient supplies of suitable target material on the ground, such a goal could be achieved in about twenty-five years. Alternatively, a 1 millibar oxygen content could be added to the atmosphere in about a century through the action of photosynthetic bacteria. Either way, once an initial supply of oxygen is available, and with a temperate climate, a thickened carbon dioxide atmosphere to supply pressure and greatly reduce the space radiation dose, and a good deal of water in circulation, plants that have been genetically engineered to tolerate Martian regoliths and to perform photosynthesis at high efficiency could be released together with their bacterial symbiotes. Assuming that global coverage could be achieved in a few decades and that such plants could be engineered to be 1 percent efficient (rather high, but not unheard of among terrestrial plants) then they would represent an equivalent oxygen-producing power source of about 200 TW. By combining the efforts of such biological systems with perhaps 90 TW of space-based reflectors and 10 TW of installed power on the surface (terrestrial civilization today uses about 15 TW) the required 120 millibars of oxygen needed to support humans and other advanced animals in the open could be produced in about nine hundred years. If more powerful artificial energy sources or still more efficient plants (or perhaps truly artificial self-replicating photosynthetic machines) were engineered, then this schedule could be accelerated accordingly, a fact that may well prove a driver in bringing such technologies into being. It may be noted that thermonuclear fusion power on the scale required for the acceleration of terraforming also represents the key technology for enabling piloted interstellar flight. If terraforming Mars were to produce such a spinoff, then the ultimate result of the project will be to confer upon humanity not only one new world for habitation, but myriads.

A GIFT TO THE FUTURE

> Witness this new-made World, another Heav'n
> From Heaven Gate not far, founded in view
> On the clear Hyaline, the Glassie Sea;
> Of amplitude almost immense, with Starr's
> Numerous, and every Starr perhaps a World
> Of destined habitation . . .
>
> —*John Milton,* Paradise Lost

The theoretical calculations are quite clear in their verdict: The Red Planet can be terraformed. But only human explorers operating on Mars can learn enough about the planet and the methods of utilizing its resources to transform such a dream into reality. Yet the game certainly is worth the candle, for what is at stake is an entire world.

In a sense, the discussion of humanity's potential to terraform Mars brings us full circle. Are we first-class citizens of the cosmos, or are we beings of lesser order? Kepler proved that the laws of the heavens were *understandable* by the human mind. The first astronauts to reach Mars will prove that the worlds of the heavens are *accessible* to human life. But if we can terraform Mars, it will show that the worlds of the heavens themselves are *subject to* the human intelligent will.

Mars could become a second home for life, all life; not only for humans, nor even just "the fish of the sea . . . the fowl of the air, and every living thing that moveth upon the Earth," but for a plenitude of species yet unborn. New worlds invite new forms, and in the novel habitats that a terraformed Mars would provide life brought from Earth could go forth and multiply into realms of diversity yet unknown.

This is the wondrous heritage that we can begin for future generations—not only a new world for life and civilization, but an example of what men and women of intelligence, daring, and vision can accomplish when acting upon their highest ideals. Gods we'll never be. But the humanity that terraforms Mars will have shown that humans are more than just animals, that we are in fact creatures who carry a unique spark that is worthy of respect. No one will be able to look at the new Mars without feeling prouder to be human. No one will be able to hear its story without being inspired to rise to the tasks that will lie ahead among the stars.

EQUATIONS FOR MODELING THE MARTIAN SYSTEM

We can estimate the average temperature on Mars as a function of the CO_2 atmospheric pressure and the solar constant using the following equation:

$$T_{mean} = 213.5(S^{0.25}) + 20(1+S)P^{0.5} \quad (1)$$

where T_{mean} is the mean planetary temperature in Kelvins, S is the amount of solar output where that of the present-day Sun equals 1, and P is the atmospheric pressure at Mars' mean surface elevation, given in bars. (One bar is what flatlanders believe is normal atmospheric pressure, 14.7 pounds per square inch. Since people living in the fetid swamps near such major capitals as Washington, London, and Paris are influential in such things, this odd unit has become a standard.)

Since the atmosphere is an effective means of heat transport from the equator to the pole, Chris McKay and I estimated:

$$T_{pole} = T_{mean} - 75(S^{0.25})/(1 + 5P) \quad (2)$$

It is also reasonable to assume, based upon a rough approximation to observed data, that:

$$T_{max} = T_{equator} = 1.1T_{mean} \quad (3)$$

and that the global temperature distribution is given by:

$$T(\theta) = T_{max} - (T_{max} - T_{pole})\sin^{1.5}\theta \quad (4)$$

where θ is the latitude (north or south).

Equations (1) through (4) give the temperature on Mars as a function of carbon dioxide pressure. However, as mentioned above, the carbon dioxide pressure on Mars is itself a function of the temperature. There are three reservoirs of carbon dioxide on Mars: the atmosphere, the dry ice in the polar caps, and gas adsorbed in the regolith. The interaction of the polar cap reservoirs with the atmosphere is well understood and is given simply by the relationship between the vapor pressure of carbon dioxide and the temperature at the poles. This is given by the vapor pressure curve for carbon dioxide, which is approximated by:

$$P = 1.23 \times 10^7 \{\exp(-3170/T_{pole})\} \quad (5)$$

So long as there is carbon dioxide in both the atmosphere and the cap, equation (5) gives an exact answer to what the carbon dioxide atmospheric pressure will be as a function of polar temperature. However, if the polar temperature should rise to a point where the vapor pressure is much greater than that which can be produced by the mass in the cap reservoir (between 50 and 100 mbar) then the cap will disappear and the atmosphere will be regulated by the regolith reservoir.

The relationship between the regolith reservoir, the atmosphere, and the temperature is not known with precision. An educated guess developed by McKay[51] is:

$$P = \{CM_a \exp(T/T_d)\}^{3.64} \quad (6)$$

where M_a is the amount of gas adsorbed in the regolith in bars, C is a constant adjusted so that equation (6) will reflect known Martian conditions, and T_d is the characteristic energy required for release of gas from the regolith (the "desorption temperature"). Equation (6) is essentially a variation of a well-known law for the change in chemical equilibrium with temperature, and so there is fair confidence that its general form is correct. However, the value of T_d is unknown and probably will remain so until after the human exploration of Mars. While we don't know what the right value for T_d is, we can bracket the problem by varying T_d from 15° to 40°K (the lower the value of T_d, the easier things are for prospective terraformers). Then we use the global temperature distribution given by equation (4) to integrate equation (6) over the surface of the planet to give us a global "regolith pressure." This gives a reasonably accurate quasi two-dimensional view of the atmosphere/regolith equilibrium problem in which most of the adsorbed carbon dioxide is distributed to the planet's colder regions. Thus, in our model, regional (in the sense of latitude) temperature changes, especially in the near-polar regions, can have as important a bearing on the atmosphere/regolith interaction as changes in the planet's mean temperature.

The results of this model, displayed graphically throughout this chapter, give strong reason to believe that Mars can be terraformed.

10: THE VIEW FROM EARTH

No bucks, no Buck Rogers.

—Anonymous

In the past nine chapters I've sketched out the technical possibilities and the vision of what we can accomplish by launching a humans-to-Mars program. Now it's time to come back to Earth. The greatest obstacle to gaining a foothold on Mars won't be found in the engineering details of a human Mars mission. It won't be found in the rigors of the journey to Mars and the long days exploring a new world. It won't be found on Mars. The greatest obstacle to sending humans to Mars resides here on our present home planet in the guise of Earthly politics. How can we get the money needed to get the program off the ground?

Some people think it can't be done. They point to the failure of the two Bush presidential space exploration initiatives (SEIs) as proof of the fact that the American political process will not support a humans-to-Mars program. The logic behind this "proof," though, is intrinsically flawed, as it is based upon the notion that because something happened a certain way in the past, it has to happen that way again. Bush Senior and Junior each tried to repeat John F. Kennedy's successful launching of the Apollo race to the Moon, they say, but in the post Cold War environment the clarion call fell flat. The Bush SEIs failed, therefore all future SEIs will fail. Q.E.D.

All very tidy, but completely inaccurate. Bush Senior did not do for his SEI what J.F.K. did for Apollo. Rather Bush did for SEI what Bush Senior did for the Kurds: announce that the hour had struck, toss the ball in the air, and walk off the field. As Dwayne Day of the Space Policy Institute has pointed out, "Bush was an advocate of space exploration in the same way he was 'the environmental president' or 'the education president'—weakly, and in name only." It is also true that NASA's 90-Day Report with its $450 billion price tag and thirty-year timeline did not help the situation, but the real problem was not the 90-Day Report, but a leadership willing to tolerate the flaws inherent in that report.

Let me make clear exactly what I mean. In June of 1990, when Senior's SEI was still in the initial stages of its downward plunge, I attended a large NASA-sponsored SEI propulsion conference at Pennsylvania State University. Addressing the plenary session of that conference, Congressman Robert Walker (R-PA) openly told representatives of the aerospace industry and aerospace press that the reason SEI funding had just been voted down in Congress was because the top NASA brass, led at that time by administrator Richard Truly, had told Congress that if NASA got what it wanted for the Space Shuttle and Space Station programs, Congress was welcome to vote however it pleased on funds for SEI. In other words, the NASA leadership refused to advocate the program that President Bush had called a national priority. Plenty of people thought this was veritable sabotage and that Truly should have been fired. The National Space Council's leaders at the time, Mark Albrecht and Pete Worden, tried to deal with the situation, but due to presidential apathy it was two years before Truly was replaced. By that time, SEI was virtually dead.

Bush Senior's lack of involvement, combined with the NASA leadership's opposition, left his SEI an orphan to be advocated by some Space Council staffers allied to a few friendly congressmen. Without any real clout, they were forced to try to fund SEI by sneaking a few small appropriations through Congress. When the administration's political opponents saw this weakness, they pounced upon it as a way to humiliate Bush and his Space Council chief, Dan Quayle. Kevin Kelly, an aide to Senator Barbara Mikulski (D-MD), led the massacre to seek out and systematically eliminate any NASA appropriation, no matter how small, that could be linked to SEI. By the time Dan Goldin became NASA administrator in 1992, the best way left to save those technology programs needed for a Mars program that still survived was to break the damning link by abolishing SEI, and, after a year or so of attempts to salvage SEI, that is what he did.

Bush Junior had a more favorable situation to work with than his dad, in that at the time he launched his SEI, (dubbed the Vision for Space Exploration, or VSE), his party had control of both houses of congress. Despite possessing this important advantage, however, Junior still managed to blow it, by allowing his first NASA administrator, the consummate bureaucrat Sean O'Keefe, who did not even want to risk sending astronauts to the Hubble Space Telescope, let alone Mars, to arrange the Moon-Mars program in such a way that it would not really begin until *after* Bush left office—even assuming his reelection to a second term. Thus, while the VSE was announced in January 2004, under the plan set forth, the overwhelming majority of NASA's human spaceflight budget would remain devoted to operating the Space Shuttle and supporting the Space Station through at least 2010. *In other words; Enjoy the Vision, suckers, but it's going to be business as usual until we're out of here.*

After letting NASA waste a year doing paper studies of completely unworkable concepts for conducting human Lunar missions without the use of a heavy lift launch vehicle (which he did not care to develop), O'Keefe eventually left amid criticism (primarily over his obstinate refusal to authorize the use of the Shuttle for the critically necessary Hubble repair mission). Unfortunately, his replacement, Dr. Mike Griffin, while a very competent engineer, and who did, in fact, save Hubble, accepted the O'Keefe action-deferred VSE strategy as a presidential policy that duty bound him to implement, rather than criticize.

I met with Griffin in his office at NASA headquarters in June 2005 and pressed him in the strongest terms to break from the O'Keefe plan. "Everything may seem aligned for the VSE, right now," I said, "with a supportive president and a supportive congress, but all that can change. On January 21, 2009, there will be a new president who had nothing to do with formulating this policy. Unless the VSE is practically approaching at least the milestone of returning to the Moon at that point, all bets will be off. For a successful VSE, speed is everything." I therefore urged him to fly just one more Shuttle flight, to save Hubble, and then shut the program down and pour all of its $4 billion/year budget into developing the heavy lift booster and other flight hardware needed for the Moon-Mars initiative. With the aid of the heavy lifter made available by such accelerated development, all the Space Station payloads scheduled for eventual Shuttle launch could get delivered soon enough. But more importantly, unless this was done, there would be virtually no chance that the embryonic VSE would be sufficiently developed by the next inauguration day to avoid abortion by the incoming administration.

Griffin listened to what I had to say, then shook his head. "I know all this," he said. "But you don't understand the constraints I am operating under. I am not the leader of the American space program. I am the administrator of the American space program. I don't make the policy. I administer the policy made by the president."

And so he did. And sure enough, things did change. Less than two years after our meeting, the GOP lost control of congress, and two years after that, the White House as well. Griffin, a Republican, wasn't kept on by the new administration, and the VSE, whose snail-paced progress had not yet produced any useful flight hardware, became highly vulnerable to cancellation.

The Obama White House assault on the VSE was delayed for one year, due to the administration's desire during early 2009 to spend money on anything and everything as part of its economic stimulus program. However, by 2010, it was ready to strike, and when it did, it hit hard. Acting under the guidance of the President's science advisor, Dr. John Holdren, an individual who had in the 1970s written that the United States is an "overdeveloped country" which needs to be deindustrialized, the administration announced its intention to cancel NASA's plan to return to the Moon (to which the VSE had devolved by this point) without putting any alternative in its place. In addition, the Constellation program, under whose banner the medium and heavy lift Ares I and V boosters and Orion crew capsule needed for the VSE were beginning to be designed, was also cancelled, leaving the post-Shuttle agency without any capability to even continue to fly humans to Earth orbit, let alone go beyond. Instead, apparently to provide the sugar necessary to make the cyanide go down, the agency would be given a set of make-work projects, such as refurbishing the Shuttle launch pads after the Shuttles stop flying, and a variety of meaningless showboat projects, such as doing demonstration battery-fired firings of the allegedly revolutionary VASIMR electric thruster (see chapter 8 focus section), without developing the large space nuclear power reactors that would be needed to make them useful in any capacity on real piloted missions.

This original Holdren policy, set forth by the administration in February 2010, would have doomed the American human spaceflight program to a decade of zero achievement, and remarkably, would have done so without saving any money. Duly outraged, a bipartisan group in the House and the Senate mobilized to push back, ultimately succeeding in getting the administration to moderate its stand. As a result, the Orion capsule program was preserved (although at this writing only equipped for flying *down* from orbit, rather than *up*), funding was appropriated to proceed with *design* of a heavy-lift launch vehicle, and a nominal goal was proclaimed of a human mission to a near Earth asteroid by 2025.

These were useful steps, but insufficient. A capsule that only flies down still leaves us without the ability to go anywhere. A goal of reaching a near Earth asteroid is acceptable, if uncompelling, but setting it for 2025 frees anyone of any obligation to do anything toward its attainment between now and at least 2016. This makes the reality of the heavy lift booster program questionable as well, since no one is going to proceed very far in developing a booster which has no missions or payloads to launch for the foreseeable future.

As a result, the U.S. human spaceflight program is now completely adrift.

Comparing the brilliant military/political strategist Napoleon Bonaparte to his dissolute grandnephew Napoleon III, Karl Marx once commented, "All historical events occur twice, the first time as tragedy and the second time as farce." The comparison holds equally well for JFK and the Bushes. It is said that while his army was being annihilated at Sedan, Napoleon III whiled away the time by playing billiards. It might be said that Bush Senior lost Mars while yachting at Kennebunkport, and Bush Junior lost the Moon while mountain-biking at Crawford. The failure of their SEIs proves nothing, except that armies don't win battles if their generals are playing billiards.

There is plenty of latent political support in this country for a humans-to-Mars program. I have experienced this firsthand when speaking on the subject before numerous public groups of every description, from Rotary Clubs to plumbers' conventions, groups with no vested interest in a Mars program as such. And the central recurring question I get is "How come we're not doing this?" "I remember Apollo," people in my audiences tell me. "Weren't we supposed to go to Mars after that? How come there was no follow-through? This is the sort of thing this country ought to be doing!"

That is what I hear, over and over again. The main public complaint about the space program isn't that it costs too much; it's that the program is not going anywhere. People feel betrayed, not by NASA, but by the politicians. The future they expected to see in the 1960s has been aborted. What's happened? How come we stopped moving? The Beltway policy wonks may tell the politicians that people in the heartland don't care about space, but everything I see firsthand tells me there is a massive disconnect here.

Some might challenge the evidence for my assertion as purely anecdotal. But if you require scientific polls, they've been produced in plenty. In a poll *Newsweek* sponsored in connection with its cover story on Mars Direct in 1994, more than half of those queried supported a piloted mission to Mars. In 2002, America Online conducted a poll in which 76 percent of respondents said that taxpayer dollars should be spent on missions to Mars, and 63 percent said they would want to go. A CBS News poll conducted in 2003 found that 80 percent of Americans think the US space program contributes to scientific advances they can use, and 85 percent say that it contributes a lot to Americans' pride and patriotism. Another CBS poll done in 2009 reported 51 percent of Americans favoring sending humans to Mars. In the same poll, performed 40 years after the Apollo Moon landings, 71 percent said that the Apollo missions had been worth the cost, dramatically more than the 47 percent who approved of the Moon program shortly after its completion in the 1970s.

Other forms of statistical data have been collected as well. For a number of years, Jon D. Miller of the Chicago Academy of Sciences has reported on the public understanding of science and technology in the United States.[52] Included in his reports are examinations of that sector of the population termed the "attentive public" to various science and technology issues. The attentive public is that group of individuals who are interested in a specific issue, who feel they are well informed about the issue, and stay current by regularly reading newspapers or magazines. They are informed enough to feel comfortable and confident in perhaps contacting a policymaker on the issue. In other words, the attentive public for an issue is that portion of the public that might be most prone to take action to support, or oppose, an issue. Those who are interested in an issue, but don't believe they are terribly well-informed, Miller classifies as the "interested public." From data collected in 1992, Miller concluded that 6 percent of the American public was attentive to space exploration, and an additional 16 percent were interested in the subject. According to Miller, the large majority of those 22 percent were found to believe that the benefits of space exploration outweigh its costs. Twenty-two percent is still a minority. But Miller also found that his "attentive" group had the highest proportion of scientifically literate individuals among its members, and was one of the best educated groups in his study of the U.S. population as a whole.

The numbers have held over time. In recent data from 2008, there appears still to be a robust 22 percent of the population very interested in space exploration, representing nearly 50 million potential adult voters.

In short, I believe, and there is every reason to believe, that if an American leader stood up (like J.F.K. *re* the Moon) and called for a humans-to-Mars program, and then stood by his guns to fight for the program and rally support, he would find himself leading a growing political juggernaut, just as Kennedy did in the early 1960s. The Mars program proposed would have to be technically and politically sound. A $450 billion price tag coupled to a thirty-year timeline can turn any proposal into a political albatross, but as we've seen, using a plan like Mars Direct, we can get to the Red Planet much cheaper and much quicker.

That said, there are at least three very different models on how a humans-to-Mars program could be accomplished. I call these models the J.F.K. model, the Carl Sagan model, and the Newt Gingrich model. Each of these has their strengths and weaknesses. Let's talk about them in turn.

THE J.F.K. MODEL

The first, and most widely understood, of all the three main approaches for a humans-to-Mars program is what I term the J.F.K. model. It's the most widely understood because it's the only one that has been done—it's how we reached the Moon. In the J.F.K. model, the president of the United States stands before the people and calls upon the nation to meet the challenge of the future. When I reread Kennedy's Apollo speeches, a sense of his greatness comes through that is unmatched by any twentieth-century orator, except perhaps Winston Churchill.

"We choose to go to the Moon!" Kennedy said, his voice ringing with destiny. "We choose to go to the Moon in this decade and do the other things, not because they are easy but because they are hard. . . . Because that goal will serve to organize and measure the best of our abilities and skills, because that challenge is one we are willing to accept, one we are unwilling to postpone, and one which we intend to win!" J.F.K. was openly visionary. While it would create new technologies, new jobs, and new knowledge, going to the Moon was fundamentally "an act of faith and vision, for we do not know what benefits await us." Few people hearing these speeches could fail to recognize they were standing in the presence of history being made.

John Kennedy's Apollo program did more than just land men on the Moon—it set in place a paradigm, both political and technical, of how space programs should be launched. First and foremost, vigorous, clear, and visionary presidential support will lead any successful effort. Kennedy did not try to sneak his program through the political process. Instead, he stood in the well of the House of Representatives and announced his intentions to a special joint session of Congress. Secondly, it was an American program. Coming during the height of the Cold War, Apollo was a grand way for America to flex its political, social, and scientific muscle on the world stage. Going to the Moon, landing men there, and bringing them back seemed on par with climbing Olympus to share the nectar with the gods. Lastly, there was money. Kennedy was forthright about the amount it would cost and, together with Lyndon Johnson, a political powerhouse in his own right, did what was required to fund it, and then some.

Can the J.F.K. approach be taken again, this time to Mars? While the foreign policy imperatives of the Cold War are now just a memory, a successful American program of Mars exploration would still have enormous impact around the world. The first nation to set foot on Mars will undoubtedly go down in history as the nation, the people, that opened the door to humanity's next great step into the future. It would show to the world and, perhaps more importantly, to us, to each and every citizen of the United States, that we still have "the right stuff," that we are still a nation of pioneers, a people who defy all limits. Is that worth $50 billion? I think so, and more.

To hear some people talk, you would think that a $50 billion space program translates into a booster simply rocketing $50 billion in large bills to the Sun's interior—big money going nowhere. The fact is, though, that funds spent taking us to Mars remain tightly bound to communities here on Earth. They are an engineer's paycheck, a welder's take-home pay, a scientist's research funds, a graduate student's stipend; they pay for innovations and inventions that will remain part of the nation's intellectual capital and that may lead to new businesses or products for earthly use; they pay for all the mission hardware required from the lowliest rivet to the latest high-tech electronics. Beyond all this, the funds spent on a humans-to-Mars program would pay for an invitation to every youth in the nation to join in a great adventure by developing their minds—the true source of all our future wealth.

In fact, the downturn of U.S. space spending at the end of Apollo was followed by a slowdown of the U.S. economy, which has remained comparatively sluggish ever since. During the 1960s, NASA spending averaged a bit more than 2.25 percent of federal spending (it peaked at nearly 4 percent of federal spending in 1964). During the same years, the U.S. economy in GDP constant dollars grew on average about 4.6 percent a year. During the early 1970s, NASAs share of the federal budget dropped to below 1 percent of federal spending, where it has remained ever since. Simultaneously, the GDP growth rate has dropped to less than 2 percent.

In contrasting the brilliant accomplishments of NASA during the Apollo era with its mediocre performance today, much is frequently made of the fact that, at its peak, NASA's budget during the 1960s comprised close to 4 percent of all federal outlays, compared to its 0.7 percent fraction today. Yet, if we take NASA's total expenditures over the Apollo program span, from 1961 when J.F.K. announced the program, to 1973 when the final Apollo-derived Skylab missions were flown, add them up, divide the sum by the 13 years in question, and translate the money into today's dollars we find that *the average NASA budget over the Apollo period was $19 billion per year, exactly the same as NASA's budget today.* And it should be pointed out that during that time NASA not only flew the piloted *Mercury, Gemini, Apollo*, and *Skylab* missions, but also the robotic *Ranger, Surveyor*, and *Mariner* probes, while doing all the development for the *Pioneer, Viking*, and *Voyager* programs as well. In addition, the space agency developed hydrogen/oxygen rocket engines, multistaged heavy lift launch vehicles, nuclear rocket engines, space nuclear reactors and radioisotope generators, spacesuits, in-space life support systems, orbital rendezvous techniques, interplanetary navigation technologies, deep-space data transmission techniques, reentry technologies, soft-landing rocket technologies, a space station, and more. In other words, virtually the entire bag of tricks that enables space exploration today was developed during the 1961–1973 period, at the same time that we flew astronauts to the Moon. Despite continued comparable expenditures, very little of importance has been developed since. So, far from entailing increased costs or the sacrifice of technology development, the focused destination-driven approach to spaceflight imposed on NASA by J.F.K.'s imperative resulted in by far the most cost-effective, inventive, and productive period in the agency's history.

The J.F.K. model is a proven success; successful both at realizing an impossible dream of getting humans to the Moon, advancing space technology, and in generating the greatest period of economic growth in the United States' postwar economic history. Today, however, the question may well be asked if the nationalistic foundations that supported Apollo exist in the current era. Instead of demonstrating American superiority, wouldn't it perhaps be better for a humans-to-Mars program to promote international cooperation? This brings us to an alternative approach to human missions to Mars, which I call *the Sagan model*, after its most consistent, eloquent, and vocal spokesman.

THE SAGAN MODEL

Carl Sagan was perhaps one of the strongest and most public voices promoting an international approach to Mars exploration, advocating such an approach in one manner or another for several decades. His original call for international human Mars exploration focused on United States/Soviet Union collaboration. He saw an American/Soviet Mars program as a way to bind two adversarial nations together in a common, historic undertaking. The energies of both nations' top engineers and scientists would be directed toward developing the aerospace, electronic, and rocket technologies necessary for an expedition to Mars and, in so doing, would use the scientific talents of both nations toward ends other than enlarging their nuclear weapon stockpiles. A mixed crew journeying to Mars was to be seen as a microcosm of the home planet, a small world where the two great powers of the world worked together.

Sagan was certainly not alone in his calls for international partnerships in space exploration. Nearly every blue-ribbon panel anointed by NASA or the president (and there have been many) over the past twenty years has called for collaborative projects in space. While Sagan's original notion may have been outpaced by political events, there remains an obvious economic benefit: More partners means more pockets. What one nation might not be able to afford, perhaps two or more, together, can. The European Space Agency's collaborative efforts have not only built a robust European program of space science, but have developed one of the most successful launch vehicles currently flying, the Ariane. Technologies as well as costs can be shared, with potentially great benefits for all. Currently, the United States lacks a heavy-lift launch vehicle with enough power to launch a Mars Direct–style initiative. Russia, on the other hand, fielded one as recently as the late 1980s, in the form of the Energia. With its 100-tonne-to-LEO capability, a revived Energia would currently be the most powerful rocket on the planet. Energia flew just twice, in part because it lacked a mission. A humans-to-Mars program would fit the bill nicely. Likewise, the international Space Station uses several Russian modules as core components of the orbital lab.

While there are obvious benefits to undertaking a Mars initiative with an international cast, there can be serious costs as well. By definition, any time a nation joins a collaborative effort, it loses control of a project. It can maintain partial control, perhaps, one vote among equals, but if the effort is truly collaborative, it will not be able to flatly declare what will or won't be. The United States' European and Japanese partners on the Space Station have had to sweat through several redesigns of the station ordered by the U.S. Congress. There was little any of the partners could do, as Congress was driving the downsizing of the project. Likewise, over the course of the ISS program, NASA has frequently had to face the possibility that our major partner in the Space Station, Russia, might not be able to honor its commitments. In general, decision making is slowed down in international programs, which tends to increase costs.

Politics aside, there are plenty of technical obstacles that can arise when undertaking an enormous collaborative effort: What if one of the partners signs on to develop a technology, and then, for whatever reason, fails to do so? What if a key partner drops out of the program entirely? What if international relations change, and a friendly partner nation should become a foe? Such events can completely destabilize the program schedule, and delays in projects as large as Apollo, the Space Station, or a Mars program tend to cascade, creating potentially disastrous consequences for the project.

When I first heard about Sagan's scheme for a joint American/Soviet mission to Mars back in the 1980s, I did not think the proposal practical. The United States was in the middle of its "Star Wars" and Pershing missile buildups, the Soviets were waging war in Afghanistan, and both were waging proxy wars with each other in El Salvador, Nicaragua, and elsewhere. In 1980, the United States and Soviet Union had been unable to participate jointly in Olympic sports competitions. The idea that we could conduct a joint Mars program over a period of years seemed chimerical. Furthermore, from a crew selection point of view, there could be few worse choices than Sagan's proposal for a mixed crew of U.S. astronauts and Soviet cosmonauts—two groups of ex-fighter pilots with years of prior training in methods of killing each other and indoctrinated with reasons for doing so. While Sagan argued that the process of collaboration itself would serve to bring the opposing nations together, I thought it more likely that conflicts between the two nations would tear the collaboration apart.

Today, however, there is a new rationale for U.S./Russian collaboration in space. Instead of using it to try to make peace with a foe, it could be used to stabilize relations with a nation that could become a true friend. Russia today is a defeated superpower with an unstable economy and a dangerous growing revanchist movement, all in a nation possessing 10,000 nuclear warheads that could be misappropriated by illegal operators or retargeted at the United States if nationalists or extremists came into power. It is thus in the self-interest of the United States to help stabilize Russia both politically and economically. Propping up the Russian economy with cash payments for space hardware is one way of doing this. Of course, proceeding in such a manner means that much of the cost-sharing justification for collaboration falls by the wayside, but from the U.S. taxpayer's point of view, money could still be saved because Russian space hardware is much cheaper than Western material.

Some say supporting the Russian space infrastructure in this way would be a mistake, since the capabilities we save may be turned against us should that country's nascent democracy collapse. This argument overlooks the fact that most of the space industries supported by a joint Mars program would be those manufacturing hardware like liquid-fuel propulsion systems, heavy-lift launch vehicles, and in-space life-support systems, all of which are of comparatively limited military utility.

Even more important, however, is this: The shared effort, risk, and adventure of a joint program could bring the U.S. and Russia together with ties of real friendship, in which case the reemergence of a strong Russia—which could occur in any case—would become a positive development for America, rather than the reverse. Indeed, after the difficult period of the past decade, there is a need today to re-cement the great Western alliance of all nations committed to the ideals of progress, reason, and freedom. A humans to Mars program could be a way to do just that.

In other words, in the context of today's global political realities, Sagan's proposal for a joint U.S./Russian, or better yet, a fully international humans-to-Mars program may have a good deal of merit. Its fundamental problem remains the programmatic risk involved—it would make the Mars program a hostage to stability in Russia, Europe, or elsewhere. But perhaps the gains are worth the chance.

THE GINGRICH APPROACH

There is a third approach to getting humans to Mars, which is more speculative than those already mentioned, but holds great promise. I call it the Gingrich approach because I came up with it under the prodding of the former Speaker of the House, and because it is in line with principles that Mr. Gingrich has supported for some time.

Here is the story behind the idea. In the summer of 1994 I was invited to come and have dinner with then-Congressman Newt Gingrich (R-GA) and some of his staff, to tell them about my ideas about Mars exploration. I explained the Mars Direct plan for a near-term, low-cost humans-to-Mars program. Gingrich was enthusiastic about the possibilities. "I want to support this with legislation," he told me, but he wanted to do it "in a more free-enterprise kind of way than just gearing up the NASA budget to go to Mars." He invited me to come on his TV show to talk more about it, which I did, and then directed me to get together with Jeff Eisenach, president of the Progress and Freedom Foundation, Gingrich's D.C. think tank.

I met with Eisenach several times, and what we came up with was the idea of a Mars Prize bill. Here's how it would work. The U.S. government would post a $20 billion award to be given to the first private organization to successfully land a crew on Mars and return them to Earth, as well as several prizes of a few billion dollars each for various milestone technical accomplishments along the way.

This is, to say the least, a novel approach to human space exploration, which up till now has been entirely government run. It also has a number of remarkable advantages. In the first place, this approach renders cost overruns impossible. Not a penny will be spent unless the desired results are achieved, and not a penny more will be spent beyond the award sum agreed upon at the start. Success or failure with this approach depends solely upon the ingenuity of the American people and the workings of the free enterprise system, not upon political wrangling. The tactic not only guarantees economic results, but it also promotes quick results and smart results. When people have their own money at stake, it's a lot easier to find and settle on practical, no-nonsense solutions to engineering problems than is ever the case in the complex and endless deliberations of a government bureaucracy. Readers may recall that when Charles Lindbergh flew the Atlantic, he did not do it as part of a government-funded program, but in pursuit of a privately posted prize. There were many such prizes offered for breakthrough technical accomplishments in aviation's early years, and collectively they played a major role in raising the art of flight from its infancy to a globe-spanning transportation network. This tradition has continued to the present day. Its most recent feat, the winning of the $50 million X-Prize for a flight to 100 km by Burt Rutan's *Spaceship One* rocketplane in October 2004, has now taken aviation into space. As a result of this success, a commercial version of Rutan's vehicle, dubbed *Spaceship Two*, will reportedly soon make opportunities for suborbital flights available to paying passengers from the general public. These accomplishments were achieved on a schedule and within a budget that would have been inconceivable within the current NASA.

There are other advantages to this approach to Mars as well. Economic growth would be spurred, prior to any government expenditure. Moreover, posting multibillion dollar prizes for breakthrough accomplishments in space would call into being not only a private space race, but a new kind of aerospace industry, one based on minimum-cost production methods. The existing aerospace industry does not work that way. Rather, the major aerospace companies contract with the government to do a job on a "cost plus" basis, which means that whatever it costs them to do the job, they charge the government a certain percentage more, usually 8 to 12 percent. Therefore, the more it costs the major aerospace companies to do a job for the government, the more money they make. For this reason their staffs are top heavy with layer after layer of useless, high-priced "matrix managers" (who manage nothing), "marketeers" (who do no marketing), and "planners" (whose plans are never used), and whose sole apparent function is to add to company overhead. Of course, since the government needs proof that the expenses claimed by the aerospace companies are actually being incurred, vast numbers of accounting personnel are also employed, to keep track of how many labor hours are spent on each and every separate contract. To give you an idea of how bad it is, at Lockheed Martin's main plant in Denver, where I used to work, and where the Titan and Atlas launch vehicles are produced, only a small minority of all personnel actually work in the factory. The fact that Lockheed Martin is cost competitive with the other large aerospace companies indicates that the rest of them are operating with similar overhead burdens as well.

The prize system would change all that, because the company's profit would be the value of the prize, minus their costs, period. They would have no incentive to run costs up. Quite the contrary, they would have every reason to drive costs down. Furthermore, their actual base costs would be lower, since their accounting and documentation burden would be much less. By creating new aerospace companies based on these principles, or forcing the existing ones to drastically reform themselves, the Mars Prize would end up saving the government and the commercial satellite industry billions of dollars, as they soon would be able to get all of their required space and launch system hardware much cheaper.

But how could the Mars Prize be set as low as $20 billion? Didn't

I say that a Mars Direct program would probably cost the government more like $30 billion? Even with another $10 to $20 billion of secondary prizes thrown in, that's hardly a deal that a private organization is likely to chase.

But my estimate of $30 billion for Mars Direct is based upon a J.F.K.-like model for the program—NASA funding existing major aerospace contractors to get the job done within their existing overhead structures, and NASA spending a lot on itself for "program management" costs in the process. If the Mars Direct or Semi-Direct missions were done on a truly private basis, with the people undertaking the effort being free to buy whatever they want from whomever they want to build whatever they want, I believe that the cost of the effort would be in the $4 to $6 billion range. This sounds incredible when compared to the $30 billion estimate for Mars Direct, to say nothing of the $450 billion estimate for the 90-Day Report, but if you look at what is actually needed to fly the mission and total it up, taking advantage of things like cheap Russian launch vehicles and other cost-saving measures, it's hard to see why the program should cost more than $4 billion or so. In the real world, you can buy an awful lot with $4 billion.

Consider the following: As a general standard, aerospace engineers will figure a $5,000 per pound cost for developing a new, high-performance jet fighter. Thus, the cost per tonne of hardware for development of such aerospace systems, which are every bit as complex as the habitats, Mars ascent vehicles, entry capsules, and other Mars Direct hardware, is about $10 million per tonne. (The McDonnell Douglas DC-X single-stage-to-orbit test vehicle came in at $6 million per tonne.) The total amount of dry hardware needed for Mars Direct or Semi-Direct, excluding launch vehicles, is certainly less than 100 tonnes. So, budget $1 billion for that. To launch everything you need to Mars, you would need to lift about 300 tonnes (this latter figure includes a lot of trans-Mars injection propellant, which is cheap stuff, less than $1,000 per tonne). Three hundred tonnes could be lifted to low-Earth orbit (LEO) by three Russian Energias, which would cost about $300 million each,[53] for a total of $900 million in launch costs, plus maybe $500 million in startup expenses to revive the Energia production line. The total cost for all hardware development and launch would thus be $2.4 billion. Toss in another $600 million for mission operations, program management, legal costs, and other odds and ends and you've got yourself a $3 billion Mars program. Even if Energia or other Russian launch vehicles (such as Proton, which is in production and whose launch cost is about $4 million per tonne to LEO) are not available or allowed, the mission still doesn't have to cost that much. Current launch costs using recent or current U.S. boosters such as Titans, Atlases, or Deltas is about $10,000 per kilogram, or $10 million per tonne to LEO. At those rates, which are extremely conservative since a heavy-lift launch vehicle, such as a Shuttle C or an Ares, would show major economies of scale relative to these medium-lift boosters, launching the required 300 tonnes would cost $3 billion. Add that to the $1 billion for hardware development and, once again, we find a total program cost of less than $5 billion.

So, if the real cost of the mission is $4 to $6 billion, a $20 billion prize ought to be able to mobilize the required capital from the private sector. No doubt there would be any number of people who would be skeptical that a manned Mars mission could be flown for $5 billion, but that wouldn't matter. If the Mars Prize bill were passed, the only thing that would matter was whether a few investors think it could. We wouldn't have to convince a majority of Congress that a humans-to-Mars program can be done cheaply; we would only have to convince a Bill Gates or an Elon Musk. This is important: The private sector is often vastly more innovative than the government because a consensus is not necessary to start something new. Rather, all it takes is one innovator and an investor who is willing to take a chance.

But, if nobody takes up the challenge, what then? In that case, the whole exercise would have cost the taxpayers absolutely nothing.

Would offering the Mars Prize damage NASA? I don't think so. Rather, it would result in an infusion of capital into the best groups at the various NASA centers, as the private consortia chasing the prize sought to subcontract expertise in particular areas of interest. This would have a very healthy influence on the technologists at NASA, as they would then be driven to develop technologies that those seeking to fly a Mars mission actually want, instead of indulging themselves with research into technologies that are not relevant.

Here's the series of prizes to drive the development of a humans-to-Mars program that I worked up for Gingrich. Note that though these prizes can be viewed as a series of steps toward an ultimate goal, a single organization need not undertake all of the challenges. A company could choose to accept one challenge and leave it at that, it could take the challenges on in succession, or it could skip the easier challenges entirely in an effort to get a crew to Mars first and win the grand prize.

CHALLENGE 1: Accomplishment of a Mars orbiter imaging mission.

The Prize: $500 million

The Conditions: The mission must successfully image at least 10 percent of the planet with resolutions of 20 centimeters per pixel or better. All images must be made available to the U.S. government, which will publish them.

The Bonus: An additional $1 million for imaging (with 90 percent coverage or better) each of the 200 sites of interest selected by NASA's Mars Science Working Group.

(Note: This challenge was subsequently achieved by NASA, through its successful *Mars Reconnaissance Orbiter* (*MRO*). Built for the agency by Lockheed Martin at a cost of $720 million, *MRO* arrived in Mars orbit in 2006 and has been mapping the Red Planet at high resolution ever since.)

CHALLENGE 2: With a robotic lander, collect a sample of Martian soil and transport the sample to Earth using propellants of Martian origin for the return flight.

The Prize: $ 1 billion

The Conditions: The soil sample size must be at least three kilograms. At least 70 percent (by weight) of the propellant mixture used on the Mars-ascent and Earth-return legs of the mission must be produced on Mars from Martian resources.

The Bonus: $10 million for each distinct rock type returned, up to a maximum of $300 million.

CHALLENGE 3: To demonstrate a long-term life-support system in space.

The Prize: $1 billion

The Conditions: A crew of three or more must be sustained in space for at least two years without resupply from Earth.

CHALLENGE 4: To deliver a pressurized rover to Mars.

The Prize: $ 1 billion

The Conditions: The vehicle must be proven capable of sustaining two humans on Mars for one week by means of a one-week test conducted on Earth during which it is driven 1,000 kilometers over unimproved terrain. The vehicle must travel at least 100 kilometers on Mars. Cabin pressure during the Mars excursion must be maintained between 3 and 15 psi. The cabin temperature must be maintained between 10° and 30° centigrade.

CHALLENGE 5: To demonstrate the first system that uses propellants of Martian origin to lift a 5-tonne payload from the surface of Mars to Mars orbit.

The Prize: $1 billion

The Conditions: At least 70 percent (by weight) of the propellant mixture used must be produced on Mars from Martian resources.

CHALLENGE 6: To demonstrate the first system that can produce more than 20 tonnes of propellant on the Martian surface in the course of a 500-day stay.

The Prize: $1 billion

The Conditions: At least 70 percent (by weight) of the propellant mixture must be produced on Mars from Martian resources.

CHALLENGE 7: To demonstrate the first system that can generate at least 15 kilowatts power (day/night average) for 500 days on Mars.

The Prize: $1 billion

The Conditions: A minimum of 2 kWe must be available at all times.

CHALLENGE 8: To demonstrate the first system that can deliver 10 tonnes of payload to the Martian surface.

The Prize: $2 billion

The Conditions: The demonstrator must provide a soft landing exerting no more than 8-gravities deceleration on the payload during any part of the trip.

CHALLENGE 9: To be the first to demonstrate a system that can lift at least 120 tonnes to low Earth orbit.

The Prize: $2 billion

The Conditions: The booster must launch from U.S. territory. Past history of the Saturn V is not eligible. A revived Saturn V system is eligible.

CHALLENGE 10: To demonstrate the first system that can put 50 tonnes onto a trans-Mars trajectory

The Prize: $3 billion

The Conditions: Hyperbolic velocity for Earth departure must be at least 4 km/s. The system must be launched by a booster or boosters with a capacity of at least 120 tonnes to low Earth orbit per launch. The booster must be launched from U.S. territory.

CHALLENGE 11: To demonstrate the first system that can deliver 30 tonnes of payload to the Martian surface.

The Prize: $5 billion

The Conditions: The demonstrator must make a soft landing exerting no more than 8-gravities deceleration on the payload during any part of the trip.

CHALLENGE 12: To be the first to send a crew to Mars and return the crew members safely to Earth.

The Prize: $20 billion

The Conditions: A majority of the crew must be Americans. At least three crew members must reach the Martian surface and remain on the planet for at least 100 days. One or more of the crew members must make at least three overland trips of at least 50 kilometers from the landing site.

The Bonus: In addition to the $20 billion, crew members will receive $1 million per person for each day spent on the Martian surface, up to a maximum bonus of $5 billion.

Some general conditions would apply to all prizes, since some challenges encompass accomplishments covered by other tasks listed. For example, any system that can deliver 30 tonnes of payload to the Martian surface can also deliver 10 tonnes. If the more difficult mission is accomplished before the less difficult one, the organization performing the mission wins both prizes. To ensure that home-grown flight systems are developed, it would be required that at least 51 percent of the cash value of all hardware used to win any of the prizes must be manufactured in the United States. This does not mean that every subsystem must be 51 percent U.S.-made. For example, a successful mission to Mars using a Russian heavy-lift booster can still claim the $20 billion mission prize, provided that 51 percent of the total mission hardware is made in the United States, but that mission will not be eligible for the heavy-lift launch system prize. Finally, the winner of any prize would be required, at the government's option, and at a cost per copy no greater than 20 percent that of the prize, to sell up to three additional copies of the winning flight system to the U.S. government. The U.S. government, in turn, would support all missions competing for the prizes with the communication services of the Deep Space Tracking Network's 34-meter-diameter dishes provided at cost, and would also provide support for all launches with the ground support and tracking systems available at the Kennedy Space Center and other potential launch sites, and make properties at these sites available at reasonable costs for launch pad construction.

If the Mars Prize bill were passed, it would be success in the field, rather than committee judgments, that would decide what architectures and technologies are best. The system of prizes would provide not only the needed incentives to get humans to Mars, but a financial "runway" as well, that will allow private organizations to accumulate the capital required to finance such a venture. For example, an organization could start by focusing on winning prize 9, the development of the heavy-lift launcher. The $2 billion prize for this is not much better than a break-even proposition, but once it is in hand, the organization would be in an excellent position to win prize 10, $3 billion for hurling 50 tonnes onto trans-Mars injection. This second prize would put the organization heavily in the black, and set it up to win prize 11, $5 billion for the first soft landing of 30 tonnes on Mars. Once that was accomplished, the organization would have in hand the primary Earth-Mars transportation system needed to fly the Mars Direct mission, and plenty of working capital, and could then launch an assault on the grand $20 billion prize for the round-trip human mission. Groups with smaller initial capital could start out by chasing some of the lesser prizes for precursor missions, and thus get into the game through the side door, so to speak. Thus, entering the contest via various routes, organizations would accumulate both capital and experience as they compete for and win prizes that rely on demonstrations of the critical technologies and accomplishments of key precursor missions required to meet the major program challenges. But the prize system does not prescribe the design of the mission—no one is obligated to go for all or any of the lesser prizes in their quest for the grand prize. The "runway" boasts multiple entrances. Each competing organization will be able to use its own creativity to determine the most efficient path to Mars, in the process creating a set of cheap transportation systems that would not merely make possible a flags and footprints mission, but the systematic exploration and settlement of the Red Planet.

After being elected Speaker of the House, Gingrich was overwhelmed with the demands of tax activists, abortion activists, balanced budgeteers, and other interest groups. He requested the Mars Prize workup, and according to Eisenach he was delighted with it. But he didn't do a thing with it, and neither did his successors in the Republican congresses that followed. The same can be said for Al Gore, who until being elected vice president, frequently intimated that he supported the Sagan plan for a joint U.S./Russia mission to Mars. As soon as he was elected, however, he lost all interest and hasn't said a word about it since. Nor can we expect any better from the political cast on stage today unless they see some evidence of political support for the idea out in the field. If we are ever going to see any action out of any of these people, we're going to have to show some strength. This brings me to my next point.

WHAT YOU CAN DO

If you want humans to get to Mars, then you need to become a space activist.

As we saw, Miller's study of the space-interested public identified a group of close to 40 million. Yet, the three main domestic space activist organizations, the National Space Society, the Planetary Society, and the Mars Society, boast a total of perhaps 100,000 members. We have immense latent support for space exploration in this country, but only a tiny fraction of it is organized. The harvest is plentiful, but the gatherers are few. Permanent organizations allied with large memberships are needed to generate the kind of political muscle required. In a nutshell, Mars needs you. It's not enough to wish the space program well; if you believe in a future that is not limited by Earth's horizons, you need to join with other like-minded individuals and make your voice heard. Joining a space activist organization is probably the best way to do that.

There are basically four organizations to choose from. I'm a bit prejudiced here because I happen to be a leader of one of them, the Mars Society. But I'll try to give you an accurate enough picture to decide where you should center your efforts. The fees and addresses listed below are accurate as of early 2011; but you should check with them directly before sending in any membership fee.

The Planetary Society is the largest of the four, with perhaps 50,000 members. Founded by Carl Sagan, Louis Friedman, and former Jet Propulsion Lab director Bruce Murray, it is led today by Friedman, astronomer Jim Bell, and TV science educator Bill Nye. The Planetary Society is primarily interested in promoting the robotic exploration of the solar system, but it is supportive of a humans-to-Mars program, provided it is done along the lines of the Sagan international collaboration model. You can join the Planetary Society by sending a check for $37 to: The Planetary Society, 85 South Grand, Pasadena, CA 91105. www.planetary.org

The National Space Society is the second largest, with 20,000 members. Founded by Wernher von Braun and Princeton space visionary Professor Gerard O'Neill, it is led today by Mark Hopkins, Kirby Ikin, and Executive Director Gary Barnhard. The primary interest of the National Space Society (NSS) is to promote the human settlement of space, including the Moon, Mars, the asteroids, and free-floating space colonies. The NSS would be equally happy supporting a Mars (or Moon) program based on any of the J.F.K., Sagan, or Gingrich models. The NSS is organized into about a hundred local chapters, which organize local and regional events, as well as a national conference once a year. You can join the NSS by sending a check for $20 to: National Space Society, 1155 15th Street NW, Suite 500, Washington, DC 20005. Membership benefits include a bimonthly glossy magazine and frequent mobilization bulletins concerning the space program. www.nss.org

The Space Frontier Foundation is the smallest organization, with about 500 members. Founded by Rick Tumlinson and Jim Muncy, and led today by Muncy, Bob Werb, and William Watson, the Space Frontier Foundation has a very strong free-enterprise tilt. Of the three approaches to Mars discussed in this chapter, it would favor only the Gingrich model. If opening space with maximum free enterprise and minimum government involvement is fundamental to your principles, then consider joining this group. The Space Frontier Foundation sponsors one national conference per year. You can join the Space Frontier Foundation by sending $25 to: The Space Frontier Foundation, 16 First Avenue, Nyack, NY 10960. www.spacefrontier.org

The Mars Society is the newest of the space organizations. Together with many other members of the Mars Underground, including Chris McKay, Carol Stoker, and Tom Meyer, as well as science fiction authors Greg Benford and Kim Stanley Robinson, I founded the Mars Society with the purpose of furthering the exploration and settlement of Mars by both public and private means. Our founding convention in Boulder, Colorado, in August 1998 drew 700 people from 40 countries, featured 180 papers and talks on everything from Mars mission strategies to the ethics of terraforming, and attracted international press coverage. As of this writing, we have around 7,000 members, divided into 80 chapters, of which 50 are in the United States and 30 span the globe. Our activities include broad public outreach, political lobbying, and the operation of two human Mars exploration simulation bases; one in the polar desert on Canada's Devon Island, and the other in the desert of southern Utah. To date, over a hundred crews of six people each have gone to these stations and undertaken simulated Mars missions ranging from two weeks to four months in duration. During these missions, crews are tasked to perform sustained programs of field exploration in geology and microbiology, while operating under many of the same constraints that human explorers would face on Mars. By doing so, we are learning a great deal about what field tactics and technologies will prove most useful when humans finally journey to the Red Planet. At the same time, press reportage of these missions, which has appeared in the world's leading media ranging from the *New York Times*, CNN, and the Discovery Channel, to the BBC and Russian and Japa-

FIGURE 10.1
The Mars Society's Flashline Mars Arctic Research Station (FMARS) is located on the rim of a giant meteor impact crater in the polar desert of Canada's Devon Island, 900 miles from the North Pole. (Photo by author)

nese national television, has helped make the vision of human exploration of our neighbor world sensuous to hundreds of millions of people around the globe.

The Mars Society holds its international convention every August. You can join either through our web site at:www.marssociety.org or by sending $50 ($25 for students) to Mars Society, 11111 W. 8th Ave. Unit A, Lakewood, CO, 80215.

If you want to reach me, you can write in care of the Mars Society address. If you want to help, sign up at the website so we can put you on the Mars Society electronic mailing list. If you join the Mars Society, you will also get access to our online library. You can get a fair number of my technical papers there, as well as those of many other authors, covering topics ranging from interplanetary propulsion technologies to the ethics of terraforming.

Making history is not a spectator sport. It's your turn at the plate.

FIGURE 10.2
Mars Society volunteer crews use the FMARS base to test exploration techniques on Devon Island, one of the most Mars-like environments on Earth. (Photo by author)

A QUESTION OF HISTORY

Establishing the first human outpost on Mars would be the most historic act of our age. People everywhere today remember Ferdinand and Isabella only because they are associated with the voyage of Christopher Columbus. In contrast, the number of people who can name predecessors and successors of Ferdinand and Isabella are few and far between, and all the wars, atrocities, palace coups, scandals, booms, and bankruptcies that must have seemed so important to people at that time are today nearly forgotten. Similarly, almost no one five hundred years from now will know what Operation Odyssey Dawn was, let alone the Whitewater scandal; they will never have heard of the wars in Iraq or Afghanistan; and they will neither know nor care whether the present United States had national health care or a balanced budget. But they will remember those who first got to and settled Mars, and the nation who made it possible.

When I was a boy, I used to read a lot of classical history. I still remember quite well one speech that Pericles, the Athenian leader, gave for the Athenian war dead at the end of the second year of Athens' desperate struggle with Sparta. To the assembled relatives he intoned: These men, your sons and husbands, are dead—and I understand you are sad. But look at what they died for: They died for Athens. And what is that, but a city which uniquely calls upon its people to be citizens, not subjects; which celebrates philosophy, science, and reason; and which allows its people to live well while exercising both their duty and right to be fully human. And then Pericles said: *Future ages will wonder at us, even as the present age wonders at us now.*

Although Athens would soon be destroyed as a major power, Pericles was correct: It's been over two thousand years and despite all the technological and literary accomplishments since, people wonder at her still. Well, if we do our job and open up the first new world for humanity on Mars, then two thousand years from now, there will probably be people living not only on Earth and Mars, but on numerous other planets throughout this region of our galaxy. Those people will have technologies and abilities that would seem as magical to us as ours would to a resident of Periclean Athens. Yet, despite all their wondrous powers, if we are the people who make it possible, then those billions of advanced beings living on worlds orbiting multitudes of civilized stars will look back at our time, and they will wonder at us.

EPILOGUE: THE SIGNIFICANCE OF THE MARTIAN FRONTIER

A little more than one hundred years ago, a young professor of history from the then-relatively obscure University of Wisconsin got up to speak at the annual conference of the American Historical Association. Frederick Jackson Turner's talk was scheduled as the last one in the evening session. A long series of obscure papers preceded his address, yet the majority of the conference participants remained to hear him. Perhaps a rumor had gotten afoot that something important was about to be said. If so, it was correct; for in one bold sweep Turner presented a brilliant insight into the basis of American society and the American character. It was not legal theories, precedents, traditions, national or racial stock that was the source of America's egalitarian democracy, individualism, and spirit of innovation, he said. It was the existence of the frontier.

"To the frontier the American intellect owes its striking characteristics," Turner said, "That coarseness of strength combined with acuteness and inquisitiveness; that practical, inventive turn of mind, quick to find expedients; that masterful grasp of material things, lacking in the artistic but powerful to effect great ends; that restless, nervous energy; that dominant individualism, working for good and evil, and withal that buoyancy and exuberance that comes from freedom—these are the traits of the frontier, or traits called out elsewhere because of the existence of the frontier."

Turner went on, driving his points home, "For a moment, at the frontier, the bonds of custom are broken and unrestraint is triumphant. There is no tabula rasa. The stubborn American environment is there with its imperious summons to accept its conditions; the inherited ways of doing things are also there; and yet, in spite of the environment, and in spite of custom, each frontier did indeed furnish a new opportunity, a gate of escape from the bondage of the past; and freshness, and confidence, and scorn of older society, impatience of its restraints and its ideas, and indifference to its lessons, have accompanied the frontier.

"What the Mediterranean Sea was to the Greeks, breaking the bonds of custom, offering new experiences, calling out new institutions and activities, that and more the ever retreating frontier has been to the United States"[54]

The Turner thesis was an intellectual bombshell, which within a few years created an entire school of historians who demonstrated that not only American culture, but the progressive humanist civilization that America generally represented resulted primarily from the great frontier of global settlement opened to Europe by the Age of Exploration.

Turner presented his paper in 1893. Just three years earlier, in 1890, the American frontier had been declared closed: The line of settlement that had always defined the furthermost existence of western expansion had actually met the line of settlement coming east from California. Today, a century later, we face the question that Turner himself posed—what if the frontier is truly gone? What happens to America and all it has stood for? Can a free, egalitarian, innovating society survive in the absence of room to grow?

Perhaps the question was premature in Turner's time, but not now. Currently we see around us an ever more apparent loss of vigor of our society: increasing fixity of the power structure and bureaucratization of all levels of life; impotence of political institutions to carry off great projects; the proliferation of regulations affecting all aspects of public, private, and commercial life; the spread of irrationalism; the banalization of popular culture; the loss of willingness by individuals to take risks, to fend for themselves or think for themselves; economic stagnation and decline; the deceleration of the rate of technological innovation. . . . Everywhere you look, the writing is on the wall.

Without a frontier from which to breathe new life, the spirit that gave rise to the progressive humanistic culture that America has represented for the past two centuries is fading. The issue is not just one of national loss—human progress needs a vanguard, and no replacement is in sight.

The creation of a new frontier thus presents itself as America's and humanity's greatest social need. Nothing is more important: Apply what palliatives you will, without a frontier to grow in, not only American society, but the entire global civilization based upon values of humanism, science, freedom, and progress will ultimately die.

I believe that humanity's new frontier can only be on Mars.

But why not on Earth, under the oceans or in a remote region such as Antarctica? It is true that settlements on or under the sea or in Antarctica are entirely possible, and their establishment and access would be much easier than that of Martian colonies. Nevertheless, at this point in history such terrestrial developments cannot meet an essential requirement for a frontier—to wit, they are insufficiently remote to allow for the free development of a new society. In this day and age, with modern terrestrial communication and transportation systems, no matter how remote or hostile the spot on Earth, the cops are too close. If people are to have the dignity that comes with making their own world, they must be free of the old.

Mars has what it takes. It's far enough away to free its colonists from intellectual or cultural domination by the old world, and unlike the Moon, rich enough in resources to give birth to a new branch of human civilization. As we've seen, though the Red Planet may appear at first glance to be frozen desert, it harbors resources in abundance that can enable the creation of an advanced technological civilization. Mars is remote and can be settled. The fact that Mars can be settled and altered defines it as the New World that can create the basis for a positive future for terrestrial humanity for the next several centuries.

WHY HUMANITY NEEDS MARS

Everything has tended to regenerate them; new laws, a new mode of living, a new social system; here they are become men.

—Jean de Crevecoeur, *Letters from an American Farmer*, 1782

The essence of humanist society is that it values human beings—human life and human rights are held precious beyond price. Such notions have been for several thousand years the core philosophical values of Western civilization, dating back to the Greeks and the Judeo-Christian ideas of the divine nature of the human spirit. Yet they could never be implemented as a practical basis for the organization of society until the great explorers of the age of discovery threw open a New World in which the dormant seed of humanism contained within medieval Christendom could grow and blossom forth.

The problem with Christendom was that it was fixed—it was a play for which the script had been written and the leading roles both chosen and assigned. The problem was not that there were insufficient natural resources to go around—medieval Europe was not heavily populated, and there were plenty of forests and other wild areas—the problem was that all the resources were owned. A ruling class had been selected and a set of ruling institutions, ideas, and customs had been selected with them, and by the law of "Survival of the Firstest," none could be displaced. Furthermore, not only had the leading roles been chosen, but so had those of the supporting cast and chorus, and there were only so many parts to go around. If you wanted to keep your part, you had to keep your place, and there was no place for someone without a part.

The New World changed all that by supplying a place in which there were no established ruling institutions. On such an improvisational stage, the players are not limited to the conventional role of actors, they become playwrights and directors as well. The unleashing of creative talent that such a novel situation allows is not only a great adventure for those lucky enough to be involved, it changes the opinion of the spectators as to the capabilities of actors in general. People who had no role in the old society could define their role in the new. People who did not "fit in" in the Old World could discover and demonstrate that far from being worthless, they were invaluable in the new, whether they journeyed there or not.

The New World destroyed the basis of aristocracy and created the basis for democracy. It allowed the development of diversity by allowing escape from those institutions that imposed uniformity. It destroyed a closed intellectual world by importing unsanctioned data and experience. It made progress possible by escaping the hold of those institutions whose continued rule required continued stagnation, and it drove progress by defining a situation in which innovation to maximize the capabilities of the limited population available was desperately needed. It raised the dignity of workers by raising the price of labor and by demonstrating for all to see that human beings can be the creators of their world. In America, from colonial times through the nineteenth century when cities were rapidly being built, people understood that America was not something one simply lived in—it was a place one helped build. People were not simply inhabitants of their world. They were makers of it.

A TALE OF TWO WORLDS

Consider the probable fate of humanity in the twenty-first century under two conditions: with a Martian frontier and without it.

In the twenty-first century, without a Martian frontier, there is no question that human cultural diversity will decline severely. Already, since the late twentieth century, advanced communication and transportation technologies have eroded the healthy diversity of human cultures on Earth. As technology allows us to come closer together, so we come to be more alike. Finding a McDonald's in Beijing, country-and-western music in Tokyo, or a Michael Jordan T-shirt on the back of an Amazon native is no longer a great surprise.

Bringing together diverse cultures can be healthy, as it sometimes results in fusions that produce temporary flowerings in the arts or other areas. It can also result in very unpleasant increases in ethnic tensions. But however the energy released in the cultural merger is expended in the short term, the important thing in the long term is that it is expended. An analogy to cultural homogenization is that of connecting a wire between the terminals of a battery. A lot of heat can be generated for a while, but when all the potentials have been leveled, a condition of maximum entropy is reached and the battery is dead. The classic example of such a phenomenon in human history is the Roman Empire.[55] The golden age produced by unification is frequently followed by stagnation and decline.

The tendency toward cultural homogenization on Earth can only accelerate in the twenty-first century. Furthermore, because of rapid communication and transportation technologies shorting out intercultural barriers, it will become increasingly impossible to obtain the degree of separation required to develop new and different cultures on Earth. If the Martian frontier is opened, however, this same process of technological advance will also enable us to establish a new, distinct, and dynamic branch of human culture on Mars and eventually more on worlds beyond. The precious diversity of humanity can thus be preserved on a broader field, but only on a broader field. One world will be just too small a domain to allow the preservation and continued generation of the diversity needed not just to keep life interesting, but to assure the survival of the human race.

Without the opening of a new frontier on Mars, continued Western civilization also faces the risk of technological stagnation. To some this may appear to be an odd statement, as the present age is frequently cited as one of technological wonders. In fact, however, the rate of progress within our society has been decreasing and at an alarming rate. To see this, it is only necessary to step back and compare the changes that have occurred in the past thirty-five years with those that occurred in the preceding thirty-five years and the thirty-five years before that. Between 1905 and 1940 the world was revolutionized: Cities were electrified; washing machines and refrigerators appeared; telephones and broadcast radio became common; home stereos were born; talking motion pictures blossomed into a grand new art form; automobiles became practical; and aviation progressed from the Wright Flyer to the DC-3 and Hawker Hurricane. Between 1940 and 1975 the world changed again, with the introduction of computers, television, antibiotics, nuclear power, Boeing 727s, SR-71s, Atlas, Titan, and Saturn rockets, communication satellites, interplanetary spacecraft, and piloted voyages to the Moon. Compared to these changes, the technological innovations from 1975 to the present seem insignificant. Immense changes should have occurred during this period, but did not. Had we been following the previous seventy years' technological trajectory, we today would have flying cars, maglev (magnetic levitation) trains, robots, fusion reactors, hypersonic intercontinental travel, reliable and inexpensive transportation to Earth orbit, undersea cities, open-sea mariculture, and human settlements on the Moon and Mars. Instead, today we see important technological developments, such as nuclear power and biotechnology, being blocked or enmeshed in controversy—we are slowing down.

Now, consider a nascent Martian civilization: Its future will depend critically upon the progress of science and technology. Just as the inventions produced by the necessities of frontier America were a powerful driving force on worldwide human progress in the nineteenth century, so the "Martian ingenuity" born in a culture that puts the utmost premium on intelligence, practical education, and the determination required to make real contributions will make much more than its fair share of the scientific and technological breakthroughs, which will dramatically advance the human condition in the twenty-first century.

A prime example of the Martian frontier driving new technology will undoubtedly be found in the arena of energy production. As on Earth, an ample supply of energy will be crucial to the success of Mars settlements. The Red Planet does have one major energy resource that we currently know about: deuterium, which can be used as the fuel in nearly waste-free thermonuclear fusion reactors. Earth has large amounts of deuterium too, but with all of the existing investments in other, more polluting forms of energy production, the research that would make possible practical fusion power reactors has been allowed to stagnate. The Martian colonists are certain to be much more determined to get fusion online, and in doing so will massively benefit the mother planet as well.

The parallel between the Martian frontier and that of nineteenth-century America as technology drivers is, if anything, vastly understated. America drove technological progress in the last century because its western frontier created a perpetual labor shortage back East, thus forcing the development of labor-saving machinery and providing a strong incentive for improvement of public education so that the skills of the limited work force could be maximized. This condition no longer holds true in America. In fact, far from prizing each additional citizen, anti-immigrant attitudes are on the rise, and a vast service sector of bureaucrats and menials is being created to absorb the energies of those parts of the population whose participation in the productive parts of the economy is no longer needed. Thus since the late twentieth century, and increasingly in the twenty-first, each additional citizen is and will be regarded as a burden.

On twenty-first-century Mars, on the other hand, conditions of labor shortage will apply with a vengeance. Indeed, it can be safely said that no commodity on twenty-first-century Mars will be more precious and more highly valued than human labor time. Workers on Mars will be paid more and treated better than their counterparts on Earth, and education will be driven to a higher standard than ever seen on the home planet. Just as the example of nineteenth-century America changed the way the common man was regarded and treated in Europe, so the impact of progressive Martian social conditions may be felt on Earth as well as on Mars. A new standard may be set for a higher form of humanist civilization on Mars, and, viewing it from afar, the citizens of Earth will rightly demand nothing less for themselves.

The frontier drove the development of democracy in America by creating a self-reliant population that insisted on the right to self-government. It is doubtful that democracy can persist without such people. True, the trappings of democracy exist in abundance in America today, but meaningful public participation in the process is deeply wanting. Consider that no representative of a new political party has been elected president of the United States since 1860. Likewise, neighborhood political clubs and ward structures that once allowed citizen participation in party deliberations have vanished. And with an average reelection rate of 90 percent, the U.S. Congress is hardly a barometer of people's will. Furthermore, regardless of the will of Congress, the real laws, covering ever broader areas of economic and social life, are increasingly being made by a plethora of regulatory agencies whose officials do not even pretend to have been elected by anyone.

Democracy in America and elsewhere in Western civilization needs a shot in the arm. That boost can only come from the example of a frontier people whose civilization incorporates the ethos that breathed the spirit into democracy in America in the first place. As Americans showed Europe in the last century, so in the next the Martians can show us the path away from oligarchy and stagnation.

There are greater threats that a humanist society faces in a closed world than the return of oligarchy, and if the frontier remains closed, we are certain to face them in the twenty-first century. These threats are the spread of various sorts of anti-human ideologies and the development of political institutions that incorporate the notions that spring from them as a basis of operation. At the top of the list of such destructive ideas that tend to spread naturally in a closed society is the Malthus theory, which holds that since the world's resources are more or less fixed, population growth and living standards must be restricted or all of us will descend into bottomless misery.

Malthusianism is scientifically bankrupt—all predictions made upon it have been wrong, because human beings are not mere consumers of resources. Rather, we create resources by the development of new technologies that find use for them. The more people, the faster the rate of innovation. This is why (contrary to Malthus) as the world's population has increased, the standard of living has increased, and at an accelerating rate. Nevertheless, in a closed society Malthusianism has the appearance of self-evident truth, and herein lies the danger. It is not enough to argue against Malthusianism in the abstract—such debates are not settled in academic journals. Unless people can see broad vistas of unused resources in front of them, the belief in limited resources tends to follow as a matter of course. And if the idea is accepted that the world's resources are fixed, then each person is ultimately the enemy of every other person, and each race or nation is the enemy of every other race or nation. The extreme result is tyranny, war, and even genocide. Only in a universe of unlimited resources can all men be brothers.

MARS BECKONS

> We have come recently to boast of a global economy without thinking of its implications, of how unfortunate we are in finding it. It would be more cheering if news should come that by some freak of the solar system another world had swung gently into our orbit and moved so close that a bridge could be built over which people could pass to new continents untenanted and new seas uncharted. Would those eager immigrants repeat the process they followed when they had that opportunity, or would they redress the grievances of the old earth by a new bill of rights . . . ? The availability of such a new planet, at any rate, would prolong, if it did not save, a civilization based on dynamism, and in the prolongation the individual would again enjoy a spell of freedom. . . .
>
> It would be very interesting to speculate on what the human imagination is going to do with a frontierless world where it must seek its inspiration in uniformity rather than variety, in sameness rather than contrast, in safety rather than peril, in probing the harmless nuances of the known rather than the thundering uncertainties of unknown seas or continents. The dreamers, the poets, and the philosophers are after all but instruments which make vocal and articulate the hopes and aspirations and the fears of a people.
>
> The people are going to miss the frontier more than words can express. For four centuries they heard its call, listened to its promises, and bet their lives and fortunes on its outcome. It calls no more. . . .
>
> —Walter Prescott Webb, *The Great Frontier*, 1951

Western humanist civilization as we know and value it today was born in expansion, grew in expansion, and can only exist in a dynamic expanding state. While some form of human society might persist in a nonexpanding world, that society will not foster freedom, creativity, individuality, or progress. Such a dismal future might seem an outrageous prediction, except for the fact that for nearly all of its history most of humanity has been forced to endure such static modes of social organization, and the experience has not been a happy one. Free societies are the exception in human history—aside from isolated pockets, they have only existed during the four centuries of frontier expansion of the West. That history is now over. The frontier opened by the voyage of Christopher Columbus is now closed. If the era of Western humanist society is not to be seen by future historians as some kind of transitory golden age, a brief shining moment in an otherwise endless chronicle of human misery, then a new frontier must be opened. Mars beckons.

But Mars is only one planet, and with humanity's powers rising as they would in an age of an open Martian frontier, the job of transforming and settling it is unlikely to occupy our energies for more than three or four centuries. Does the settling of Mars then simply represent an opportunity to "prolong, but not save, a civilization based upon dynamism?" Isn't it the case that humanist civilization is ultimately doomed anyway? I think not.

The universe is vast. Its resources, if we can access them, truly are infinite. During the four centuries of the open frontier on Earth, science and technology have advanced at an astonishing pace. The technological capabilities achieved during the twentieth century would dwarf the expectations of any observer from the nineteenth, exceed the dreams of one from the eighteenth, and appear outright magical to someone from the seventeenth. The nearest stars are incredibly distant, about 100,000 times as far away as Mars. Yet, Mars itself is about 100,000 times as far from Earth as America is from Europe. If the past four centuries of progress have multiplied our reach by so great a ratio, might not four more centuries of freedom do the same again? There is ample reason to believe that they would.

Settling the Red Planet will drive the development of ever faster modes of space transportation; terraforming Mars will drive the development of new and more powerful sources of energy. Both of these capabilities in turn will open up new frontiers ever deeper into the outer solar system, and the harder challenges posed by these new environments will drive the two key technologies of power and propulsion ever more forcefully. The key is not to let the process stop. If it is allowed to stop for any length of time, society will crystallize into a static form that is inimical to progress. That is what defines the present age as one of crisis. Our old frontier is closed. The first signs of social stagnation are clearly visible. Yet progress, while slowing, is still extant: Our people still believe in it and our ruling institutions are not yet incompatible with it.

We still possess the greatest gift of the inheritance of a four-hundred-year-long Renaissance: To wit, the capacity to initiate another by opening the Martian frontier. If we fail to do so, our culture will not have that capacity long. Mars is harsh. Its settlers will need not only technology, but the scientific outlook, creativity, and free-thinking inventiveness that stand behind it. Mars will not allow itself to be settled by people from a static society—those people won't have what it takes. We still do. Mars today waits for the children of the old frontier. But Mars will not wait forever.

SPECIAL ADDENDUM: THE MARS METEORITE DISCOVERIES OF 1996

We don't need to wait until the Mars Sample Return mission to study Martian geology at first hand. A few pieces of the Red Planet have taken the trouble to travel here. In 1996, one of them, a 2-kilogram rock known as ALH84001 caused quite a stir.

The history of ALH84001 is as follows. The rock was formed a kilometer or two underground on Mars about 4.5 billion years ago, not long after the formation of the planet itself. About 3.6 billion years ago it was fractured, probably by a meteor impact not too far away from its location. Then, about 26 million years ago, another impact ejected it from Mars, causing it to fly around in space, until as chance would have it, it encountered the Earth 13,000 years ago and landed in Antarctica. All of these facts are known through various kinds of chemical analysis and isotopic dating techniques. For example, the rock's Martian origin is proved by its oxygen isotope ratios and the discovery of pockets of encased gas whose composition matches that of the Mars atmosphere as measured by *Viking*. Its formation date is supported by both samarium-neodymium and rubidium-strontium ratios, which allow the determination of age because these are mother-daughter radioactive decay pairs. Similar mother-daughter isotope ratios indicate the date of the fracturing impact, although with less certainty. The time it spent on Earth after landing can be determined by conventional carbon 14 dating, and the time spent during spaceflight is known from the amount of isotope changes induced by cosmic rays. Adding these two latter amounts of time together gives us the date of the planetary ejection event. Thus the basic chronology of ALH001's career is not in dispute.[56]

National Science Foundation Antarctic Meteorite program geologist Roberta Score found the rock in Antarctica's Allan Hills in early 1984 (thus the name, ALH84001). Despite being identified by Score immediately as anomalous, the rock was placed in cold storage at NASA Johnson Space Center and more or less ignored. There it remained until 1993, when a sample of it was mistakenly delivered to meteorite researcher David Mittlefehldt, who had ordered a different rock for examination as part of his program of studying diogenites. Mittlefehldt saw zoned carbonates in the rock and realized ALH84001 was no ordinary meteorite; it was a member of the one-in-a-thousand class that comes to Earth from Mars.

Thus rescued from obscurity by a process somewhat analogous to the workings of the judicial system of the old Austro-Hungarian Empire ("despotism tempered by inefficiency"), ALH84001 was made the subject of special study by a team of researchers.

By August 1996, the team, consisting of Johnson Space Center scientists David McKay, Everett Gibson, Kathie Thomas-Keprta, and Chris Romanek, and Stanford University Chemist Richard Zare, had written up their study in a *Science* magazine[57,58] paper showing remarkable results. According to the paper, ALH84001 showed evidence indicating the presence of bacteria on Mars 3.5 billion years ago. A discovery of this magnitude had obvious political implications, and so the team briefed NASA Administrator Dan Goldin. Goldin thereupon briefed Vice President Al Gore, and the preprinted paper ended up in the hands of a White House political strategist, whose mistress leaked it to the press. Thus a secret that had been kept by a fairly substantial group within the scientific community for over a year (I, who had no involvement whatsoever in this work, heard a little about it as early as the summer of 1995), was leaked within a week after it became known to the

White House. To preempt distorted press coverage, the team had no choice but to call a press conference explaining their discovery prior to formal publication.

The press conference was held at NASA headquarters on August 6, 1996. Introduced by NASA Administrator Goldin, the presenting panel included McKay, Gibson, Thomas-Keprta, and Zare, as well as noted paleobiologist Professor J. William Schopf, the director of the Center for the Study of Evolution and the Origin of Life at UCLA, who was there to provide skeptical counterpoint.

As presented, the teams' lines of evidence arguing for past biological activity included the presence in the rock of carbonate globules, organic polycyclic aromatic hydrocarbons (PAHs), photographs of small structures resembling fossil bacteria, and crystals of minerals including pyrrhotite, greigite, and magnetite that are frequently created biogenically. I briefly address each of these types of evidence in turn.

Carbonates

Carbonates are formed by the reaction between carbon-dioxide bearing water and rocks. There is no doubt that carbonates exist in ALH84001. This is evidence not for life *per se*, but for the existence of an aqueous environment that could have supported life. The team argued that their evidence showed that the carbonates formed at temperatures below 80°C (176°F), which would be acceptable for life. Skeptics have proposed alternative mechanisms where the carbonates could have formed at temperatures as high as 450°C (842°F), perhaps during one of the shock events. However the greigite and PAHs in the rock would have been destroyed at such temperatures, so this high-temperature explanation is inconsistent with the data. Furthermore, discounting the carbonates in ALH84001 seems beside the point. We know from orbital imaging of water erosion features that there was liquid water on Mars 3.5 billion years ago, and such water was almost certainly well loaded with carbon dioxide. It would thus be surprising if a Mars rock from that period did not contain carbonates. In any case, the existence of an aqueous environment on ancient Mars that the carbonate demonstrates is beyond dispute.

PAHs

PAHs are organic molecules, but that does not mean they were created by life. They have been found in ordinary meteorites, and few would argue for a biological origin in those cases. In addition, some have dismissed the PAHs in ALH84001, arguing that they could be the result of terrestrial contamination. The PAHs in ALH84001 do not represent the full spectrum of PAHs found in terrestrial pollution, and they are present at about a thousand times the concentration ordinarily found in Arctic or Antarctic samples that have been contaminated by Earth's atmosphere. Moreover, the concentration of PAHs in ALH84001 increases as one moves from the exterior to the interior, which is the exact opposite of what one would observe if the source of the PAHs were terrestrial contamination. So the PAHs are from Mars, but they don't prove life. They do show, however, that there was organic chemistry going on underground on Mars at that time, which is very interesting.

Fossil-like structures

The team displayed electron microscope photographs of things that certainly looked like fossils. One even had an uncanny resemblance to a segmented worm. There are two problems with this line of evidence, however. In the first place, as Schopf pointed out at the press conference, inorganic processes can frequently produce little rock structures, called "foolers," that look like fossils but are not. The second problem is that the fossil-looking things in ALH84001 are about an order of magnitude smaller than any known bacteria. There is thus a basis for arguing that they cannot be fossils, because there is no way that all the biochemical complexity that goes into making a bacterium can be packaged in a volume that small. I don't think that this argument is valid; it's like saying that an unprecedentedly large fish cannot be a fish because no previous fish was ever that big. Moreover, for reasons that I will explain below, I think that searching for lifeforms, or fossils of lifeforms, smaller and simpler than known bacteria is exactly the most important form of exobiological research to be done on Mars. So ruling such things out if we see them is a bad idea. That said, however, there is no proof that the fossil-like formations observed in ALH84001 are in fact fossils.

Possible biogenic minerals

As noted, the team showed that ALH84001 contained a number of sets of tiny crystals of various minerals, including magnetite and pyrrhotite, that generally are found on Earth as products of bacterial activity. Unfortunately, they can also be produced by inorganic processes. The team showed that some of the carbonate in the area where the minerals were found appeared to have been eroded by acids, which would indicate the prevalence of pH conditions incompatible with the basic chemistry required to precipitate magnetite and pyrrhotite by inorganic means. Schopf, however, was unconvinced. The two events could have happened at different times. If you want to prove that such minerals are of biologic origin, he said, you need to show that they have been laid down in the kind of linear formations, or chains, that are unique to life. The team had not done that.

Thus, while they had identified four intriguing types of phenomenon friendly to biology, none of these individually proved the existence of past life in ALH84001. The team argued however, that taken in combination, the simplest explanation for the ensemble was biological activity. But preponderance of evidence is a subjective criterion, and speaking for many scientists of similar temperament, Schopf was unconvinced. Quoting a famous dictum attributed to Carl Sagan, he said, "Extraordinary claims require extraordinary evidence."

So there the battle lines were drawn, with the team and others adhering to their side attempting to expand or defend the evidence, and large numbers of skeptics attempting to debunk it. Much of this argument generated more heat than light. For example, at the Mars Society convention in Toronto in August 2000, there was a one-on-one debate between Simon Clemett, a coworker of chemist Richard Zare and a member of the team, and Prof. Ralph Harvey, of Case Western Reserve University, an outspoken skeptic. Harvey was the far more capable debater, and appeared to get the better of Clemett, but on examination, many of his points lacked merit. For example, Harvey derided the fact that the team had only chosen to publish the photographs of the rock that contained the apparent microfossils, but had not published the much larger number that failed to show any such objects. This was nonsense. If you want to show that deer live in the woods you only need to show the photos in which deer appear. Discarding the null pictures is entirely legitimate.

One of the positive aspects of the dispute however, was that it resulted in the most intense study of any single meteorite, Martian or otherwise. As part of these investigations, Professor Joseph Kirschvink of Cal Tech made an important discovery; he found chemical evidence showing that in the course of its journey from Mars ejection to Earth landing, that large parts of the rock had never exceeded 40°C (104°F).[59] This confirmed an earlier theoretical calculation by Professor Jay Melosh of the University of Arizona, who had predicted based on the mathematics of shock interactions, that rocks could be ejected from planets without being excessively heated.[60] Kirschvink's experimental confirmation of Melosh's math is very important because it means that material can be transferred between planets *without being sterilized.* Thus if there had been bacteria in ALH84001 at the time of its ejection, they would have survived the trip to Earth.

Of course, ALH84001 arrived on Earth only 13,000 years ago, and any Marsbugs riding on it would have been greeted by hordes of native Earthling bacteria, fully adapted to their homeworld's environment and eager to eat the awkward newcomers. But what of the more distant past? Mars rocks have been landing on Earth (and Earth rocks on Mars) since the birth of the solar system. Mars, being smaller, cooled from its initial molten state before the Earth did. Life would have had its chance to get started on Mars before it did on Earth. So what if Martian bacteria were delivered to Earth before there were any natives to challenge them? *What if life on Earth came from Mars?* Kirschvink's discovery threw that possibility wide open.

Then, in the fall of 2000, a bombshell was dropped into the debate. The ordnance in question came from famed astrobiologist Imre Friedmann. In 1974, Friedmann had essentially founded the field of astrobiology as a science, by discovering the first known "cryptoendolithic" organisms—bacteria capable of surviving extremely cold and dry environments by hiding inside of rocks (thus the name crypto-endo-lith)—within rock samples included among the effects of scientist-explorer Wolf Vishniac, who had fallen to his death while searching for life amid the cliffs of Antarctica during December 1973.[61,62] By the late 1990's Friedmann was getting on in years, but as events would show, still kicking. Hard.

Schopf had dismissed the original ALH84001 team's magnetite claims because they had found no linear formations, or chains. Magnetite crystals created by magnetotactic bacteria are laid down in chains. In contrast, magnetite crystals created by nonbiological processes have no such geometrical organization.

Friedmann found the magnetite chains. They were there, all right, and in a paper published in February 2001 he and his collaborators showed them for all the world to see.[63] Not only that, the Friedmann team (consisting of Friedmann, Jacek Wierzchos of the University of Lleida, Spain, Carmen Ascaso of the Centro de Ciencias Medioambientales, Madrid, and Michael Winklhofer of the Geophysics Institute of the University of Munich) showed they met a set of criterion that very forcefully indicated a biological origin. These criterion "which could not be present in abiotically formed chains of magnetite crystals (no such chains have ever been observed in nature)" were: (i) Uniform crystal size and shape within chains, (ii) gaps between crystals, (iii) orientation of elongated crystals along the chain axis, (iv) Halo traces of membrane around chains, and (v) flexibility (bending) of chains. Friedmann et al. didn't mince words: "We conclude that the chains of electron-opaque particles in ALH84001 are magnetofossils, as no other consistent explanation would account for these findings."

Despite denials from the other side, Friedmann et al. really did nail the lid on the Allan Hills meteorite controversy. If extraordinary claims require extraordinary evidence, they had certainly supplied it. In light of Kirschvink's proof of the possibility of interplanetary bacterial transfer, however, by 2001 showing the existence of life in ALH84001 was really no extraordinary claim at all. After all, we know there was life on Earth 3.6 billion years ago, at a time when there was liquid water on Mars. Furthermore, with the higher rate of asteroidal impacts typical of the early solar system there was certainly plenty of natural transport available. So there *must* have been some bacteria on Mars at that time, if from no other source than the *Earth*. The real question is what was the source of the bacteria. We shall return to the significance of this question shortly.

But by validating the magnetofossils in ALH84001, the Friedmann team did much more than show the presence of life, they showed the presence of a particular form of life, magnetotactic microbes. Now on Earth, magnetotactic bacteria use their little compasses to allow them to navigate up and down to reach places where the oxygen concentration in their fluid medium suits them best. Therefore, one does not find magnetotactic bacteria on Earth until the oxygen level in the atmosphere had built up to significant concentrations roughly 2.3 billion years ago. Readers who know some geological history may wonder about this; after all, it is well known that photosynthetic cyanobacteria appeared on Earth close to 3.5 billion years ago. Why did it take so long for our planet to start to become oxygenated? The reason is that the limited amount of photosynthesis the primitive cyanobacteria could perform could not overpower the capacity of the Earth's plate tectonics to recycle fixed carbon back into the atmosphere as carbon dioxide.

Friedmann's findings thus imply the presence of free oxygen on Mars in significant quantities more than a billion years before it was available on Earth. This is not too surprising. Because it is a smaller planet, Mars' tectonic activity is much weaker than the Earth, in fact it is essentially nonexistent today. Therefore, the Red Planet would not recycle biologically fixed carbon as effectively as the Earth, and this might give primitive cyanobacteria a chance to oxygenate the place much faster.

But here is where things get wild. On Earth, there is significant evidence that the rate of evolution is correlated with the concentration of oxygen in the atmosphere. There is a clear statistical correlation between these factors, but there is a logical causal relationship as well. The presence of oxygen allows for more energetic chemical reactions, and thus more energetic and complex organisms. If we consider the development of animals, for example, these complex oxygen breathing creatures are each composed of multicelluar organizations of oxygen-respiring nucleated cells. These nucleated cells, or eukaryotes, are themselves complex organizations of subsystems, such as mitochondria (cellular energy generators), which in the distant past were once free-living bacteria. According to the now generally accepted theory known as symbiogeneis, developed originally by biologist Lynn Margulis of Boston University, it is believed that the complex nucleated cells that compose all higher animals and plants themselves originated as colonies of bacteria whose different members evolved to specialize in various activities. Thus bacteria have an analogous relationship to animal (or plant) cells as single-celled animals do to multicellular animals.[64]

Examining the fossil and geologic record, it has been determined that the appearance of cells using mitochondria correlates with the rise of atmospheric oxygen concentration to between 1 percent and 2 percent of present atmospheric levels (PAL). Chloroplasts (specialized cellular photosynthesis units) appeared roughly 2 billion years ago, when oxygen levels rose to 5 percent PAL. Around 600 million years ago, with oxygen levels rising to about 20 percent PAL, multicelluar animals burst upon the scene with a suddenness that has caused their advent to be called "the Cambrian Explosion."

The significance of the relationship between atmospheric oxygen levels and Martian evolution was first identified by astrobiologist Chris McKay in a series of daring papers published in 1996[65,66]. In these papers, McKay argued that we should not take the pace of evolution on Earth as a necessarily representative model. The Earth took 3.2 billion years from the end of the heavy asteroidal bombardment (which presumably precluded any life before its conclusion) period 3.8 billion years ago to generate multicelluar life, but since the rate of evolution is conditioned by the presence of oxygen, on Mars things conceivably could have occurred much faster. Mars' warm-wet juvenile period only lasted about 1 billion years, before the CO_2 atmosphere thinned out and the planet lost its beneficial greenhouse effect. On Earth, evolution acting over such a span was only able to yield bacteria. On Mars, with greater amounts of free oxygen present, it might have produced much more. It might have produced nucleated cells. It might even have produced complex multicelluar animals and plants.

In 1996, when McKay proposed these ideas, many people, including me, regarded them as fanciful speculations. But Friedmann's demonstration of Martian magnetotactic bacteria changed that. He had shown that on Mars, just 200 million years after the end of the heavy bombardment, there were critters of a type that took 1600 million years to appear on Earth. Suddenly McKay's ideas didn't seem to strange after all.

MARS AND THE ORIGIN OF LIFE ON EARTH

The origin of life on Earth is a mystery. Despite centuries of investigation by innumerable researchers, no evidence has ever been produced for the presence on Earth, either now or at any time in the past, of any free-living microorganisms simpler than bacteria. This is a striking fact: While bacteria are frequently thought of as simple life forms, they are actually extremely complex molecular machines, using highly evolved mechanisms to enable survival, metabolism, growth, reproduction, mobility, and innumerable other functions. It is thus inconceivable that bacteria could actually represent the earliest life to emerge from chemistry. A period of prior evolution had to occur, starting with much simpler forms, developing by degrees into the elaborate organisms we call bacteria. Yet we have good fossils of cyanobacteria, apparently similar to current forms, existing on Earth 3.5 billion years ago. This is only 300 million years after the close of the heavy bombardment period that would have made life impossible here prior to 3.8 billion years ago. This is an extraordinarily short time to expect native bacteria to evolve from chemistry, especially if we consider that the fossil record shows that for the next 2 billion years, the pace of evolution on this planet was extremely slow.

From the mathematical point of view, it is apparent that the general pace of evolutionary change in the biosphere is fastest at the present, and the farther back we go in the past as illuminated by the known fossil record, the slower evolution occurs. Thus, it took 2 billion years for bacteria to evolve sufficiently to produce nucleated single celled organisms ("eucaria"), but only another 900 million years to produce the first true multicelluar plants and animals. In another 400 million years we see complex vascular plants, fish, amphibians, reptiles, and protomammals, and in the next 200 million years, we see trees with seeds, grasses, flowering plants, dinosaurs, birds, mammals, and man. As we have seen, this rate is correlated with the concentration of oxygen in the atmosphere. But the general pattern observed also is that the more evolved life becomes, the greater its capability develops for ever more rapid evolution. It thus appears to be rather a stretch to argue that the simplest life forms that preceded bacteria should have been able to span the huge evolutionary chasm separating organic chemistry from complex bacteria in a geological blink of an eye, but then have evolution slam on the brakes for the next 2 billion years. If anything, it is to be expected that such simple prebacteria operating in an oxygen deprived environment should have accomplished their evolutionary ascent in the most tedious fashion.

Furthermore, as mentioned above, there are no surviving examples of this class of organisms on Earth. This seems extraordinary, and is not well explained by the supposition that such creatures were driven into extinction by the more highly evolved bacteria. After all, despite the development of more complex eucaria, bacteria are still very much around, and the single-celled eucaria survive quite nicely despite the evolution of still more complex animals and plants. Complexity always comes at an evolutionary cost, leaving plenty of room for the simpler folk who preceded the more complex forms.

Thus, while bacteria could not have been the first life, both the fossil record and current biological surveys strongly support the assertion that bacteria were in fact the first life *on Earth*. The only way to resolve this situation is to hypothesize that bacteria did not evolve on Earth, but arrived here fully evolved from space. This hypothesis, known generally as "panspermia," is further supported by the observation that many varieties of bacteria have adaptations that allow them to survive in dormancy for long periods in hard vacuum, ultracold, and radiation environments that can only be found in outer space. In general, in biology, all adaptations have costs, and organisms do not support adaptations which have no utility. If we found a species of land animal with vestigial aquatic adaptations, we would assume that its ancestors came from the sea. Similarly, it can be argued that the presence of astronautical adaptations among bacteria strongly supports the conjecture that their ancestors came from space.

The panspermia hypothesis is unpopular among origin-of-life researchers, because it completely ducks the central question of interest in their field, the origin of living things from nonliving chemistry. Actually, panspermia is not irrelevant to these concerns, because it opens up the possibility that life may have originated in more favorable environments than the early Earth, for example on a planet offering the chemically reduced environment found so favorable for the development of amino acids in the 1950's experiments by Miller and Urey.[67] In these experiments, Professor Harold Urey's graduate student Stanley Miller achieved scientific immortality by mixing methane, ammonia, and water vapor in a flask and zapping the mixture with electric sparks to produce a large number of the amino acids that are considered fundamental (and, prior to Miller, unique) to biology. These experiments have been criticized as irrelevant to the origin of life, because the early Earth offered a more oxidizing environment in which the Miller-Urey reactions would not readily occur. If panspermia is possible, however, then these criticisms are moot. Whatever its relevance to questions of the origin of life, it is clear that the panspermia hypothesis bears heavily on the issues of the prevalence and distribution of life in the universe.

TRANSPORT OF LIFE BETWEEN EARTH AND MARS

As discussed above, it is now well established that throughout its history, the Earth has been the target of numerous impacts by asteroids and comets that have had the capability to eject large amounts of unshocked, and therefore unsterilized, material into interplanetary and interstellar space. Collaborators of University of Arizona Professor Jay Melosh, such as Swedish biologist Curt Mileikowski, have published calculations showing that significant fractions of this material can find their way to nearby planets such as Mars within time scales that are very short compared to the demonstrated viable lifespans of dormant bacteria.[60] Thus, throughout geological history, innumerable bacteria have almost certainly been transported from Earth to Mars. Furthermore, if there is or ever was bacterial life on Mars, natural transport of these organisms has occurred from Mars to Earth as well. Indeed, 500 kg/

year of unsterilized Martian rocks are estimated to fall on Earth every year. This observation indicates that the current very expensive "planetary protection" programs instituted by various space agencies to prevent the transportation of microorganism between planets on artificial spacecraft are without rational foundation; the microbes already have plenty of their own spacecraft and have been making the trip regularly and in large numbers for the past 3.5 billion years.

The ease of natural transport of bacteria between Earth and Mars makes it unlikely, albeit not impossible, that any past or present Martian life could have a separate origin from terrestrial life. For two separate origins to occur, they would have to be nearly simultaneous, as otherwise the planetary life which originated first would preemptively spread. Rather, the most realistic possibilities are that both Earth and Mars were seeded simultaneously from an outside, presumably interstellar source, or that life developed indigenously from chemistry on either Earth or Mars, and then rapidly seeded the other as soon as forms, such as bacteria, which could survive spaceflight on the planet of origin evolved. We have seen from the previous discussion that the lack of prebacterial organisms on Earth undermines the supposition that life developed indigenously on Earth. Therefore, the most probable alternatives are either: a) Life originated on Mars and then seeded the Earth, or b) The Earth and Mars were seeded simultaneously from an interstellar source.

It will be observed that I do not include as a likely possibility that only Earth, and not Mars, could have been seeded from the outside. This is because it is now clear that Mars had liquid water on its surface for hundreds of millions of years of its early history. Thus, if an interstellar source of bacteria were available, Mars certainly would have been seeded, too.

The narrowing of possibilities to (a) and (b) above, recasts the issues surrounding the search for life on Mars in a decisive way. The key point is not whether bacteria ever existed on Mars; they almost certainly did. Rather the key issue is whether prebacterial organisms (prebacteria) either do or did exist on Mars. If we can find evidence for such prebacteria on the Red Planet, then we should conclude that (a) is true. If we don't, then we are driven to conclude (b).

The implications of either of these results would be spectacular. For example, if prebacteria can be found on Mars, then we would finally gain an understanding of the fundamental steps involved in the transition from chemistry to life. Furthermore, since much of Mars's surface is fairly well preserved going back 3.8 billion years, we would have the opportunity to read the history of life's development from nonliving chemistry directly from the fossil record. In essence, we would be gaining the opportunity to read the book of life itself.

On the other hand, if the search for life on Mars only reveals evidence for the same sort of well-developed, spaceflight-capable bacteria we see as the earliest inhabitants of the Earth, then that would say that both planets were seeded from interstellar sources. That would prove the validity of interstellar panspermia, and therefore imply that microbial life should be expected on nearly all of the many billions of microbe-suitable planets that are now believed to exist throughout our galaxy. As microbes are the source for evolution to higher forms, this would greatly increase the odds for the widespread existence for complex and intelligent life as well.

The search for both living and fossilized prebacteria on Mars thus emerges as one central to an understanding of the place of both life and humanity in the universe. It seems pretty clear that this can only be undertaken competently by human explorers operating on the planet's surface. The reason for this is that prebacterial fossils, in the absence of distracting and more obvious bacterial fossils, are likely to be very old, and thus rare. Furthermore, if McKay is right—and it now appears possible that he might be—then we need to be looking for fossils of macroscopic animals and plants as well. Since Mars' wet period ended 3 billion years ago, any fossils of these creatures will have to be at least that old, and considerably more rare than dinosaur bones are on Earth. It would thus be wildly unreasonable to hope that any such things would be found in a kilogram of material returned to Earth as part of a robotic sample return mission.

But the real scientific bonanza on Mars would come from the recovery of live organisms, either prebacteria, or, if we are very lucky, bacteria or even eucaria of a separate origin. We need live organisms if we are to examine their structure, and learn in detail what were the steps by which prebacteria made the transition from chemistry to life. We need live organisms if we are going to be able to determine for sure whether Martian bacteria represent a common or separate origin from those of the Earth. And if Mars life does represent a separate genesis, only living samples will tell us in what ways it was able to choose an alternative path as that followed by the life of the Earth.

To obtain such living samples, we will need to set up drilling rigs capable of penetrating a kilometer or more below the Martian surface to reach liquid ground water and the active biosphere it probably hosts. That is no task for small robotic probes.

If we are to find out the truth about the nature of life, human explorers will have to go to Mars.

APPENDIX I: FOUNDING DECLARATION OF THE MARS SOCIETY

The time has come for humanity to journey to Mars

We're ready. Though Mars is distant, we are far better prepared today to send humans to Mars than we were to travel to the Moon at the commencement of the space age. Given the will, we could have our first teams on Mars within a decade.

The reasons for going to Mars are powerful

We must go for the knowledge of Mars. Our robotic probes have revealed that Mars was once a warm and wet planet, suitable for hosting life's origin. But did it? A search for fossils on the Martian surface or microbes in groundwater below could provide the answer. If found, they would show that the origin of life is not unique to the Earth, and, by implication, reveal a universe that is filled with life and probably intelligence as well. From the point of view learning our true place in the universe, this would be the most important scientific enlightenment since Copernicus.

We must go for the knowledge of Earth. As we begin the twenty-first century, we have evidence that we are changing the Earth's atmosphere and environment in significant ways. It has become a critical matter for us better to understand all aspects of our environment. In this project, comparative planetology is a very powerful tool, a fact already shown by the role Venusian atmospheric studies played in our discovery of the potential threat of global warming by greenhouse gases. Mars, the planet most like Earth, will have even more to teach us about our home world. The knowledge we gain could be key to our survival.

We must go for the challenge. Civilizations, like people, thrive on challenge and decay without it. The time is past for human societies to use war as a driving stress for technological progress. As the world moves towards unity, we must join together, not in mutual passivity, but in common enterprise, facing outward to embrace a greater and nobler challenge than that which we previously posed to each other. Pioneering Mars will provide such a challenge. Furthermore, a cooperative international exploration of Mars would serve as an example of how the same joint action could work on Earth in other ventures.

We must go for the youth. The spirit of youth demands adventure. A humans-to-Mars program would challenge young people everywhere to develop their minds to participate in the pioneering of a new world. If a Mars program were to inspire just a single extra percent of today's youth to scientific educations, the net result would be tens of millions more scientists, engineers, inventors, medical researchers, and doctors. These people will make innovations that create new industries, find new medical cures, increase income, and benefit the world in innumerable ways to provide a return that will utterly dwarf the expenditures of the Mars program.

We must go for the opportunity. The settling of the Martian New World is an opportunity for a noble experiment in which humanity has another chance to shed old baggage and begin the world anew; carrying forward as much of the best of our heritage as possible and leaving the worst behind. Such chances do not come often, and are not to be disdained lightly.

We must go for our humanity. Human beings are more than merely another kind of animal; we are life's messenger. Alone of the creatures of the Earth, we have the ability to continue the work of creation by bringing life to Mars, and Mars to life. In doing so, we shall make a profound statement as to the precious worth of the human race and every member of it.

We must go for the future. Mars is not just a scientific curiosity; it is a world with a surface area equal to all the continents of Earth combined, possessing all the elements that are needed to support not only life, but technological society. It is a New World, filled with history waiting to be made by a new and youthful branch of human civilization that is waiting to be born. We must go to Mars to make that potential a reality. We must go, not for us, but for a people who are yet to be. We must do it for the Martians.

Believing therefore that the exploration and settlement of Mars is one of the greatest human endeavors possible in our time, we have gathered to found this Mars Society, understanding that even the best ideas for human action are never inevitable, but must be planned, advocated, and achieved by hard work. We call upon all other individuals and organizations of like-minded people to join with us in furthering this great enterprise. No nobler cause has ever been. We shall not rest until it succeeds.

The above declaration was signed and ratified by the 700 attendees at the Founding Convention of the Mars Society, held August 13–16, 1998 at the University of Colorado at Boulder, Colorado. If you agree, I invite you to join. Further information is available at www.marssociety.org or by writing the Mars Society, 11111 W. 8th Ave, unit A, Lakewood, CO, 80215.

APPENDIX II: CONTINUING THE FIGHT FOR MARS DIRECT, 2001–2011

Between 2001 and 2011 I wrote a large number of articles continuing my advocacy for Mars Direct. Collectively, they represent a kind of chronicle of the debate within and around the space community concerning human exploration and the many issues that were encountered and had to be dealt with if the prospects for human Mars exploration were to advance. Parts of them are necessarily repetitious (such as the summary description of the Mars Direct plan that is included in many of them), but a great deal of new material is presented as well. For example, they include extensive discussions of the lunar base initiative proposed by the Bush administration, the fight to save the Hubble Space Telescope, and the controversy surrounding the decision by the Obama administration to redirect NASA back to a nondestination driven mode of operation.

I had originally planned to include these articles as an appendix to this edition. However, as even the abridged collection is more than 200 pages in length, that proved to be impossible. Instead I have posted them on my company website at www.pioneerastro.com.

The articles exhibited there include:

"Victory from Space." *Space News*, September 24, 2001.

"Osama Bin Laden Found, On Mars!" Special to the *Weekly World News*, (submitted October 8, 2002, declined by the publisher).

"NASA NExT Program Needs a Destination." *Space News*, October 2002.

"Forward with Space Nuclear Power." *Space News*, February 3, 2003.

"No Time to Cut and Run." *St. Petersburg Times*, February 9, 2003.

"AP Falsely Reports Mars Radiation Data." Mars Society internet bulletin, March 14, 2003.

Testimony of Dr. Robert Zubrin at Senate Commerce, Science, and Transportation Committee Hearings: "Future of NASA." Wednesday, October 29, 2003.

"Mars Is Our Goal." Letter to *The New York Times*, December 7, 2003.

"The Choice for Kitty Hawk." *The Washington Times*, December 15, 2003.

"Don't Desert Hubble." *Space News*, February 9, 2004.

"Hubble Honorable Discharge?" Letter to *The Washington Times*, February 24, 2004.

"Tighten the Exploration Initiative." *Space News*, April 2004.

"Review of NASA Lunar Program Requirements Documents." Study for NASA Exploration Systems Mission Directorate (ESMD), October 18, 2004.

"How to Build a Lunar Base, Part 1: The Launch Issue." *Space News*, February 21, 2005.

"How to Build a Lunar Base, Part 2: The Mission Plan." *Space News*, February 28, 2005.

"How to Build a Lunar Base, Part 3: Evolution to Mars." *Space News*, March 7, 2005.

"Getting Space Exploration Right." *The New Atlantis*, Spring 2005.

"The Case for a Small CEV." *Space News*, July 4, 2005.

"Where Is NASA Going?" *Space News*, September 26, 2005.

"The Vision at Risk." *Space News*, March 27, 2006.

"Hubble Decision a Victory for Reason." *Space News*, November 2006.

"Don't Wreck the Mars Program." *Space News*, August 1, 2007.

"To the Stars! (But Stay on Budget?)." Letter to *The New York Times*, November 25, 2008.

"Augustine's Pathway to Nowhere." *Space News*, August 24, 2009.

"NASA Needs a Destination." *Space News*, February 22, 2010.

"Obama's Fake Space Program." *New York Daily News*, April 16, 2010.

"Will Obama Wreck NASA?" *Commentary*, June 2010.

"Opening Space with a 'Transorbital Railroad.'" *The New Atlantis*, Fall 2010.

"The New Sputnik." *Space News*, December 13, 2010.

Unfortunately, except for the successful efforts to save Hubble and the Mars Science Lab, the story related by the articles is not a happy one. Indeed, while the science-driven robotic Mars exploration program accomplished much over the period in question, NASA's human space exploration program, operating without (or in willful denial of) any rational plan, is no closer today to sending humans to Mars than it was in 2001 (or, arguably, 1971 for that matter).

Still, there is much to be learned from studying any battle, and perhaps more from a defeat than from a victory. We came close, during the past decade, to getting a humans to Mars program launched, but, at the end of the day, the opportunity was blown. As the German writer Friedrich Schiller once famously said of the French Revolution, "A great moment found a little people." Hopefully we will do better next time, because there will certainly be a next time. For—to paraphrase the most celebrated speech of a Frenchman of a more recent time, who may have been many things but certainly not little, Charles De Gaulle—*Mars has lost the battle, but Mars has not lost the war.*

On to Mars.

GLOSSARY

Aerobraking: A spacecraft maneuver using friction with a planetary atmosphere to decelerate from an interplanetary orbit to one about a planet.

Aeroshell: A heat shield used to protect a spacecraft from atmospheric heating during aerobraking.

Apogee: The highest point in an orbit about a planet.

Atmospheric pressure: The pressure an atmosphere exerts. On Earth at sea level, the atmospheric pressure is 14.7 pounds per square inch. This amount of pressure is therefore known as one "atmosphere" or one "bar."

BEIR: Biological effects of ionizing radiation.

Bipropellant: A rocket propellant combination including both a fuel and oxidizer. Examples include methane/oxygen, hydrogen/oxygen, kerosene/hydrogen peroxide, and so on.

Buffer gas: An effectively inert gas that is used to dilute the oxygen required to support breathing or combustion. On Earth, the 80 percent nitrogen found in air serves as a buffer gas.

Conjunction: The position of a planet behind the Sun as seen from another planet. When Earth and Mars are in conjunction, they are on opposite sides of the Sun.

Conjunction mission: A mission that flies about half of the way around the Sun to travel from one planet to another. Conjunction missions have the lowest propulsion requirements.

Cosmic ray: A particle, such as an atomic nucleus, traveling through space at very high velocity. Cosmic rays originate outside of our solar system. They typically have energies of billions of volts and require meters of solid shielding to stop.

Cryogenic: Ultracold. Liquid oxygen and hydrogen are both cryogenic fluids as they require temperatures of −180° and −250°C, respectively, for storage.

Delta 2: An expendable launch vehicle manufactured originally by McDonell Douglas and now by Boeing, capable of throwing 1,000 kg on a direct trajectory from Earth to Mars.

Delta-V (also written ΔV): The velocity change required to move a spacecraft from one orbit to another. A typical delta-V required to go from low Earth orbit to a trans-Mars trajectory would be about 4 km/s.

Departure velocity: The velocity of a spacecraft relative to a planet after effectively leaving the planet's gravitational field. Also known as ***hyperbolic velocity***.

Direct entry: A maneuver in which a spacecraft enters a planet's atmosphere and uses it to decelerate and land without going into orbit.

Direct launch: A maneuver in which a spacecraft is launched directly from one planet to another without being assembled in orbit.

Electrolysis: The use of electricity to split a chemical compound into its elemental components. Electrolysis of water splits it into hydrogen and oxygen.

Electron density: The number of electrons per cubic centimeter. The higher the electron density of an ionosphere, the better it reflects radio waves.

Endothermic: A chemical reaction requiring the addition of energy to occur.

Epicycle: A small circle whose center travels along the path of a larger circle. Ancient and medieval astronomers described the motion of the planets by envisioning that each planet traveled in a circle—the epicycle—whose center moved along a larger circle centered on the Earth.

Equilibrium constant: A number that characterizes the degree to which a chemical reaction will proceed to completion. A very high equilibrium constant implies near complete reaction.

ERV: Earth return vehicle.

ET: External tank.

EVA: Extravehicular activity.

Exhaust velocity: The speed of the gases emitted from a rocket nozzle.

Exothermic: A chemical reaction that releases energy when it occurs.

Fairing: The protective streamlined shell containing a payload that sits on top of a launch vehicle.

Fast conjunction mission: A conjunction-type mission (see above) in which some extra propellant is used to shorten the flight time.

Free-return trajectory: A trajectory which, after departing Earth, will eventually return to the Earth without any additional propulsive maneuvers.

GCMS: Gas chromatograph mass spectrometer.

Geothermal energy: Energy produced by using naturally hot underground materials to heat a fluid, which can then be expanded in a turbine-generator to produce electricity.

Gravity assist: A maneuver in which a spacecraft flying by a planet uses that planet's gravity to create a slingshot effect which adds to the spacecraft's velocity without any requirement for the use of rocket propellant.

Heliocentric: Centered about the Sun. A heliocentric orbit is one that transverses interplanetary space and is not bound to the Earth or any other planet.

Hohmann transfer orbit: An elliptical orbit, one of whose ends is tangent to the orbit of the planet of departure, and whose other end is tangent to the orbit of the planet of destination. The Hohmann transfer orbit is the purest incarnation of the conjunction-class orbit, and as such is the lowest energy path from one planet to another.

Hydrazine: A rocket propellant whose formula is N_2H_4. Hydrazine is a "monopropellant," which means that it can release energy by decomposing, without any additional oxidizer required for combustion.

Hyperbolic velocity: The velocity of a spacecraft relative to a planet before entering, or after effectively leaving, the planet's gravitational field. Also known as "approach" or "departure" velocity.

Hypersonic: A speed many times the speed of sound; in common usage Mach 5 or greater.

Ionosphere: The upper layer of a planet's atmosphere in which a significant fraction of the gas atoms have split into free positively charged ions and negatively charged electrons. Because of the presence of freely moving charged particles, an ionosphere can reflect radio waves.

Isp: A commonly used abbreviation for specific impulse (see below).

ISPP: In-situ propellant production.

JSC: Johnson Space Center, Houston, Texas.

kb/s: Kilobits per second.

Kelvin degrees: The Kelvin or "absolute" scale is a method of measuring temperature which starts with its zero point set at "absolute zero," the temperature at which a body in fact possesses no heat. Thus, 273 degrees Kelvin is the same temperature as 0° centigrade, the freezing point of water. Each additional degree Kelvin corresponds to one additional degree centigrade.

kHz: Kilohertz, a measure of frequency used in radio. One kHz equals 1,000 cycles per second.

km/s: Kilometers per second.

kW: Kilowatts.

kWe: Kilowatts of electricity.

kWe-hr: The total amount of energy associated with the use of one kilowatt of electricity for one hour.

kWh: The total amount of energy associated with the use of one kilowatt for one hour.

LEO: Low Earth orbit.

LOR: Lunar orbit rendezvous.

LOX: Liquid oxygen.

MAV: Mars ascent vehicle.

Methanation reaction: A chemical reaction forming methane. In the Mars Direct mission, the methanation reaction is the Sabatier reaction, in which hydrogen is combined with carbon dioxide to produce methane and water.

MHz: Megahertz, a measure of frequency used in radio. One MHz equals 1,000,000 cycles per second.

millirem: 1/1,000th of a rem (*see* below).

Minimum energy trajectory: The trajectory between two planets requiring the least amount of rocket propellant to attain (see Hohmann transfer).

m/s: Meters per second.

MOR: Mars orbit rendezvous.

MSR: Mars sample return.

MSR-ISPP: Mars sample return employing in-situ propellant production.

MWe: Megawatts of electricity.

MWt: Megawatts of heat. One megawatt equals 1,000 kilowatts.

NEP: Nuclear electric propulsion.

NIMF: Nuclear rocket using indigenous Martian fuel.

NTR: Nuclear thermal rocket.

Opposition: The position of a planet in the opposite direction from the Sun as seen from another planet. When Earth and Mars are in opposition, they are on the same side of the Sun, and thus closest to each other.

Opposition mission: A mission that flies most or all of the way around the Sun (~360 degrees) to travel from one planet to another, swinging into the inner solar system in the process in order to increase speeds. Opposition missions have the highest propulsion requirements.

Perigee: The lowest point in an orbit around a planet.

Pyrolyze: The use of heat to split a compound into its elemental constituents.

Regolith: What most commonly refer to as dirt.

rem: The measure of radiation dose most commonly used in the United States. One hundred rems equals one Sievert, the European unit. It is estimated that radiation doses of about 60 to 80 rem are sufficient to increase a person's probability of fatal cancer at some time later in life by 1 percent. Typical background radiation on Earth is about 0.2 rem/year.

RWGS: Reverse water-gas shift reaction.

RTG: Radioisotope thermoelectric generator.

Sabatier reaction: A reaction in which hydrogen and carbon dioxide are combined to produce methane and water. The Sabatier reaction is exothermic, with a high equilibrium constant (see above).

Saturn V: The heavy-lift launch vehicle used to send the Apollo astronauts to the Moon. The Saturn V could lift about 140 tonnes to LEO.

SEI: Space Exploration Initiative.

SNC meteorites: Named for the locations where the first three were found (Shergotty, Nakhla, and Chassigny), SNC meteorites are believed on the basis of very strong chemical, geologic, and isotopic evidence to be debris thrown off of Mars by impacting meteorites.

Sol: One Martian day, 24.6 hours long.

Solar flare: A sudden eruption on the surface of the Sun that can deliver immense amounts of radiation across vast stretches of space.

SPE: Solid polymer electrolyte.

Specific impulse: The specific impulse of a rocket engine is the number of seconds it can make a pound of propellant deliver a pound of thrust. If you multiply the specific impulse of a rocket engine, given in seconds, by 9.8, you will obtain the engine's exhaust velocity in units of meters/second. Specific impulse is generally viewed as the most important factor in judging a rocket engine's performance. Frequently abbreviated "Isp."

SRB: Solid rocket booster.

SSME: Space Shuttle main engine.

SSTO: Single-stage-to-orbit.

Stable equilibrium: An equilibrium condition, which, if displaced by some external force, will return on its own to its original state. A ball on top of a hill is in unstable equilibrium, because if pushed in either direction it will roll away, accelerating itself from its original position. A ball on a flat surface in the bottom of a bowl is in stable equilibrium, because if pushed, it will roll back to its starting point.

STR: Solar thermal rocket.

Telerobotic operation: Remote control of some device, such as a small Mars rover equipped with TV cameras, by human operators at a significant distance away.

Thrust: The amount of force a rocket engine can exert to accelerate a spacecraft.

Titan IV: An expendable launch vehicle formerly manufactured by the Lockheed Martin Corporation, capable of delivering 20,000 kg to LEO or 5,000 kg to a minimum energy trans-Mars trajectory.

TMI: Trans-Mars injection, a maneuver which places a payload or spacecraft on a trajectory to Mars.

TW: Terrawatt, one terrawatt equals 1,000,000 megawatts. Human civilization today uses about 15 TW.

TW-year: The total amount of energy associated with the use of one terrawatt for one year.

Unstable equilibrium: See stable equilibrium, above.

Vapor pressure: The pressure exerted by the gas emitted by a substance at a certain temperature. At 100°C, the vapor pressure of water is greater than the Earth's atmospheric pressure and so it will boil.

W/kg: Watts per kilogram.

NOTES

1. P. Berton, *The Arctic Grail*, Penguin Books, 1989.

2. G. Levin, "A Reappraisal of Life on Mars," D. B. Reiber, ed., *The NASA Mars Conference*, Volume 71, Science and Technology Series of the American Astronautical Society, Univelt, San Diego, CA, 1988.

3. N. Horowitz, "The Biological Question of Mars," D. B. Reiber, ed., *The NASA Mars Conference*, Volume 71, Science and Technology Series of the American Astronautical Society, Univelt, San Diego, CA, 1988.

4. J. Postgate, *The Outer Reaches of Life*, Cambridge University Press, Cambridge, UK, 1994.

5. J. W. Head et al.; "Possible ancient oceans on Mars: evidence from Mars Orbit Laser Altimeter data," *Science*, 286, 2134–2137, 1999

6. Katharine Sanderson, "Water could be flowing on Mars today," *Nature News*, December 6, 2006, http://www.nature.com/news/2006/061206/full/news061204-7.html (Accessed December 11, 2010). For the full scientific paper, see M. C. Malin, et al. "Present-Day Impact Cratering Rate and Contemporary Gully Activity on Mars, *Science*, 314.1573–1577 (2006).

7. Henry Bortman, "Odyssey Finds Large Concentrations of Water on Mars," *Astrobiology Magazine*, March 4, 2002, http://www.astrobio.net/exclusive/47/odyssey-finds-large-concentrations-of-water-on-mars (accessed December 11, 2010). "Odyssey Finds Water Ice in Abundance Under Martian Surface," Jet Propulsion Lab press release, May 28, 2002, http://mars.jpl.nasa.gov/odyssey/newsroom/pressreleases/20020528a.html (accessed December 11, 2010).

8. Steve Squyres, *Roving Mars: Spirit, Opportunity, and the Exploration of the Red Planet*, Hyperion, 2006.

9. "Mars Express Confirms Methane in Martian Atmosphere," ESA Press Release, March 30, 2004, http://www.esa.int/SPECIALS/Mars_Express/SEMZ 0B57ESD_0.html (accessed December 11, 2010).

10. "Discovery of Methane Reveals Mars Is Not a Dead Planet," NASA press release, January 15, 2009. http://www.nasa.gov/home/hqnews/2009/jan/ HQ_09-006_Mars_Methane.html (accessed December 11, 2010).

11. A. Cohen et al., *The 90 Day Study on the Human Exploration of the Moon and Mars*, U.S. Government Printing Office, Washington, DC, 1989.

12. R. Zubrin, D. Baker, and O. Gwynne, "Mars Direct: A Simple, Robust, and Cost-Effective Architecture for the Space Exploration Initiative," AIAA 91-0326, 29th Aerospace Science Conference, Reno, NY, January 1991.

13. T. Stafford et al., *America at the Threshold: Report of the Synthesis Group on America's Space Exploration Initiative*, U.S. Government Printing Office, Washington, DC, May 1991.

14. R. Zubrin and D. Weaver, "Practical Methods for Near-Term Piloted Mars Missions," AIAA 93-2089, 29th AIAA/ASME Joint Propulsion Conference, Monterey, CA, June 28–30, 1993. Republished in Journal of the British Interplanetary Society, July 1995.

15. M. Goldman, "Cancer Risk of Low Level Exposure," *Science*, March 29, 1996.

16. S. Kondo, *Health Effects of Low Level Radiation*, Kinki University Press, Osaka, Japan, 1993.

17. C. Comar et al., "The Effects on Populations of Exposure to Low Levels of Ionizing Radiation: Report of the Advisory Committee on the Biological Effects of Ionizing Radiations (BEIR)," Division of Medical Sciences, National Academy of Sciences and National Research Council, Washington, DC, 1972.

18. B. Clark and L. Mason, "The Radiation Show Stopper to Mars Missions: A Solution," presented to the AIAA Space Programs and Technologies Conference, Huntsville, AL, September 1990.

19. L. Simonson, J. Nealy, L. Townsend, and J. Wilson, "Radiation Exposure for Manned Mars Surface Missions," NASA Technical Publication-2979, Washington, DC, 1990.

20. J. Letaw, R. Silverberg, and C. Tsao, "Radiation Hazards of Space Missions," *Nature*, 330, no. 24 (1987):709–10.

21. A. Thompson, "Artificial Gravity for Long Duration Space Missions," presentation to Martin Marietta Scenario Development Team, February 1990.

22. M. Carr, *Water on Mars*, Oxford University Press, New York, 1996, pp. 24–29.

23. J. Gooding, "2005 Sample Return: Martian Meteorites and Curatorial Plans," presentation to the Mars Exploration Long-Term Strategy Working Group, Johnson Space Center, Houston, TX, September 20, 1995.

24. R. Zubrin, S. Price, L. Mason, and L. Clark, "Report on the Construction and Operation of a Mars In-Situ Propellant Production Plant," AIAA-94-2844, 30th AIAA Joint Propulsion Conference, Indianapolis, IN, June 1994. Republished in *Journal of the British Interplanetary Society*, August 1995.

25. R. Zubrin, S. Price, L. Mason, and L. Clark, "An End to End Demonstration of Mars In-Situ Propellant Production," AIAA-95-2798, 31st AIAA/ASME Joint Propulsion Conference, San Diego, CA, July 10–12, 1995.

26. B. Clark, "A Day in the Life of Mars Base 1," *Journal of the British Interplanetary Society*, November 1990.

27. B. Mackenzie, "Metric Time for Mars," AAS 87-269, in C. Stoker, ed., *The Case for Mars III*, Volume 75, Science and Technology Series of the American Astronautical Society, Univelt, San Diego, CA, 1989.

28. B. Mackenzie, "Building Mars Habitats Using Local Materials," AAS 87-216, in C. Stoker, ed., *The Case for Mars III*, Volume 74, Science and Technology Series of the American Astronautical Society, Univelt, San Diego, CA, 1989.

29. R. Boyd, P Thompson, and B. Clark, "Duricrete and Composites Construction on Mars," AAS 87-213, in C. Stoker, ed., *The Case for Mars III*, Volume 74, Science and Technology Series of the American Astronautical Society, Univelt, San Diego, CA, 1989.

30. B. Jakowsky and A. Zent, "Water on Mars: Its History and Availability as a Resource," in J. Lewis, M. Mathews, and M. Guerreri, eds., *Resources of Near-Earth Space*, University of Arizona Press, Tucson, 1993.

31. "Water Ice in Crater at Martian North Pole," ESA press release, July 28, 2005 http://www.esa.int/SPECIALS/Mars_Express/SEMGKA808BE_0.html (accessed December 11, 2010)

32. C. Stoker et al., "The Physical and Chemical Properties and Resource Potentials of Martian Surface Soils," in J. Lewis, M. Mathews, and M. Guerreri, eds., *Resources of Near-Earth Space*, University of Arizona Press, Tucson, 1993.

33. T. Meyer and C. McKay, "The Atmosphere of Mars—Resources for the Exploration and Settlement of Mars," AAS 81-244, in P. Boston, ed., *The Case for Mars*, Volume 57, Science and Technology Series of the American Astronautical Society, Univelt, San Diego, CA, 1984.

34. J. Williams, S. Coons, and A. Bruckner, "Design of a Water Vapor Adsorption Reactor for Martian In situ Resource Utilization," *Journal of the British Interplanetary Society*, August 1995.

35. G. O'Neill, *The High Frontier*, William Morrow, New York, 1977.

36. J. Lewis and R. Lewis, *Space Resources: Breaking the Bonds of Earth*, Chapter 9, Columbia University Press, New York, 1987.

37. R. Zubrin, "Diborane/CO_2 Engines for Mars Ascent Vehicles," AIAA 95-2640, 31st AIAA Joint Propulsion Conference, San Diego, CA, July 10, 1995. Republished in *Journal of the British Interplanetary Society*, September 1995.

38. S. Geels, J. Miller, and B. Clark, "Feasibility of Using Solar Power on Mars: Effects of Dust Storms on Incident Solar Radiation," AAS-87-266, in C. Stoker, ed., *The Case for Mars III*, Volume 75, Science and Technology Series of the American Astronautical Society, Univelt, San Diego, CA, 1989.

39. R. Haberle et al., "Atmospheric Effects on the Utility of Solar Power on Mars," in J. Lewis, M. Mathews, and M. Guerreri, eds., *Resources of Near-Earth Space*, University of Arizona Press, Tucson, 1993.

40. M. Fogg, "Geothermal Power on Mars," *Journal of the British Interplanetary Society*, November 1996.

41. R. Zubrin, "Nuclear Thermal Rockets Using Indigenous Martian Propellants," AIAA-89-2768, AIAA/ASME 25th Joint Propulsion Conference, Monterey, CA, July 1989.

42. R. Zubrin, "Long Range Mobility on Mars," *Journal of the British Interplanetary Society*, 45 (May 1992), pp. 203–210.

43. B. Cordell, "A Preliminary Assessment of Martian Natural Resource Potential," AAS 84-185, in C. McKay, ed., *The Case for Mars II*, Volume 62, Science and Technology Series of the American Astronautical Society, Univelt, San Diego, CA, 1985.

44. R. Zubrin and D. Baker, "Mars Direct, Humans to the Red Planet by 1999," IAF-90-672, 41st Congress of the International Astronautical Federation, Dresden, Germany, October 1990. Republished in *Acta Astronautica*, 26, no. 12 (1992): pp. 899–912.

45. R. Zubrin and D. Andrews, "Magnetic Sails and Interplanetary Travel," AIAA-89-2441, AIAA/ASME, 25th Joint Propulsion Conference, Monterey, CA, July 1989. Published in *Journal of Spacecraft and Rockets*, April 1991.

46. A. Clarke, *The Snows of Olympus: A Garden on Mars*, W. W. Norton, New York, 1995.

47. M. Fogg, *Terraforming: Engineering Planetary Environments*, Society of Automotive Engineers, Warrendale, PA, 1995.

48. R. Forward, "The Statite: A Non-Orbiting Spacecraft," AIAA 89-2546, AIAA/ASME, 25th Joint Propulsion Conference, Monterey, CA, July 1989.

49. C. Sagan, "The Planet Venus," *Science*, 133 (1961):849–858.

50. J. Pollack and C. Sagan, "Planetary Engineering," in J. Lewis, M. Mathews, and M. Guerreri, eds., *Resources of Near-Earth Space*, University of Arizona Press, Tucson, 1993.

51. C. McKay, J. Kastings, and O. Toon, "Making Mars Habitable," *Nature* 352 (1991):489–496

52. J. Miller, "The Information Needs of the Public Concerning Space Exploration," Special report to the National Aeronautics and Space Administration, 1994.

53. B. Lusignan et al., "The Stanford US–USSR Mars Exploration Initiative, Final Report," Stanford University School of Engineering, Stanford, CA, July 1992.

54. F. J. Turner, *The Frontier in American History*, H. Holt & Co., New York, 1920.

55. C. Quigley, *The Evolution of Civilizations*, Liberty Fund, Indianapolis, IN, 1961.

56. K. Sawyer, *The Rock from Mars: A Detective Story on Two Planets*, Random House, New York, 2006.

57. D. McKay et al., "Search for Past Life on Mars: Possible Relic Biogenic Activity in Martian Meteorite ALH84001." *Science*, 273:924–930, 1996.

58. E. Gibson, D. McKay, K. Thomas-Keprta, and C. Romanek, "The Case for Relic Life on Mars." *Scientific American*, 277:36–41, 1997.

59. Ben Weiss and Joseph Kirschvink, "Life from Space." *The Planetary Report*, Nov/Dec 2000.

60. C. Mileikowsky et al., "Natural Transfer of Viable Microbes in Space," *Icarus*, 145, 391–427, July 2000

61. R. Zubrin, *Mars on Earth*, Tarcher Penguin, New York, 2003, pp. 18–21.

62. E. I. Friedmann, "Endolithic Microbial Life in Hot and Cold Deserts," *Origins of Life*, vol. 10, p.223.

63. E. I. Friedmann, J. Wierzchos, C. Ascaso, and M. Winklhofer, "Chains of Magnetite Crystals in the Meteorite ALH84001: Evidence of Biological Ori-

gin." *Proceedings of the National Academy of Sciences*, vol 98, no. 5, 2176–2181, February 27, 2001.

64. L. Margulis and D. Sagan, *Microcosmos; Four Billion Years of Evolution from Our Microbial Ancestors*, University of California Press, 1997.

65. C. McKay, "Oxygen and the Rapid Evolution of Life on Mars," in J. Chela-Flores and F. Raulin (eds.), *Chemical Evolution: Physics of the Origin and Evolution of Life*, 177–184, Kluwer Academic Publishers, printed in the Netherlands, 1996.

66. C. McKay, "Time for Intelligence on Other Planets," in L.R. Doyle, editor, *Circumstellar Habitable Zones*, Travis House Publications, Menlo Park, pp. 405–419, 1996.

67. S. Miller, *The Origins of Life on Earth*," Prentice Hall, 1974.

REFERENCES

On Mars as a Planet

M. Carr, *The Surface of Mars*, Yale University Press, New Haven, 1981. Updated with a new edition from Cambridge University Press, 2007. The best introduction to Mars yet written.

M. Carr, *Water on Mars*, Oxford University Press, New York, 1996. A very readable book based on all the data of its time, focusing on the central issue of water on Mars, past and present.

H. Kieffer, B. Jakowsky, C. Snyder, and M. Mathews, *Mars*, University of Arizona Press, Tucson, 1992. A collection of 114 papers by virtually the entire community of Mars science specialists. Rather technical, but quite complete relative to its date of publication.

J. Bell, *The Martian Surface: Composition, Mineralogy, and Physical Properties*, Cambridge University Press, Cambridge, 2008. Some 650 pages long, this huge collection of papers is a major comprehensive update to the Kieffer, Jakowski, Snyder, and Mathews *Mars* volume published in 1992.

N. Barlow, *Mars: An Introduction to Its Surface, Interior, and Atmosphere*, Cambridge Planetary Science Series, Cambridge University Press, 2008. A manageable summary of our knowledge about Mars, including data through 2006.

W. K. Hartmann, *A Traveler's Guide to Mars*, Workman Publishers, New York, 2003. Written as an illustrated tour guide, this book provides a lighter introduction to the Red Planet.

J. Kargel, *Mars: A Warmer, Wetter, Planet*, Springer-Praxis, New York, 2004. Written in 2004, this book made a polemical case for a water-rich Mars that has since been largely validated by subsequent discoveries.

On Robotic Exploration of Mars

F. Taylor, *The Scientific Exploration of Mars*, Cambridge University Press, Cambridge, 2010. Contains the single best history of all robotic Mars missions to date, as well as a good short summary of our current knowledge of the planet.

P. Raeburn and M. Golombek, *Mars: Uncovering the Secrets of the Red Planet*, National Geographic Society, Washington, DC, 1998. Heavily illustrated, with some 3-D panoramas included, this book provides a great report on the *Pathfinder* mission.

S. Squyres, *Roving Mars: Spirit, Opportunity, and the Exploration of the Red Planet*, Hyperion, 2006. The principal investigator of the Mars Exploration Rover mission tells its story.

J. Bell, *Postcards from Mars: The First Photographer on the Red Planet*, Plume, New York, 2010. A compendium of wonderful images returned by the Spirit and Opportunity rovers.

On Human Missions to Mars

D. Reiber, *The NASA Mars Conference*, Volume 71, Science and Technology Series of the American Astronautical Society, Univelt, San Diego, 1988.

P. Boston, *The Case for Mars*, Volume 57, Science and Technology Series of the American Astronautical Society, Univelt, San Diego, 1984.

C. McKay, *The Case for Mars II*, Volume 62, Science and Technology Series of the American Astronautical Society, Univelt, San Diego, 1985.

C. Stoker, *The Case for Mars III*, Volumes 74 and 75, Science and Technology Series of the American Astronautical Society, Univelt, San Diego, 1989.

C. Stoker and C. Emmart, *Strategies for Mars: A Guide to Human Exploration*, Volume 86, Science and Technology Series of the American Astronautical Society, Univelt, San Diego, 1996.

T. Meyer, *The Case for Mars IV, Proceedings of the Fourth Case for Mars Conference Held at the University of Colorado in 1990*, Univelt, San Diego, 1996.

R. Zubrin, *From Imagination to Reality: Mars Exploration Studies of the Journal of the British Interplanetary Society*, Univelt, San Diego, 1997.

R. Zubrin and M. Zubrin, *Proceedings of the Founding Convention of the Mars Society Held August 13–16, 1998, Boulder, Colorado,* Univelt, San Diego, 1999.

P. Boston, *The Case for Mars V, Proceedings of the Fifth Case for Mars Conference Held at the University of Colorado in 1993,* Univelt, San Diego, 2000.

K. McMillen, *The Case for Mars VI, Proceedings of the Sixth Case for Mars Conference Held at the University of Colorado in 1996.* Univelt, San Diego, 2000.

F. Crossman and R. Zubrin, *On to Mars: Colonizing a New World*, Apogee Books, 2002.

F. Crossman and R. Zubrin, *On to Mars 2: Exploring and Settling a New World*, Collector's Guide Publishing, 2005.

On the Folklore of Mars

J. N. Wilford, *Mars Beckons*, Alfred Knopf, New York, 1990.

O. Morton, *Mapping Mars: Science Imagination, and the Birth of a New World*, Picador USA, New York, 2002.

R. Zubrin, *Mars on Earth: The Adventures of Space Pioneers in the High Arctic*, Tarcher Penguin, New York, 2003.

K. Sawyer, *The Rock from Mars: A Detective Story on Two Planets*, Random House, New York, 2006.

M. Roach, *Packing for Mars: The Curious Science of Life in the Void*, W. W. Norton, New York, 2010.

INDEX

Page numbers in *italics* refer to figures.

ABOUT THE AUTHORS

Robert Zubrin is president of Pioneer Astronautics, an aerospace R&D company, and the founder and president of the Mars Society, an international organization dedicated to furthering the exploration and settlement of Mars by both public and private means. He lives with his wife, Hope, a science teacher, in Golden, Colorado.

Richard Wagner is the former editor of *Ad Astra*, the journal of the National Space Society. He lives in Northampton, Massachusetts.